Hurricane Jim Crow

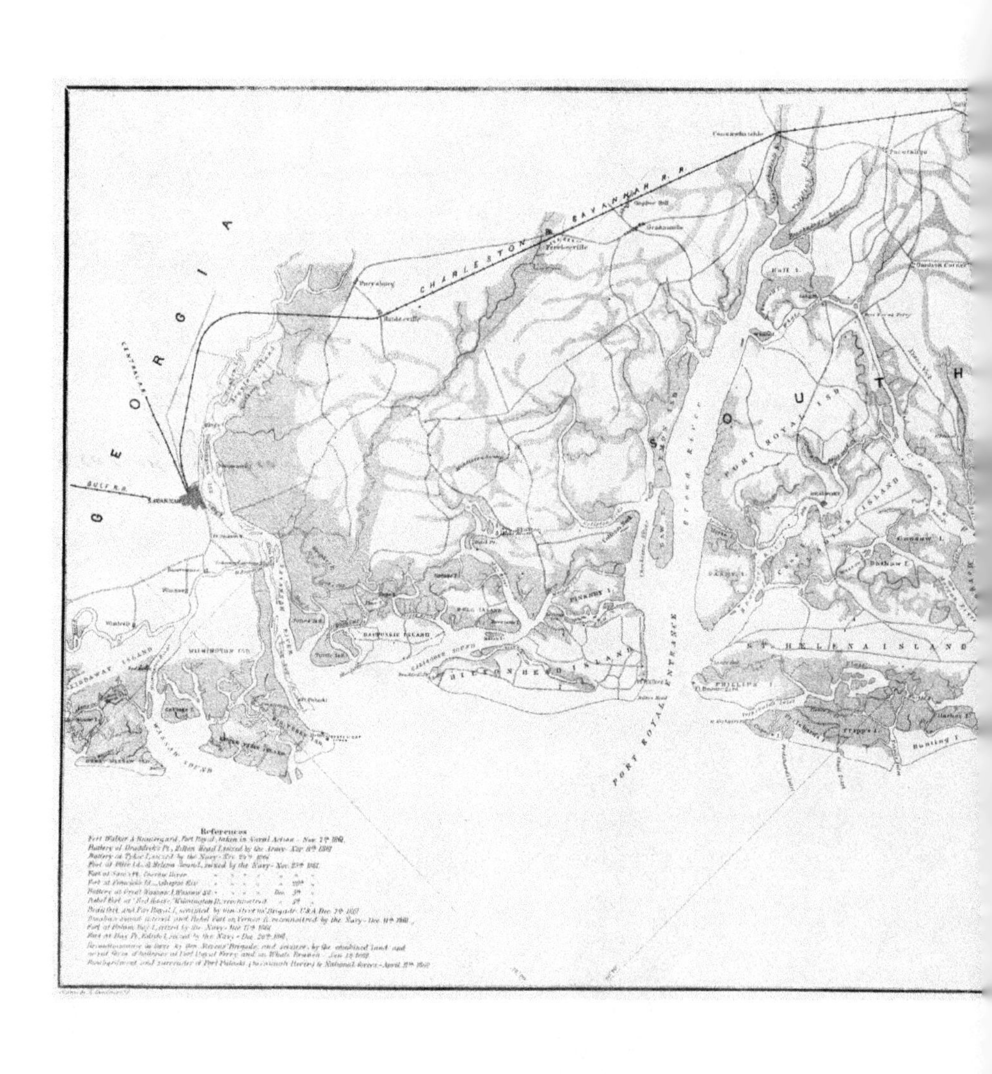
GEORGIA
SOUTH
CHARLESTON & SAVANNAH R.R.
HILTON HEAD ISLAND
ST. HELENA ISLAND
PORT ROYAL ISLAND
PORT ROYAL ENTRANCE
References

Hurricane Jim Crow

How the Great Sea Island Storm of 1893 Shaped the Lowcountry South

CAROLINE GREGO

The University of North Carolina Press
Chapel Hill

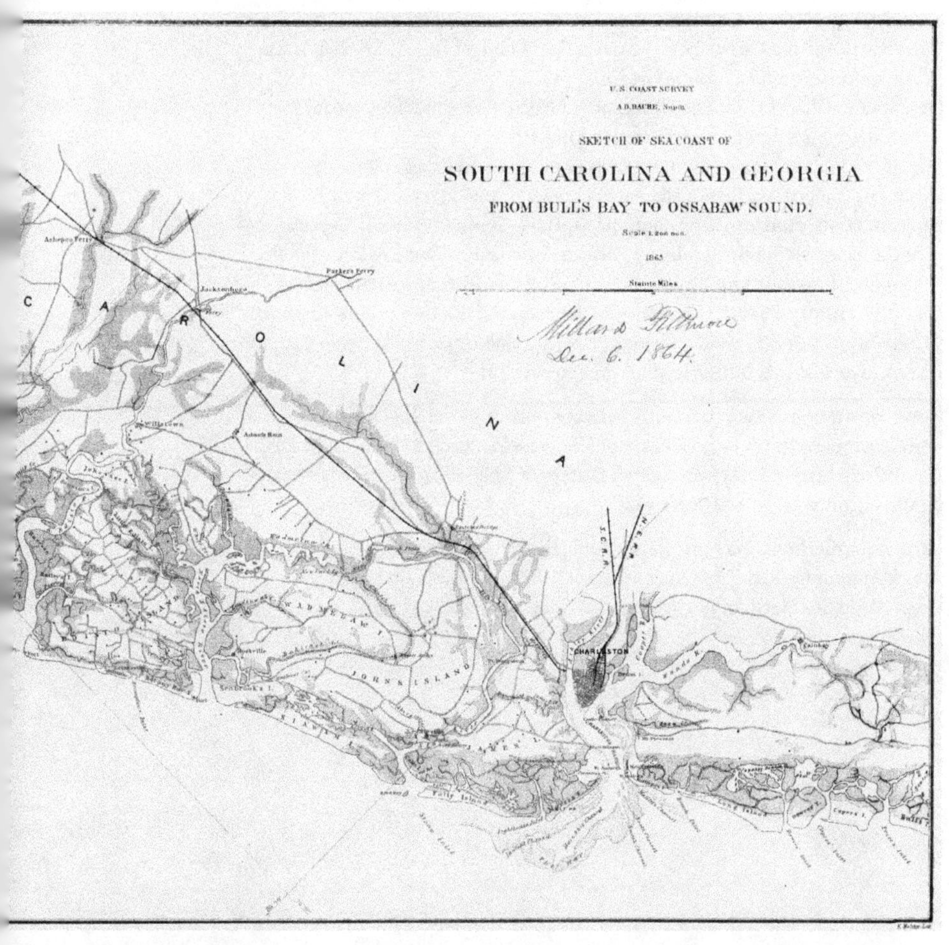

The Lowcountry, from Ossabaw Sound to Bull's Bay. A. Lindenkohl, E. Molitor, and U.S. Coast Survey, *Sketch of Sea Coast of South Carolina and Georgia from Bull's Bay to Ossabaw Sound* (U.S. Coast Survey, 1863), Library of Congress Geography and Map Division (LCCN 2007630406).

This book was published with the assistance of the Fred W. Morrison Fund of the University of North Carolina Press.

Set in Minion Pro by Westchester Publishing Services

Library of Congress Cataloging-in-Publication Data
Names: Grego, Caroline, author.
Title: Hurricane Jim Crow : how the Great Sea Island Storm of 1893 shaped the Lowcountry South / Caroline Grego.
Description: Chapel Hill : The University of North Carolina Press, [2022] | Includes bibliographical references and index.
Identifiers: LCCN 2022023741 | ISBN 9781469671345 (cloth ; alk. paper) | ISBN 9781469671352 (pbk. ; alk. paper) | ISBN 9781469671369 (ebook)
Subjects: LCSH: Hurricanes—Economic aspects—South Carolina. | Hurricanes—Social aspects—South Carolina. | African Americans—Segregation—South Carolina. | African Americans—South Carolina—History. | Atlantic Coast (S.C.)—History. | South Carolina—Race relations. | South Carolina—History.
Classification: LCC F277.A86 G74 2022 | DDC 305.896/0730757—dc23/eng/20220623
LC record available at https://lccn.loc.gov/2022023741

Cover illustrations: *Inset*, African Americans outside of the Sea Island Red Cross relief headquarters, 1893–94 (Library of Congress Prints and Photographs Division, LC-USZ62-122100); *background*, High Battery of Charleston after the hurricane (South Carolina Historical Society).

Portions of this book were previously published in a different form as "Black Autonomy, Red Cross Recovery, and White Backlash after the Great Island Sea Storm of 1893," *Journal of Southern History* 85, no. 4 (November 2019): 803–40.

For my parents, my sister, and Graeme

Contents

Graphs, Illustrations, and Tables

Graphs

Illustrations

Tables

Preface
The List

The list of the dead is piecemeal and relentless. Extended families. Married couples. Children bearing only the last names of their mother or father. No name, only "Unknown (Body)." Two hundred eighty-two entries line the pages of the forever-partial list of the dead, collected over two days by a group of hastily deputized, exhausted assistants to the Beaufort County coroner after the Great Sea Island Storm rolled through the Lowcountry on the night of August 27 to 28, 1893.[1] The list of named and unnamed, later compiled by archivist Grace Cordial, represents perhaps a tenth of those who perished in the powerful hurricane and its rushing storm surge. It does not include anyone who lived outside of Beaufort and its immediate environs or those who died in the months or years after of sickness, old injuries, hunger, or whatever deprivation gradually sapped the life from them.

All but a couple dozen of the thousands of hurricane dead were African American. Estimates of the storm's death toll hover frustratingly between fifteen hundred and five thousand: an unknowable number, likely around two thousand, perished in a single night, and many more passed away in the days, weeks, and even years later. The African Americans who died, whether on that one night or after, lived and worked in the part of the Lowcountry that the hurricane struck hardest, the marshy edge of the southeastern coast stretching from the mouth of the Savannah River to Winyah Bay, South Carolina. They inhabited tenement buildings in Charleston, sharecropper cabins near rice and sea island cotton fields, company-owned shacks along phosphate-rich marshes, and yeoman cottages on sea island land that they were proud to own—places that they had sometimes claimed or where they had sometimes been placed by the structural demands of labor and capital. Outside of this list, few of the names of the hurricane dead are recoverable in any physical archive, though they may persist in private family memories.

This list is valuable not simply as a record of the dead but as a tangible artifact of people who once lived. For some of these sea islanders, this list represents one of very few times, if not the only, that their name appeared in a record that yet exists. Some of these names allow us to reconstruct the contours of their lives. Abby and Phibby Hunt, in their seventies, had been married for decades; she kept house while he worked as a farm laborer, maybe in the rice fields near Sheldon.[2] Perhaps Philip Brisbane's mother, Daphne, who owned her farm on St. Helena Island, mourned his death and buried his body on her land.[3] Renty Capers, who had attended the historic Penn School on St. Helena Island, grew up in a big, rambling family who lived on adjacent farms, staying close even after growing up and getting married.[4] It is painful to see the entries for children whom the coroner only identified by surname—we are left with the grief of their parents, echoing across the generations, and the knowledge of lives cut short. Monday Washington lost seven children in one night. Joe Drayton's wife and seven children died too. Sookie Perry perished that night with her six young ones and likely her husband as well. Rose and Monday Polite, Scipio Heyward, and Peggy Johnson lost four each and other relatives besides. Like the many entries of "unnamed child" on the list, those others who are cataloged as unnamed were not nameless when they lived. Their inclusion is a poignant reminder of the limits of our knowledge and the extant historical record.

The list is part of the hurricane's archive. This archive coalesced around the hurricane, revealing the workings of politics, labor, daily lives, human relationships, and the Lowcountry environment in ways that we might not have had the opportunity to understand without it.[5] That archive is fraught with the same issues that many historians encounter when writing and learning about the South. Few of the sources in the hurricane's archive—besides a handful of petitions that the American Red Cross collected, some accounts taken at the Penn School on St. Helena Island, and a few letters and interviews scattered about—were made by African Americans, and those relatively unmediated by the white gaze are even rarer.

That said, an analytical eye and a willingness to consider the reasonable realm of possibilities are two of the most indispensable tools to a historian regardless of context, and they have been put to use here. White southerners spoke very freely of their racism, as they did not see their belief in white supremacy as something either to be ashamed of or to hide; Black southerners, in turn, had to weigh carefully what they said to white southerners. The prejudices of white southerners—both in the moment and of those who collected and curated their documents later—have indelibly altered which historical conversations the contemporary reader can be privy to. Often we must listen in on a one-sided conversation. Fortunately, the hurricane's archive contains more than passing glimpses at the lives of Black southerners in the late nineteenth and early twentieth century, and that is humbling and gratifying. The limitations of the hurricane's archive may be frustrating, but its possibilities are immense. They compel us to consider what we owe to the dead, to those who survived, and to their descendants today.

Acknowledgments

Everything about writing history—for me, on a place that I care deeply about and know well, and about a history that I feel a sense of responsibility to—is ultimately communal. There hasn't been a stage of this project that I completed without invaluable help from many people. I have to start with Paul Sutter. His incisive editorial and historical eye, his willingness to work with my specific vision for what kind of historian I wanted to be, and his ability to push me when I needed to expand what that vision was have made me a better scholar. I also must credit Peter H. Wood for helping me find this project in the first place. I was a South Carolinian excited to meet a scholar who cares about South Carolina all the way out in Colorado, and our conversations about Lowcountry history brought me to the Great Sea Island Storm of 1893. At the same time, his steadfast dedication to mentorship and his boundless generosity in helping build a new generation of historians have shaped it along the way. Taking to heart one of Peter's favorite sayings, I have tried to fight like hell for the dead in this book.

Thanks to other formative scholars in the University of Colorado Boulder History Department for their guidance, the books that they assigned, and the ethos that they encouraged: most of all Thomas Andrews, Lee Chambers, Samanthis Smalls, and Natalie Mendoza. Thanks also to other young academics I met at CU Boulder for sharing their work, advice, and kindness with me many times over the years: Julia Frankenbach (who provided valuable comments just before I began revising the manuscript), Alessandra Link, and Sara Porterfield.

I also owe so much to my writing group: Mandy Cooper, Robert Greene, Lindsay Stallones Marshall, and Steve Hausmann. We convened in 2020 amid the COVID-19 pandemic, a truly terrible year for, among so much else, building in-person scholarly community. But through online conversations and monthly writing workshops, I have grown as a historian and a writer, and their feedback and friendship have helped nourish me. I will always be grateful to Ashleigh Lawrence-Sanders, David Dangerfield, and Christopher Barr for being my Lowcountry sounding board. And to Melissa DeVelvis, thank you for being ready to discuss whatever issue—job market, archival find, or the politics of history in South Carolina—was at hand for either of us.

The small but mighty History Department at Queens University of Charlotte has been a port in a storm, at a time when the job market is worse than ever. Thanks to Sarah Griffith, Barry Robinson, and Bob Whalen for their unflagging advocacy for me; it is overwhelming to move from graduate student to (visiting assistant) professor absent the concerns of contingency and pandemic alike, and the three of them eased the transition at every turn.

Many conferences, workshops, and talks have prompted revisions and new directions in this project that have made it better. The Workshop on the History of the Environment, Agriculture, Technology, and Science; the Southern Forum on Agricultural, Rural, and Environmental History; the Boston Environmental History Seminar at the Massachusetts Historical Society; the American Society of Environmental History; the Coastal Discovery Museum on Hilton Head; the "Ranger Talks" at the Reconstruction Era National Historical Park; and the Southern Historical Association have all been wonderful venues for me to present and improve my work. I must also thank the *Journal of Southern History*'s Randal Hall and Bethany Johnson, peerless editors who helped prod my nascent work into form as a proper journal article. Thanks also to historians who have provided advice for this project over the years: Albert Way, Judith Carney, Christopher Pastore, Allison Dorsey, Tore Olsson, Evan Bennett, Robert Rouphail, William Horne, Scott Gabriel Knowles, Mart Stewart, and Keri Leigh Merritt, among others.

I've frequently returned to some core archives over the course of this process, and I have always appreciated their librarians' patience and helpfulness. The South Caroliniana Library in my hometown of Columbia, South Carolina, is one of my real archival homes, and I am grateful for its rich holdings and helpful staff, including Edward Blessing. The South Carolina Historical Society in Charleston, especially with the assistance of Virginia Ellison, and the South Carolina Department of Archives and History have also been wonderful places to return to frequently. The Beaufort District Library, with assistance from its archivist, Grace Cordial, was where my project began in the summer of 2016. The University of North Carolina at Chapel Hill's Special Collections at the Louis Round Wilson Library (thank you Julie and Nancy for hosting me on that trip!) and the Library of Congress, most of all its Clara Barton Papers, are two other institutions whose holdings made this work possible. The Lowcountry Digital Library, especially its photograph collections, have been so useful in thinking through parts of the project, with appreciation to Leah Worthington; and thanks to Dale Rosengarten for her incredible work on the Jewish Heritage Collection, also digitized there. Florence Nthiira Mugambi at the Melville J. Herskovits Library of African Studies at Northwestern University helped me arrange a "visit," albeit through a graduate student researcher, Nathan Ellstrand; I appreciate their work and patience in navigating the strictures of pandemic-inflected archival research.

I am also grateful for the institutional support that I have received in researching, writing, and presenting my work from the Rachel Hines Prize at the Carolina Lowcountry and Atlantic World Program at the College of Charleston, the Samuel P. Hayes Research Fellowship from the American Society for Environmental History, the University of Colorado Boulder History Department and Graduate School, the American Council of Learned Societies for the Mellon/ACLS Dissertation Completion Fellowship, and the Queens University of Charlotte College of Arts and Sciences.

I could not imagine a more stalwart proponent for this project than University of North Carolina Press editor Brandon Proia; his remarkable efficiency, along with his ability to grasp what matters most in a project and his skill in bringing that out, made bringing this book to production a smoother process than I ever imagined it could be. I also extend many thanks to Beatrice Burton for her fantastic work on the index.

I'd like to thank my parents and my little sister, Emily, for all their support and love over the years. In a very literal sense, this book would not exist without them because they've housed me, fed me, and put up with my post-archival rants whenever I visit for research trips. But much more importantly, they have always set examples to help guide my approach to my scholarship, my teaching, and my life outside of my work. My sister, Emily, is an incredibly talented and creative artist, dedicated to her craft with an unmatched work ethic. My mom, Dr. Rhonda Grego, is just about the smartest person I know, and I aspire to her perseverance and astuteness. My dad, Dr. John Grego, has a resolute determination and loyalty to his principles and to his work, both in the academy and his environmental advocacy. To my closest friends, whose support and care have meant so much: Alison Salisbury, Claire Sumaydeng-Bryan, Caitlin Bockman, Alana Boileau, Christina Turner, Jessica Rose, and Aly Olivierre (I am determined to have one of your maps in my next book!). And, of course, to Graeme Pente, my husband—the first person to hear my ideas and rants and the first, most patient, most dedicated editor of all my work. We met while we were both PhD students at Boulder, and he is my partner in all things.

Hurricane Jim Crow

Introduction

On the night of the storm, Quash Stevens looked out of the windows of his clapboard house, perched on a thumb of land jutting out from Kiawah Island into the Kiawah River.[1] "We Had Agraite stoim Heair," he wrote the next day, "the Tide Caime owpe in my House."[2] The ocean had washed two to eight feet deep across the long, narrow island, and Quash could see nothing but water. Quash would have despaired of the meager sea island cotton crop and for the cattle and marsh tackies (small, sure-footed horses of the old Spanish stock) that roamed Kiawah.[3] And as the manager of the island for twenty-seven years, he would have known that this storm would spell ruin for the provisions that the dozen Black families living on Kiawah had put by and the scarce fresh water on the island, where drought routinely left cattle, crops, and humans alike parched.[4] A hurricane of this magnitude did not bode well for the future of Kiawah and its inhabitants, and this realization would have distressed Quash. His life was tied to Kiawah as ancestral family land. His father, Elias Vanderhorst, was of a wealthy Lowcountry family. After the Civil War, Elias sent Quash to Kiawah to make something of the island, some portion of which had been in the Vanderhorst family since 1772.[5]

Quash, however, was not Elias's recognized heir, and he was not white. Quash was the son of Elias and an enslaved woman whose name the Vanderhorsts did not record.[6] Quash dedicated his adult life to Kiawah, but his white relations only permitted him to lease the island and subjected him to decades of alternating neglect and micromanagement. Quash's life, which straddled slavery to the early decades of Jim Crow, throws light on the fraught intimacies of Black and white relations in the Lowcountry, the intrinsic difficulties and joys of living and working on a sea island with its own volatile environment, and the meaning of a hurricane's intrusion into Black communities on the sea-girt edge of the American South.

This book began with a list of many people who died in the hurricane. It continues and will conclude with an account of one who survived—a man whose life was at once meaningfully unique and profoundly emblematic. The transformations that Quash lived through, that he bore witness to, that he wrote about, and that he died during defined the early decades of Jim Crow in the Lowcountry and thus this book. The hurricane's role in Lowcountry history can be understood through Quash's story, which reveals the slow, accumulating pressure that the storm wound around his life on Kiawah Island.

While we leave off with Quash Stevens for the time being, we now pick up the broader story of the hurricane, Jim Crow, and the Lowcountry. The history of the Lowcountry is enveloped by disaster from the late days of the Reconstruction era to the Great Depression.[7] W. E. B. Du Bois wrote about slavery as the driving disaster of U.S. history, a man-made system of labor exploitation, white supremacy, and capital accumulation that, pushed to the extremities of cruelty and depredation, collapsed in on itself when the enslaved reached their own collective breaking point.[8] In Du Bois's telling, the calamity was nothing natural—but it was material. This book traces a second disaster birthed from the first, albeit one contingent on a particular set of not only material but also environmental circumstances: the rise of Jim Crow in the Lowcountry. The Great Sea Island Storm brought forth a new iteration of an old regime.

This book takes the hurricane as a point of departure from the late years of the Lowcountry's long Reconstruction to understand what the early decades of Jim Crow in the region, from the early 1890s to the late 1920s, looked like, how it came to be, and what forces shaped it.[9] Even as the hurricane fomented crisis, it introduced a set of alterations in what was possible in the Lowcountry.[10] The hurricane became a force of contingency, in which the subsequent constraints, opportunities, and scars shaped the paths of the people, communities, and politics caught up in the storm's aftermath.[11] The stories that occupy the pages of this book are deeply human, and to live in the Lowcountry meant to struggle against, to work with, or to be subjected to the environment—dependent on who you were.

Observers have, for centuries, sensed that Lowcountry nature *mattered*, and it has preoccupied settlers, tourists, and writers alike. Descriptions of the Lowcountry from at least the last two hundred years frequently followed a poetic script. Let us begin with the weather, as it was—the heat, the humidity, an insalubrious climate in the summer and early fall that gives way to the blue beauty of winter skies and the gentle, sweet air of spring.[12] The flora—yellow jasmine climbing up the straight trunks of tall pines, Spanish moss falling in wispy gray curtains from live oaks brushed with resurrection fern. The marshes—velvety cordgrass that shifts from green in summer to gold in winter, the slicks of pluff mud, the puck-puck of oysters spitting at low tide. The creeks and rivers—wide, looping through marshes from the coastal plain to the ocean. The beaches—mild surf, white-yellow-gray sand, ghost crabs scurrying along the shore at dawn and dusk. The images are lovely and sedating.

Such a portrait is not technically inaccurate. But the Lowcountry's physical beauty, described so often and in a familiar set of tropes, feels like a butterfly trapped in a glass. Writings that fixate on Lowcountry nature can seem to be avoiding or even obfuscating the labor and the history that wrought this place, because they focus on elements of nature at the expense of environment.[13] A full understanding of the region requires recognizing that labor, race, class, and gender produced a distinctive Lowcountry environment and have long mediated experiences of the Lowcountry's nature and that settler colonialism, slavery, and white supremacy have molded the landscape that is now so routinely consumed by predominantly white tourists.[14]

The physical legacies of that history, inscribed on the landscape, are not difficult to find.[15] Shell rings, of mounded and compacted oyster shells bleached white with time, mark Indigenous villages or sites of ceremonial feasts. Rice fields, which enslaved Africans carved out of floodplain forests, have grown into sprawling, grassy riverside wetlands. Wide sinkholes pit the wet edges of marshy islands, where workers strip-mined phosphatic rocks in the late nineteenth century. The scruffy pine plantations that line country roads suggest old fields that once sprouted cotton, corn, or sweet potatoes, which enslaved and, later, free African Americans grew. These tangible vestiges of history are not limited to rural landscapes. The scars of history are embedded in the urban architecture of Charleston itself, including monuments glorifying enslavers—fewer now than before, at least. So are the more contemporary wounds of segregation, gentrification, and policies crafted to dispossess and exclude Black Charlestonians, emblemized by the disrespectfully named Septima P. Clark Expressway, a roadway that sliced apart historic Black neighborhoods and that city and state officials named after one of South Carolina's greatest civil rights leaders. The marks of this history on the land are everywhere in the Lowcountry. In examining the world that the hurricane collided with, this book seeks to reveal the fullness of the Lowcountry environment, from urban to rural communities, from marsh to ocean, and to excavate the painful and tender histories embedded in its soils.

The Lowcountry is dynamic and contradictory, fragile and mutable by nature. In the Lowcountry—which has long troubled its inhabitants with its ferocious heat and humidity, its water- and mosquito-borne diseases, its thick forests and miry marshes, and its web of waterways—it is impossible to ignore the long history of humans confronting and frequently feeling defeated by the region's difficult meteorological and physical features.[16] Hurricanes are an indelible part of the Lowcountry, which has so often confounded those who seek to harness and tame it into a fecund, productive, and profitable landscape. Whites exploited enslaved labor to achieve those ends. They accumulated massive fortunes off the backs of enslaved workers, and their wealth was only possible through constant vigilance and thousands of Black lives cut short. Once emancipation relieved African Americans of the burden of continual management of the rice fields, the system rapidly eroded under the Lowcountry's environmental conditions—including the Great Sea Island Storm.

The Lowcountry's apparent distinctiveness is the result of historical components that have origins in far-flung places, which European colonizers and enslavers, Indigenous peoples, and enslaved Africans brought to bear on the islands clinging to the coast of South Carolina.[17] African American–led efforts have pushed the Lowcountry's predominant historical narrative from white supremacist mythologies vaunting the region as a golden kingdom of rice to a complex understanding of the Lowcountry as Indigenous territory, a chain of sprawling slave labor camps, contested terrain between enslavers and the enslaved, and the homeland of the Gullah-Geechee people.[18] The region was the site of the Stono Rebellion, the abode of the wealthiest white men in the thirteen English colonies, the tinderbox of the Civil War, and the wellspring of a new birth of freedom among African Americans during the Civil War–era Port Royal Experiment.[19] This history of imperialism, enslavement, and hard-won liberation is, of course, characteristic of the history of the United States, not divergent from it.

Yet within the Lowcountry, the environment frequently played an important role in shaping a localized, idiosyncratic trajectory of those processes.[20] As the historian Mart Stewart wrote, "the natural environment was part of this convergence" of people and place, in which the "success or failure" of Lowcountry inhabitants' efforts to make a living or impose their will "was the consequence not only of their political and social relationships with one another but also of their relationship with nature."[21] The interplay of humans and nature in the Lowcountry coalesces in the environment—a word that brings together the hurricane as a weather event and as a disaster driven by social, political, and economic circumstances. The hurricane was thus an integral part of the region's history, not a singular event acting outside of that history. This book uses the hurricane to explain how the Lowcountry, whose course from Indigenous homeland to labor camp to African American homestead historians have charted well, shifted once again from a stronghold of Black yeomanry in the late nineteenth century to an impoverished and depopulated region that was a business opportunity for corporate developers by the mid-twentieth.

But this is not simply a declensionist narrative. Reconstruction lingered in the Lowcountry in meaningful ways into the 1890s because of the region's Black majority and the preponderance of African American landownership, and it took an extraordinary set of environmental and political circumstances to tear the era's gains away from the determined hands of African Americans.[22] The muddied, flooded land that the Great Sea Island Storm left behind became the ground on which coastal African Americans, wealthy white southerners, well-meaning reformers, and others met. There, they tested the limits of each other's power. African Americans sought a recovery that would bolster their defenses against the white supremacist project engulfing the rest of the state. White interlocutors from the North engaged in paternalistic humanitarian endeavors that were too temporary to affix a permanent lifeline for African Americans, especially those living on the sea islands, struggling to maintain their homes and their rights. The white elite hungered to amplify the hurricane's damage upon this troublesome population of African American landowners and workers.

The hurricane was, to be sure, a destructive meteorological event of exceptional force even among the storms that routinely thrashed the Lowcountry. However, what transpired after the winds died down and the waters receded was not predetermined. Even as the region's history of slavery and deep-rooted racism created a likely set of outcomes, Black residents of the Lowcountry never accepted those possibilities as inevitable. The hurricane was a tragedy of horrific proportions: but the long effects of the storm did not have to be disastrous, and African Americans worked hard to forestall that disaster.

Sometimes, though, the ire of white South Carolinians toward their fellow Black citizens was self-defeating, and while whites may have wanted African Americans to suffer, they did not always realize in time that any calamity that struck their workforce could also hurt their income. The hurricane thoroughly wrecked white South Carolinians' visions of economic progress of a "New South" for the Lowcountry. Once they understood the extent of the destruction, they then took steps to ensure that the Lowcountry, if it could not be prosperous for whites, would not be productive for African Americans either. As historians have long pointed out, the New South was less a reality than a white supremacist fantasy of a smoothly operational, lucrative, and semi-industrialized Jim Crow, in which a compliant and exploited Black workforce labored to deliver capital to white shareholders.[23]

And indeed, the hopes that white boosters in the Lowcountry had for economic advancement in a New South were rarely more than an illusion.[24] They had difficulty attracting investors to a region filled with resistant whites eager to cling to the agrarian activities that had brought their families wealth under slavery and African Americans who justifiably were leery of white bosses, preferring to set their own terms.[25] Though southern agricultural production overall reached prewar levels by 1880, "a series of financial and ecological calamities threatened the pillars of the embryonic New South," the historian Erin Stewart Mauldin explains.[26] This book explores the tensions between the New South—what many white southerners dreamt of—and Jim Crow, what white southerners manufactured instead. During this era, the inextricable combination of racial oppression and economic exploitation of African Americans, such an effective moneymaker for whites in antebellum times, only sometimes made white southerners as wealthy as they had expected or rendered the South as innovative and productive as they claimed it would. They settled for the exercise of power, whenever they could. To fully comprehend the instability of the chimerical New South, the environment must be accounted for. In the Lowcountry, the hurricane revealed that instability more than any other singular event because of how sprawling its effects were. The New South required an amenable environment. The Lowcountry never was.

The history that spiraled out from the hurricane has been memorialized on the land in literal and visible ways as well. A brass plaque mounted on a granite pedestal sits on Beaufort, South Carolina's, manicured waterfront promenade, which overlooks the Beaufort River. Alongside other plaques overviewing the history of the small riverside town, this plaque dwells on the Great Sea Island Storm and its long-lived impact on the town of Beaufort. Another historical marker stands on the grounds of the Penn Center, the historic African American school-turned-cultural-institution on St. Helena Island, which describes the toll the storm took on the island and the American Red Cross–led recovery effort after.[27]

Residents of the Lowcountry have made meaning of the hurricane and generated public commemorations of how it coalesced with very human forces. This book seeks to follow their example, exploring that nuanced, painful, and contradictory history, which draws together overlapping cycles of environmental change, political repression, and communal traditions of resistance, survival, and care that have circulated through the region for centuries. The book begins with a portrait of the Lowcountry as it was before August 27, 1893, and then charts the hurricane's path through the region. Next is the contentious recovery efforts in the days, weeks, and months after the storm hit—but it does not end there. Instead, the book explores the long effects of the hurricane from the mid-1890s to the late 1920s to illustrate how the disaster was made, slowly and sometimes with unintended consequences, through white supremacist policies, economic collapse, and a thousand individual choices shaped by structural inequities. Finally, at the end, the book returns to Quash Stevens and the life he led after the Great Sea Island Storm scoured Kiawah Island.

This is neither declensionist tragedy in which coastal African Americans never stood a chance against Jim Crow nor a naïve tale in which South Carolinians banded together to rebuild despite the hurricane's ravages, because neither is accurate. Instead, it takes shape as a telescoping series of narratives in which no one's actions, no matter how righteous or unjust, were ever fully triumphant or utterly futile, and in which the hurricane's effects sometimes faded and sometimes sharpened over time. This book delves into the lives of the people who lived through and with the Great Sea Island Storm of 1893 to contend with human folly, tenderness, cruelty, indifference, and dogged determination *and* the historical background, environmental context, and structural and racial injustices that shaped their experiences and their communities. This approach, to a disaster that cut short the lives of thousands of African Americans and so altered the futures of tens of thousands more, is meant to express a humanist care and dedication for those thousands that will ensure a historicized remembrance of the disaster's human toll. This focus on the lives of the South Carolinians who worked, lived, fought, and struggled on the edge of the Atlantic during the rise of Jim Crow will place the hurricane into Lowcountry history.

PART I

Hurricane

Chapter 1

The Lowcountry

Reverend Charles Cotesworth Pinckney, with centuries of wealth and privilege lending weight to his words, lamented in an 1889 lecture before the South Carolina Historical Society that "whenever [Negroes] dominate by numbers and political preponderance most of them have retrograded towards the worst phases of African life."[1] Emancipation, he argued, had proven a futile experiment. According to this descendant of one of the state's most powerful white families, the liberation of four million African Americans had failed to bear fruit. Now, a generation later, they were instead "on the verge of returning to a state of barbarism." Pinckney, a long-term rector at Charleston's Grace Episcopal Church, who had been born in Beaufort, construed liberty as a gift willingly given rather than a right that enslaved workers had seized from men like him during a bloody war.[2] In Pinckney's view, the time had come for white southerners to reclaim their birthright of white supremacy and to rescind the privileges of Black freedom.

The Lowcountry's newest traditions—of Black landownership and political participation—signaled to Pinckney and other white elites what they saw as the disaster of emancipation. By the late 1880s, African Americans in many parts of South Carolina had been swallowed into sharecropping schemes, bullied and beaten away from polls, and thrown out of the halls of power: not wholly so in the Lowcountry. African Americans there had levied their superiority of numbers and federal support to build communities along the coast that shunned white control as much as possible. This rejection stung Pinckney, a constant reminder of the Confederacy's loss and the shift in power that the Civil War enabled in the Lowcountry. Doubtless emboldened by the resurgence of white-led state governments across the South, Pinckney laid claim to the region, calling his lecture "Our Blighted Sea Islands." He appealed to his peers to reestablish enslavers and their descendants as the true owners of the sea islands, to cast out the "blight" of Black sea islanders.

The blight that Pinckney feared was in fact a remaking of the Lowcountry environment that rejected white elites' notions of productivity, racial hierarchy, and labor discipline. To maintain the Lowcountry's agricultural productivity required specialized knowledge deeply embedded in place. And that knowledge lay with African Americans, who used it to determine their own meaning of *productivity*—to redefine it in a way that benefitted them and their kin rather than the white landowners who had once enslaved them. Labor in the Lowcountry relied on a granular understanding of the environment, which varied from island to island, marsh to marsh, river to river. African Americans, as they sought to establish autonomy and safeguard their communities, turned time and time again toward their environmental sensibilities to guide how they sustained their livelihoods, their homes, and their future. They derived nourishment, spiritual healing, and cultural traditions from the land. They also drew upon their labor power to shape the earth. They habitually levied their knowledge of the Lowcountry's environments to negotiate terms with white landowners who desperately relied on Black expertise for their own enrichment.

Coastal African Americans cultivated their centuries-deep understanding of the Lowcountry in part out of necessity. The Lowcountry environment was not always beneficent. Indeed, enslavers had exploited some of its most dangerous elements to subdue enslaved workers. They combined their regimens of overwork with the Lowcountry's heat, humidity, and brackish water to inflict high death rates on children, to cripple adults' backs and legs through overwork in difficult conditions, and to force women back into the dirt of cotton fields too soon after giving birth. Through these methods, they tried to transform the coast into a regimented chain of labor camps. Even this was an imperfect and incomplete project. Along the margins of rice and cotton fields, in thick woods, sandy beaches, and salt marshes, many of the enslaved cultivated their own relationships with the plants, the animals, and the earth to which their captors were not privy.[3] The forces of nature, too, refused easy harnessing to the project of enslavement and environmental subjugation. The ocean and the atmosphere continued to shift waterways and sculpt the supple slips of land that composed the coast, insensible to the vision of enslavers.

But in the late nineteenth century, with the chains of slavery broken, African Americans undertook a remaking of the Lowcountry, a reshaping of the environment, the labor regimes, and the political relations that had kept them in bondage. They developed new relations between labor and environment that threatened what remained of the old economic order that enriched the white elite. A single generation out of slavery, African Americans along the coast had developed agile, flexible ways of making a living that eluded white control to the greatest degree feasible. Pinckney and other white South Carolinians could not bear this reclamation. Black autonomy anywhere was a threat to white supremacy everywhere. But for the time being, their rumblings stayed in the realm of the rhetorical. What African Americans had built in the Lowcountry was too resilient and too embedded for the white elite to destroy alone.

The origins of Black landed autonomy in the post-Reconstruction era had its roots in two acts of coercion, compelling enslavers to relinquish what they should never have claimed as their own. One was the Emancipation Proclamation, which African Americans on the sea islands celebrated at midnight on January 1, 1863, at the very moment that the proclamation went into effect.[4] The other was particular to the sea islands in Beaufort County and less ostentatiously thrilling: the U.S. Direct Tax Act of 1862, which ultimately allowed for the redistribution and resale of abandoned labor camps to newly emancipated African Americans.

In Beaufort County, Lincoln's Emancipation Proclamation only confirmed what Black sea islanders had already decided for themselves fourteen months earlier. In November 1861, a grand fleet of the U.S. Navy sailed down the eastern seaboard and into Port Royal Sound.[5] The U.S. government needed to secure a port on the southern Atlantic coast to successfully blockade the Confederacy, and the deep waters of the sound and its quiet location between Charleston and Savannah made it an ideal target.[6] As the artillery boomed across the blue waters of the sound, panicked enslavers fled Beaufort County.[7] They left behind 10,000 African Americans. Over the course of the war, 20,000 more from across the region would flee from their labor camps across U.S. lines into Beaufort County.[8] The sea islands became a haven for African Americans claiming their liberation.[9] The federal government relegated them all to the status of "contraband," a nebulous condition neither free nor enslaved. But African Americans understood their position differently.[10] They "stood at freedom's door," without their enslavers present to shackle them any longer.[11]

On the sea islands, each with distinct communities and traditions but also altered by the inflow of freedom-seeking migrants, the federal government attempted to instill a sense of order and discipline, according to their definitions. Starting in 1862, northern educators, policy makers, and missionaries moved to Beaufort and the nearby islands to establish schools, foremost among them the Penn School on St. Helena, and churches for African Americans. Undergirding their efforts was a racist paternalism that deemed Black sea islanders as yet ignorant and unfit for the rigors of citizenship and in need of the strong hand of federal guidance to place them on a path to virtuous, industrious behavior befitting of free workers. The U.S. government preferred to ensure stability through what they saw as a system of free labor to "preserve . . . large-scale productive units" that kept Black sea islanders concentrated in small communities based on farms.[12]

While eager for the education that white northerners offered to them, Black sea islanders resisted the efforts of federal agents to coerce them into wage labor in the cotton fields of the county. Black sea islanders did not view

The Penn School. This detail from a stereograph of the Penn School on St. Helena Island shows the original prefabricated schoolhouse. The Penn School's first classes convened in a room at the Oaks, a former labor camp, but it quickly outgrew those cramped quarters and was moved to its current location near Frogmore, St. Helena Island. The Penn School ceased to be a primary and secondary school in 1948 and adopted the name it is known by today, Penn Center. It was and continues to be a cultural and educational community hub. Hubbard & Mix (photographers), *Miss Laura Town's i.e. Towne's School, St. Helena Island, South Carolina* (taken between 1863 and June 1866), Robin G. Stanford Collection, Library of Congress Prints and Photographs Division (LCCN 2015646740).

wage labor as a true form of freedom, for it relegated them to picking cotton for white bosses, without providing any sense of self-determination over land, time, or labor.[13] The federal government balked at Black sea islanders' outright rebellions against white authorities when they imposed untenable restraints on Black labor practices.[14] But Black sea islanders used the advantage of sheer numbers and the necessity of their labor to the Lowcountry economy to push back against the efforts of the federal government and, later, white South Carolinians to subdue their emancipatory spirit. They chafed under top-down policies that sought to control their labor, preferring instead to rely upon their "extensive experience managing their own affairs."[15]

Black sea islanders' purchase of land from the U.S. government in some cases helped their resistance succeed. This outcome was unique within the former Confederacy, where less than 5 percent of all formerly enslaved people became landowners in the several years immediately following the Civil War.[16] Even in the Lowcountry, the federal government refused consistently to back Black rights to landownership. Under the authority of the 1862 tax act, the government had seized 101,930 acres of land in the Lowcountry for nonpayment of taxes.[17] General William T. Sherman's Field Order No. 15, issued in 1865, set aside that confiscated acreage on the sea islands as a reserve for newly freed African Americans. But during the winter of 1866 to 1867, the U.S. government ordered the military to evict thousands of Black sea islanders from farmlands that the order had promised to them.[18]

The federally facilitated land redistribution effort that Black southerners had anticipated was thus systematically constrained by federal authorities themselves. Even in the Lowcountry, which should have been the site of the effort's greatest success, the land reforms were a shadow of what they could have been. Some Black southerners, though, were able to purchase land, albeit under these limited circumstances. Under the 1866 Freedman's Bureau Bill, African Americans settling in St. Luke and St. Helena Parishes of Beaufort County had permission to buy confiscated land at $1.25 per acre. Nearly 2,000 families—some from the area and some who moved in to make a new life for themselves—purchased roughly 20,000 acres, in plots up to twenty acres.[19] With official titles in hand, African Americans could prevent former enslavers from later reclaiming their land. While land purchase was most common on the large sea islands within the bounds of Beaufort County like Hilton Head, St. Helena, Port Royal, Daufuskie, and Lady's Islands, African Americans on the Georgia sea islands of Sapelo and St. Simons established self-sufficient farming communities as well. On their newly acquired acreage, African Americans built their own homes, farms, schools, and local governments. Sometimes they drew upon white support in these endeavors, and sometimes they refused it.

The effects of the Port Royal Experiment lingered for decades after the Civil War. Its legacy outlasted the ostensible end of Reconstruction in South Carolina in 1876 when former Confederate general Wade Hampton III overthrew the interracial Republican government with the use of voter suppression, fraud, and a vigilante militia and replaced it with a Democratic, white supremacist regime. In Beaufort, many northern reformers and educators simply stayed put after the war, imbuing the local political scene with something approaching white solidarity with Black Republican politics. Beaufort, a riverside town chartered in 1711 on Port Royal Island, became the heart of this experiment in multiracial political alliance. Indeed, Beaufort was perhaps the most robustly Republican town in the South in the early 1890s. Its population of 3,500 included a number of Black professionals and elected officials. Beaufort town and county were awash with Black politicians: the local elected seats of sheriff, clerk of court, probate judge, school commissioner, coroner, county commissioners, one state senator, and three state representatives were all filled by African American men.[20] The port collector, a federally appointed position, was Robert Smalls, the famed African American Civil War veteran.

The contrast between Beaufort and its environs and the rest of the state was perhaps most visible in comparison to Charleston. By the early 1890s, the two towns, separated by fifty miles of coastline and a vast gulf of political differences and historical contingencies, represented two different outcomes of the Civil War. In Beaufort County, the United States could be said to have won. In Charleston, however, the white elite championed the Lost Cause and dominated the municipal government.[21] Confederate veterans and the descendants of enslavers or former enslavers themselves reigned over the city's African American population and enforced conservative politics over a population of 60,000. White Charlestonians, while begrudging of anything they viewed as federal intrusion, began to welcome economic efforts that filled their pockets and slowly and in limited fashion accepted the wage labor, industrialization, and economic modernization, all dependent on control over cheap African American labor, that other white Southerners called the New South. Black Charlestonians thus took refuge in their own institutions and made few inroads elsewhere in the city.

Black residents of Beaufort County hoped to stave off the danger that Charleston represented. For in Beaufort County, high rates of Black landownership, a nearly all-Black population, and a Republican-led government empowered African Americans to carve out a living for themselves and their families, making a malleable livelihood in which they participated in a variety of economic activities.[22] The rates of landownership among Black sea islanders remained exceptionally high for decades and facilitated a positive cycle of autonomous subsistence and community support. With plots of land to fall back on for subsistence crops and a few acres of a cash crop like cotton, Black residents of the area were able to mostly escape entrapment in a single wage-labor job or tenancy arrangements of the kind that white Southerners used elsewhere to keep African Americans under control. What white South Carolinians heralded as a New South economy based on wage labor, Black sea islanders regarded as too reminiscent of slavery. Beaufort County, while the stronghold of Black landownership and political participation in the early

1890s, was not entirely alone. In pockets of the coast, like Colleton County, and in autonomous communities outside Charleston, Black residents replicated similar patterns.

Black residents of the coast who did not own land were likely to be cash renters, often under the "two days' system," a form of renting and land tenure particular to the Lowcountry.[23] Under the two days' system, African Americans gave two days of their work week to the white landlords whose land they lived on. Seen most frequently on Edisto and Johns Islands, large sea islands nearer to Charleston than Beaufort, or on the rice lands farther inland, the two days' system granted African Americans five to seven acres of land, rent-free, and occasionally a house and fuel. Under this system, African Americans did not owe rent to the landlord nor a portion of their own crop. The two days' system was decidedly unfavorable for white landowners accustomed to full-scale exploitation of Black labor, because two days each week was hardly sufficient to make them rich. And for African Americans unable to purchase their own land, it was the next best option. Under share rentals, the most restrictive and least common form of tenancy in the Lowcountry, landowners permitted workers to live on and farm their property in exchange for a certain percentage of the crops harvested, an exploitative system that left tenants in debt and that dominated much of the rest of the South.

The Lowcountry's significant Black majority also helped bolster Black self-determination. Without a significant contingency of whites, it was difficult for the white elite to launch a successful fight against African American landed rights. Beaufort County, of course, was the most trying environment for white wheelers and dealers seeking total political control over the state. There, in 1890, 2,563 whites lived in Beaufort County, as compared to 31,883 African Americans.[24] The white population on the sea islands and also around the rice lands was small in general, with a few white families living on their properties. This was no accident of history. Lowcountry enslavers had long avoided living on the rural land that they owned for much of the year, for the heat and malaria conspired to forge an unhealthy environment that they thought was best left to enslaved workers. They therefore fled the dangerous clime of the rural Lowcountry in the summer and early fall for townhouses in Savannah, Charleston, Atlanta, Columbia, and sometimes farther to large cities in the Northeast, budget permitting. Their descendants carried on similar traditions when they could afford it.[25] After the war, especially in the region around Beaufort where the most comprehensive programs of land redistribution and resale to African Americans occurred, white landowners had a few avenues forward. If former enslavers had the money to pay their taxes, they reclaimed their land and "made a living" as landlords to Black tenants. Many maintained a seasonal residence elsewhere.

Land tenure rates by South Carolina coastal counties, 1890.

County	*Number of farms*	*Owner cultivated, %*	*Cash rental, %*	*Share rental, %*
Barnwell	4991	34.9	35.8	29.3
Beaufort	3762	72	27.3	0.6
Berkeley	5999	53.4	43.9	2.6
Charleston	632	23.7	75.9	0.3
Colleton	—	71.6	22.1	6.4
Georgetown	1208	52.3	24.1	23.4
Hampton	2542	44.7	50.2	5.1
Horry	2097	86.8	6.2	6.9
Williamsburg	3267	58.1	37.3	4.5
State wide	115,008	44.7	27.8	27.5

Source: Data from Strickland, "Traditional Culture."

Black and white population of South Carolina coastal counties, 1890.

County	*Black*	*White*	*% Black*	*% White*
Barnwell	30,602	14,010	69%	31%
Beaufort	31,553	2,563	92.50%	7.50%
Berkeley	47,766	7,661	86%	14%
Charleston	35,200	24,637	59%	41%
Colleton	26,410	13,870	66%	34%
Georgetown	16,837	4,020	81%	19%
Hampton	13,737	6,807	67%	33%
Horry	5,617	13,639	29%	71%
Williamsburg	18,525	9,250	66%	34%
Coastal county average	226,247	96,457	70%	30%
State wide	692,503	458,454	60%	40%

Source: Data from published table stapled in notebook of Ben Tillman, 1894, envelope 8, in Diaries and Notebooks, 1861–1905, series 8, Tillman Papers, Clemson University.

African American farmers, whether independent or renters, found greatest financial success and stability on Hilton Head Island, St. Helena Island, Port Royal Island, Lady's Island, and smaller nearby islands like Coosaw Island, where they planted sea island cotton and subsistence crops.[26] Cotton in particular lent itself well to small-scale farms, whereas areas of rice production, like on the Combahee River upstream of Beaufort, were less conducive to Black yeomanry because of the intensive labor and high capital costs necessary for rice cultivation. In large swaths of the Lowcountry, protected by their numbers, many African Americans lived in cottages on their own land, in repurposed cabins from the old rows of former labor camps, or along the edges of white-owned properties as renters or sharecroppers. The homes of African Americans who owned their land were typically sturdier than the houses of Black sharecroppers or renters, because they had more resources to fix up their houses as they wanted. Anne Simons Dea, a young white woman whose family owned hundreds of acres of rice lands near Georgetown, failed to recognize that the comparative poverty of Black renters on her own family's property might have had something to do with what they paid and provided to them. She tutted at the "miserable log-huts, with clay chimneys stopping short of the pitch of the roof; broken fences; dirty yards, guiltless of the least attempt at fruit or flowers; and at the doors, clusters of ragged children."[27]

Whites, whether South Carolinians or northern visitors, often dwelled on scenes of poverty among Black southerners—the former as supposed evidence of the depravity and indolence of African Americans and the latter as a condescending way of condemning systems of southern rural labor or painting a picture of an isolated, "primitive" region. Neither vision was true to life, and both said more about the people who imagined them than the material conditions of coastal African Americans. Lowcountry African Americans in the postwar period were, when possible, deliberately insulated and independent.[28] They had built lives for themselves and their families that suited their desires for autonomy and landownership, albeit sometimes within restricting or exploitative circumstances that were not of their making. They may have lived on sea islands or marshy necks of land that could require taxing journeys to reach, but that should not suggest a disconnect from the broader world or a parochial turn of mind. If anything, coastal African Americans, especially on islands like St. Helena, had formed communities that afforded them a greater degree of liberty than anywhere else in the southern United States.

To reach the sea islands, boats and the rare bridge crossed waterways, from narrow creeks slipping through the marshlands to wide, curving rivers with surprisingly dangerous undercurrents—a geography that made fast travel and easy transportation impossible. Once on the islands, cart tracks cut through the fine sandy earth for horse- and ox-drawn wagons to make their way through fields and woods of oak and pine. Shaded by fig, peach, plum, or ubiquitous pine trees, clean-swept yards of packed dirt—one described as being so clean that "it looked as if a leaf had never fallen on it"—led to homes with wood siding, often with a porch, a kitchen garden, a chicken coop, a barn for livestock and tools, and a cart and buggy behind the house.[29]

Inside, most residences had walls of pine, a cooking stove, wood furniture, and a kerosene lamp on a kitchen table.[30] A Bible was the most common book in any given home, though many African Americans complemented their Christian religious beliefs with hoodoo, a syncretic set of spiritual and

Gathering Sweet Potatoes and *A Sea Island Home*. Throughout the book are illustrations, like these two, by Daniel Smith. Smith accompanied Joel Chandler Harris as he toured the sea islands to investigate the conditions of the region a few months after the hurricane. Smith's illustrations vary in quality throughout the two articles that Harris published in *Scribner's* in 1894: some, which I have avoided, verge on racist stereotypes or seem to have an almost lurid gaze for African American suffering, and others are drawn with a more sensitive grace. In the first illustration, African American women dig up sweet potatoes, welcome provisions after a hard winter. In the second, Smith has painted a scene of a sea island homestead, with a clean-swept yard, an outdoor cooking area, and a tall spray of palmettos behind the cabin. *Left*: Daniel Smith, *Gathering Sweet Potatoes*, in Harris, "Sea Island Hurricane," 270; *right*: Daniel Smith, *An Island Home*, in Harris, 231.

medicinal practices of West African, Indigenous, and Christian origins that emerged in the Lowcountry.[31] The houses, usually of one to three rooms and an attic, were divided among the common living space and kitchen, a bedroom for the heads of the household, usually another for girls or elderly family, and the attic for growing boys. Mosquito nets swathed beds with patchwork quilts in the bedrooms to ward off the buzzing insects—though in 1893, their role in the transmission of malaria, endemic to the region, was still unknown.

Adjacent fields produced sweet potatoes, corn, cotton, and other crops, coaxed forth by Black farmers with their livestock and plows and the humid air wafting off the ocean. Many grew rice for household consumption, in poorly drained patches of land: nourishment that tied them to ancestral West African rice fields and that composed the staple starch of their diet and cuisine. Black sea islanders fed their cows, horses, and oxen with dried spartina harvested from the marsh flats and let them loose on fallow lands.[32] Black sea islanders, who counted among their numbers many talented boatmen, also turned to the water to enrich their diet.[33] They dug oysters, clams, and mussels from the pluff mud of the salt marshes, dipped crab pots into the water for blue crabs, cast nets for shrimp and baitfish, and fished from the shores and from small boats for whiting, red drum, spot, croaker, spotted sea trout, sheepshead, and flounder. If they owned a gun, they could hunt for rabbit, squirrel, deer, raccoons, possum, and ducks, though pigs, usually at various levels of domestication, were a primary source of animal-derived protein. The Lowcountry environment could be volatile and dangerous at times and could require immense amounts of labor to render it arable and productive at others, but it could also yield an abundance of game and fish. To reduce the lives of coastal African Americans to a condition of isolation would mean ignoring how they interacted with the environment and the meanings that they made from the Lowcountry. They worked, thought, and played *with* the land and the water.[34]

Over the course of the decades following emancipation, Black residents of the Lowcountry built an economy ill-suited to white southerners' capitalist-oriented assessments of economic vitality. They supplemented their yeoman farming practices with cash earned by renting out their labor to white landowners and industrialists.[35] In these ways, African Americans made a living in the decades after the Civil War in ways that dodged the vagaries of both the market and white employers. Black autonomy in the Lowcountry constituted a rejection of capital accumulation through the exploitation of Black labor and was the result of the sustained struggle to attain self-sufficiency in the decades immediately after the collapse of a national economy that was entirely reliant upon the shackling of Black labor.[36]

White commentators frequently derided the Lowcountry economy as stagnant, a poor shadow of its former wealth. But for coastal African Americans, those standards had little to do with what they valued, especially since such appraisals reduced them and their ancestors to simple economic cogs within a moneymaking machine that yielded them no coins. Operating under that understanding of the market, African Americans were wary of relying too much on a single industry for a wage, preferring instead to diversify, to draw upon their variety of skills to make a living for themselves, their families, and their loved ones. They developed their own economic ethos grounded in the Lowcountry, which centered on flexibility, autonomy, and subsistence.[37] A comfortable and self-sufficient life for the family unit was largely the goal.

But white landowners, many former enslavers or their descendants, hungered for wealth, which was represented by the people who had once been and produced their capital. Whites subsequently turned to what had once been the second greatest source of their wealth and what many still had in abundance: land. Yet that guaranteed no security. White landowners and coastal African Americans continually vied over a wide array of land tenure and work contracts.[38] As African Americans purchased plots of land and turned to maintaining their own crops, the scale of rice cultivation, once the backbone of the Lowcountry economy, broke under its own weight. White landowners could not maintain the sheer size of antebellum rice production because African Americans, freed from the obligation of slavery, did not want to work in the mud of the rice fields at the same intense pace.

As the agricultural policy maker and white landowner Harry Hammond described it, "the difficulty of obtaining labor" for white landowners dramatically reduced the price of rice lands. By 1880, properties that were at "$200 to $300/acre are now at $20 or $30, or less," with two million acres of land appropriate for rice "now lying unused . . . most of it in its original wilderness."[39] These abandoned lands, which lay along tidal rivers as they snaked from the sea islands deep into the coastal plain, were overgrown with grasses, transformed into freshwater marshlands. With emancipation's reconfiguration of the Lowcountry economy, another major change was effected in the environment of the Lowcountry as rice fields were abandoned to wildlife and vegetation.

Black tenant farmers after emancipation often refused to work in rice fields year-round.[40] Upkeep of the fields needed as much attention as the crop. Rice required fresh water pushed onto fields by the ocean's tidal pulse, which lowers and raises the water table twice daily: rice fields could not be too near the ocean for fear of saltwater incursion nor too far inland from the ocean's tidal force. Farmworkers therefore had to constantly monitor not only the crop but also the river, the tides, and the weather to ensure that the rice, positioned precariously in the intertidal zone, would not perish. While freedpeople did not have to clear new rice fields from old-growth floodplain forests as their enslaved ancestors had, the need to maintain the vast infrastructure of the rice fields and to retain an acute sensitivity to shifts in water salinization, the weather, and the tides carried over to the postwar period.

Indeed, rice fields were always at the mercy of changing environmental and meteorological conditions. Freshets, or floods from upstream caused by heavy precipitation, could waterlog fields beyond the point of proper drainage. Storm surges from the Atlantic could saturate fields with saltwater and kill the rice. Drought as well as tidal surge also threatened rice fields with salinization, a threat that increased the closer the fields were to the ocean.[41] Large-scale rice culture in South Carolina had always been a profoundly vulnerable enterprise because of the region's susceptibility to the elements, the rigors of its maintenance in the face of continual water erosion, and the back-breaking nature of the work, extracted from a workforce pressed into unwilling service. African Americans who continued to engage in rice cultivation after emancipation found the rigors of rice cultivation trying, especially as its once-forcibly maintained infrastructure crumbled through the erosion of weather and the lack of regular upkeep.[42]

While African Americans in the region obtained their own property in proportions unmatched across the South, none of them owned enough land to justify the commercial production of rice. Furthermore, the land reclaimed by the federal government for inexpensive resale to African Americans was usually on sea islands too close to the salty ocean, like St. Helena Island, for large-scale tidal rice cultivation. Instead, they raised a wide variety of crops. Black landowners left much of their land to lie fallow and planted the rest in subsistence crops and one-eighth- to several-acre plots in cotton, sometimes intermingled with corn and sweet potatoes.[43] Cotton was a handy cash crop for them to grow with minimal investment and equipment, as a few acres could yield a useful amount of cash at the end of the harvest. Sea island cotton was a regional favorite, with its silky, luxurious lint two to three times as long as that of short staple cotton; the plants thrived on the humid ocean breezes that wafted over the sea islands.[44] And because cotton did not require the specificity of timing that rice did, coastal African Americans could also leave their cotton crop alone while they worked in white landowners' rice fields for an occasional daily wage or as part of the two days' system.[45] Consequently, commercially grown rice continued in a limited fashion that suited the work patterns of African Americans balancing labor for cash and for subsistence.

The many African Americans who still labored in white-owned rice and cotton farms after emancipation were tethered by economic need to a cash wage, hemmed in by white landowners who yielded them little by way of money or land. They nonetheless found ways of negotiating more favorable working conditions *because* white landowners were strapped for labor. If a white landowner needed additional labor to tend or harvest their crops, they usually hired their Black tenants or nearby farmworkers for fifty cents a day, an average wage for fieldwork in the late nineteenth century. Harry Hammond groused that the two days' system was to the disadvantage of the white landowners, because much of their fertile land lay uncultivated due to the dearth of labor. Two days of labor each week was simply not enough to push a large property to its full capacity, and, furthermore, the system "burdened" Hammond's land "with the support of a much larger population than necessary to its cultivation."[46]

White landowners in the Lowcountry complained frequently about how African Americans defied white control and reduced this independence to racist stereotypes. Lewis Grimball, a white landlord on the Pon Pon River, bemoaned that "negroes are so careless" because they did not handle storm-damaged rice with sufficient delicacy.[47] Frances Butler Leigh, who operated her father Pierce Mease Butler's former labor camps on Butler's and St. Simons Islands in Georgia, wrote that the rice workers on Butler's Island were "so hopelessly lazy to be almost worthless as labourers."[48] Elizabeth Allston Pringle, the daughter of former South Carolina governor Robert F. W. Allston, who enslaved 630 African Americans in Georgetown County, operated the remnants of the family's rice labor camp in the postwar decades. She regularly fretted over the "idle, shambling, trifling element" that she saw in the African Americans who worked her land.[49] She also noted with frustration that the yield of rice varied immensely depending on whether the workers rented the rice fields or simply received a daily wage for working in rice. Each year, she hired Black laborers to plant between twenty and thirty acres of land in rice and rented out 100 to 150 acres. Black day laborers in the "wage fields" curiously harvested only seventeen bushels per acre, while Black renters coaxed anywhere between thirty to forty-five bushels from each acre that they cultivated for their own sale.[50]

White landowners on the coast tried to regain control over Black laborers in the postwar years, but they found that African Americans, who had sustained resistance to slavery over centuries, chafed under systems of white-owned agriculture if they had no stake in what price the harvest fetched at market. Enslaved rice workers had resisted their condition by appearing to accidentally break tools, by suffering from moderate illnesses with unusual frequency, or by other everyday acts of rebellion—and that, apparently, was not a legacy that their descendants had forgotten.[51] African Americans engaged in rice cultivation because their knowledge of it was valuable to white landowners, rice still afforded an opportunity to make a wage, and white landowners, who found it difficult to obtain labor, were more open to negotiation to terms favorable to African Americans.

To be sure, any system of bonded agricultural labor short of slavery was less profitable for white South Carolinians because they now had to pay their laborers. But for African Americans without the money to purchase their own land, the two days' system at least gave them time to tend to their own land and to pursue other opportunities for employment outside of the boundaries of the plantation. Anne Simons Dea, whose family hired Black rice workers near Georgetown through the late nineteenth century on the two days' system, referred to the system as "contract days." These had to be renegotiated at the start of each planting season, with Black rice workers sometimes waiting until the latest moment to do so. Dea grumbled one spring that "the hands have at last come up to contract for the year. Work should have begun long before February . . . [but] the hands will not contract as long as they have anything left from their last year's crop; when that is gone they will work."[52] Black rice hands on the Dea land arranged for specific "contract days" so that "they are at liberty to work elsewhere on the days that are not 'contract days.'"[53] This gave them the time to work on their own land or to hire their labor out for a cash wage elsewhere. In a white South hostile to labor organizing, the negotiation over contract days was one of the few moments of opportunity for Black workers to exert control over the terms of their labor.

Black farmworkers knew that their labor was instrumental in raising a crop and that white landowners had little choice but to capitulate to their terms. Frances Butler Leigh presented herself as a no-nonsense manager, adept at manipulating the Black workers on her family's property—but her accounts often show African Americans gaining the upper hand. A frustrated Leigh wrote that the farmworkers at Butler's Island, where they raised rice, "were quite convinced that if six days' work would raise a whole crop, three days' work would raise half a one, with which they as partners are satisfied, and so it seemed as if we should have to be too."[54] She chronicled the extensive, multiyear negotiations between white landlords and Black tenant farmers on the Altamaha River from 1866 through Reconstruction's end. Black farmworkers would not work "by the day" but instead insisted on the old task system by which the length of the workday was determined not by the number of hours in the field but by the completion of specific, predetermined tasks.[55] They also refused regular wage labor that deprived them of a stake in the harvest, rejected working without compensation detailed ahead of time, and bought land to escape white-owned farms whenever possible.[56]

Black workers were standing up for the most basic of labor rights, yet white landowners perceived their actions as intolerable insolence. One year, Leigh recounted a healthy rice harvest season in 1867, after which she paid $6,000 out to the rice workers. "The result was," she recounted, "that a number of them left me and bought land of their own"—much of it in the pine woods, "where the land was so poor they could not raise a peck of corn to the acre."[57] Any independence, no matter how impoverished, was preferable to the servitude of tenant farming.

African Americans were not interested in toiling to the point of collapse so that white landowners in the Lowcountry could profit from their labor. Their efforts to establish self-determination existed at many scales, from outright rebellion, as during a wave of strikes in Lowcountry rice fields in 1876, to the annual negotiations of contracts between white landlords and Black tenant farmers.[58] They fought to loosen the hold of white landowners over their labor and to reduce the amount of time they spent tending the rice fields of the Lowcountry. Consequently, rice production at a level that benefitted and enriched white South Carolinians had to decrease. Commercial rice cultivation may have greatly declined after emancipation, but African Americans were mostly content with the form that rice culture took in the postwar decades. From their perspective, rice cultivation had not collapsed; nor was it ruined. It now existed at a scale that better suited their wants and needs rather than those of white landowners, their former enslavers.

But even that equilibrium was delicate, for the future of rice on the coast was unclear. Its cultivation continued after the war because of the enduring nature of rice field architecture, the sunk capital of white landowners in rice lands, the cultural value associated with rice and rice growing, and the knowledge of rice culture that African Americans used to support their families. "Rice been money in dem time you know," said Ben Horry in 1937.[59] Horry, a freedman in Georgetown County, South Carolina, knew that rice had once been the currency of the Lowcountry. That was no longer true. In 1850, enslaved rice workers in South Carolina grew and harvested nearly 160 million pounds of rice, the high-water mark of rice production in South Carolina's history.[60] In 1870, African American laborers harvested 32 million pounds—a substantial number, yet still a significant drop because newly emancipated African Americans frequently refused to return en masse to rice fields and instead began to diversify their participation in the Lowcountry economy. Despite what the South Carolina naturalist and landowner James Henry Rice described as "a few hectic efforts" as white landowners after Reconstruction tried to leverage their reclaimed political power over African Americans, there would be no recovery to antebellum levels of production in rice, just a meandering decline.[61] African Americans did not mourn this slow death.

African Americans also looked to other industries for a cash wage, foremost among them the phosphate mines of the postwar coast, which were the great hope among white businessmen for a New South Lowcountry. Enslaved workers in the Lowcountry had uncovered phosphatic rocks along the rivers and in the marsh mud for generations. Phosphate rocks varied in appearance. Rocks found in marshes or along riverbanks were usually "light yellowish brown in color" and "soft and chalky" in texture, whereas rocks found along the bottom of Lowcountry rivers were generally darker, "being dark brown or gray or even jet-black," sometimes "inclosed [*sic*] by a hard lustrous enamel."[62] Francis Simmons Holmes, who owned Springfield Plantation on the west bank of the Ashley River, ordered thirty enslaved workers to dig, haul, and spread out the curious rocks for analysis in 1832.[63] Geologists, naturalists, and chemists made their way to South Carolina to investigate the rocks, which were "rarely larger than a brick and usually the size of an adult's fists" and were interspersed amid "marl, green-sand marl, marl-stone, clay, coprolites, conglomerates, fossil teeth, or fossil bones."[64] They, too, used enslaved labor to assist in their projects or to determine the feasibility of digging up the rocks. Michael Tuomey, a geologist who wrote two definitive reports on the state's natural history, attested that two enslaved workers could open a twenty-foot-square marl pit, four and a half feet deep, in five days. Chemical analysis found that the rocks varied from a content of 22–28 percent phosphoric acid, a chemical that could be processed into a rich fertilizer.[65]

The South Carolina phosphate beds, scientists agreed after decades of research into the region's geology, began their accumulation during the Oligocene and early Miocene epochs twenty-three million years ago. The Lowcountry fell below sea level and became "a great estuary" that fishes, amphibians, and reptiles made their home. Their fossilized remains settled along the ocean floor, eventually compressing into a layer five to ten feet thick.[66] As the coast elevated into dry land once more five million years ago, marshes much like those in the Lowcountry today covered this layer of bone and organic detritus. Water seeped down through the mud and dissolved the phosphate, but the lime carbonate in the fossils reprecipitated the dissolved phosphate into the knobby rocks that enslaved workers found hundreds of thousands of years later. The phosphate collected in intermittent beds buried between five and fifteen feet under coastal soils in thin, brittle seams three to fourteen inches thick.[67] These beds stretched from the Wando River south to St. Helena Island, in five primary groups named according to the nearest river: the Wando Basin, the Cooper Basin, the Ashley Basin, the Edisto Basin, and the Coosaw Basin.[68] Enslavers, interested in rejuvenating southern soils that their own use of monoculture under the exploitation of chattel slavery had exhausted, encouraged scientists in determining the potential use of these rocks.[69]

The phosphate industry, which boomed in the years after the Civil War, became another element of the Lowcountry economy that allowed coastal African Americans to exert their relative independence and collect a decent wage. African Americans recalled taking a job in phosphate as a means of additional support or on their way to landownership. Henry Brown, whose older brothers had fought for the United States during the Civil War, labored in the phosphate mines as a young man once he grew too big to clean chimneys, before settling into a job as a gardener of a house in Charleston.[70] Sam Polite, born on St. Helena Island, bought fifteen acres of land after the Civil War and then managed "five hundred head ob man in rock" at the phosphate mines while his family maintained the farm.[71] Roughly 5,000 Black South Carolinians found employment in the phosphate industry from the industry's early years in the late 1860s to its demise by the early 1900s.

African Americans could earn comparatively high wages in the phosphate mines—much higher, in fact, than pay for a day laborer on a farm or for an upstate textile worker, which imparted a powerful incentive to engage in periodic labor in the mines. Even white landowners admitted the draw of the industry because of its relatively high pay for labor. Hammond observed, "Since the war, the industries connected with the working of the phosphate rock in the rivers, and on the main lands adjacent to them, have furnished the men with employment at higher wages than could be obtained elsewhere in the State."[72] Significantly, miners worked and were paid according to the task system in which each worker was given a predetermined task to complete after which they could be done for the day. That, along with the wage, helped attract African Americans already accustomed to the patterns of task labor and who could not afford to abandon their own land for too long.

The phosphate industry, funded by northern and southern investors, gained its footing with the establishment of the first South Carolina phosphate company, the Wando Fertilizer Company, in 1867.[73] White South Carolinians pursued the possibility of finding a new use for old rice lands and new sources of profit. Although they all drew almost exclusively from the labor pool of Black South Carolinians, the industry soon split into three distinct entities, each with its own unique equipment, extraction methods, and labor needs: land mining, river mining, and fertilizer production. Land mining concentrated along the rivers near Charleston, on the Ashley and Cooper. River mining clustered around the Beaufort area, on the Coosaw River. Fertilizer companies snapped up cheap land on the swampy Charleston Neck for their factories—though these fertilizer companies were not dependent on a domestic supply of South Carolina phosphate for raw material. Phosphate executives prioritized riverfront access for shipment of the processed phosphate to ports in the northeastern United States, England, and Germany.

Though short-lived, the industry did command the national phosphate market and held a significant position in the global phosphate market. In 1889, South Carolina held a market share of 95 percent of all U.S. phosphate consumed globally in 1889.[74] Production of land and river rock topped out at 542,000 long tons (2,240 pounds per long ton) in 1889, with twenty-four land-mining companies and forty-three river-mining companies in operation in the state between 1867 and 1896.[75] The total investment in the land and river mining companies totaled $5.5 million at the time.[76]

African Americans who took jobs in phosphate found the labor grueling. It was as arduous as that involved in rice cultivation, and it bore at least some resemblance to this familiar occupation, especially in the exposure of phosphate workers to the hot sun and muck of Lowcountry work sites. Land mining occurred along wide, marshy flats that workers cross-hatched with ditches radiating out from each other like the irrigation ditches of a rice field: from a main trench sprouted lateral trenches roughly 600 feet apart, bisected by smaller line ditches, and with paired subsidiary ditches dug at right angles from the laterals. The goal was for workers to scour the land of phosphate as thoroughly as possible while not hindering the transportation of phosphate from the fields to the processing plants. Phosphate miners did this work, hand mining, with shovel and pickaxe, though larger, wealthier companies later invested in steam shovels that spared miners the grinding labor of ditch digging. Steam shovels, though costly, had advantages for management and miners alike. They could remove 700 to 800 tons of phosphate rock per acre to the 600 tons removed by hand mining.[77] A miner's daily task in phosphate constituted clearing a twelve- to fifteen-foot-long, six-foot-wide trench of its load of phosphatic rock. Miners were not under contract and did not get paid by the month, instead receiving "short-term cash wages" quickly.[78]

Because of the familiar nature of the task system and the flexibility of payment, the labor was very attractive to many Black South Carolinians, from those who owned their own farms to working-class Black Charlestonians. A cash-strapped man could undergo the brief journey to a phosphate mine, present himself for work, labor for a few days, and return home with a few dollars in his pocket. The summer, between planting and harvesting, and the winter, after harvest and before preparing for planting, were popular times for African Americans to leave their farms for the phosphate mines, where they took up residence in company-owned dorms. These dorms sometimes resembled the quarters that once housed the enslaved. Men well into adulthood—for the average age of the phosphate miner hovered above thirty for the industry's duration—were the primary workers in the mines, and all but a few were Black.[79] Though some single young men worked year-round, the majority were family men who went sporadically to the mines to bring additional money back to their farms.[80]

The final source of labor for some mining companies was involuntary and a reminder to African Americans on the relatively autonomous coast of the threat of white domination. Management looked to the state's penitentiaries for forced laborers between 1880 and 1889.[81] Cahill & Wise Phosphate Works leased nearly 10 percent of the state's imprisoned population, bringing on 58 convicted laborers and paying the state $12.50 a month for each. R. S. Pringle, a phosphate magnate, paid ten dollars each month for 120 forced convict laborers at a mine in Summerville.[82] The Charleston Mining and Manufacturing Company, the dominant land-mining company, with an investment more than its next three competitors combined, also leased convict laborers to work in its mines. The men slept in the former cabins of enslaved workers at Drayton Hall on the Ashley River, a housing arrangement that, with little subtlety, reinforced their own temporary reenslavement.

But the prisoners' condition of servitude was, far too often, not necessarily temporary. Unpaid labor in the phosphate mines frequently claimed their lives. When the South Carolina General Assembly halted the leasing of convict workers to the phosphate mines in 1889, they primly cited a new policy of only leasing imprisoned men to work in "healthful localities," and the phosphate mines did not qualify.[83] In a shielded yet unmistakably grim elaboration on the assembly's statement, the state prison superintendent remarked in his annual report in 1890 that, in the year since imprisoned men ceased working in phosphate, "the general health of the convicts has been better than last year and the death rate less."[84] And indeed, whereas 111 prisoners died in 1890, 69 perished in 1891 after the implementation of the new policy.[85] Though the reports do not list the cause of death, it is likely that these forced laborers died of the same ills that killed enslaved laborers working under similar conditions: malnutrition, exhaustion from overwork, sicknesses, poor sanitation conditions, or a combination of each. As a convict laborer, labor in the phosphate mines could be a death sentence.

River mining, the second form the phosphate industry took in South Carolina, existed within an arcane tangle of state oversight and corporate direction because of its location on public domain rivers. In the early days, African Americans were functionally independent contractors who collected phosphatic rocks that washed up along the edges of rivers around Beaufort and sold them directly to phosphate companies. Rivers, which belonged to the state, were fair game for fishing, crabbing, shrimping, hunting, and recreation, and that included their shores. As such, African Americans could freely forage phosphate rock from the rivers as they pleased, and some even dove into the murky, treacherous depths of intertidal waterways to fish phosphatic rocks

Rate and number of prisoner deaths in South Carolina, 1884–94.

Year	*Deaths*	*Number of prisoners*	*Mortality rate*
1888	100	976	10.20%
1889	101	887	11.40%
1890	111	846	13.10%
1891	69	774	8.90%
1892	42	845	4.90%
1893	55	953	5.70%
1894	51	1042	4.90%

Source: Data from Benjamin Tillman, "Message of Benjamin R. Tillman, Governor, to the General Assembly of South Carolina at the regular session commencing November 17, 1894," (Columbia: Charles A. Calvo Jr., State Printer, 1894), 34, folder 23, box 1, P U Series, Tillman Papers, Clemson University.

up from riverbeds. In 1869, at least 1,200 Black men made money by harvesting thousands of tons of rock and sometimes earning as much as five dollars a day for their catch, which they sold to phosphate companies.[86]

The commons principle that made gathering and selling phosphatic rocks accessible to African Americans, however, rendered it a bureaucratic nightmare once investors tried to formalize river mining. In 1869 and 1870, the general assembly finagled deals behind the scenes to bestow exclusive rights to a few wealthy phosphate investors for a per-ton royalty. A handful of Black Republicans pushed back with antimonopolist vigor against these efforts. They found sympathy with the Republican governor, Robert Scott, who vetoed an 1870 bill that he saw as a "naked grant . . . to a few individuals."[87] The general assembly overrode his veto and then licensed the Marine and River Phosphate Mining and Manufacturing Company of South Carolina as the corporation with exclusive rights to mine river rock for twenty-one years, at a royalty of a dollar per ton.[88] Two other river mining companies, Coosaw Mining and Oak Point Mining, subcontracted with Marine and paid royalties through the larger company, a practice of dubious legality. A few other independent operators managed to wrangle a license out of the general assembly before 1876, but none came close to the dominance of the Marine Company. Between 1880 and 1896, this company produced half of all river phosphate—with the blessing of the state government, which in 1876 passed an act that confirmed Marine's exclusive rights to river mining.[89]

River mining required a fleet of specialized boats. Dredges were each accompanied by a tugboat, a wash boat, a lighter, and two small flats.[90] River miners ate and slept on dredges and washers, which returned to shore for repairs, to switch out crews, or to resupply. Dredge and washer engineers could bring their family on board, an indication of how long these voyages could be.

Dredges sat flat upon the water, with a huge, rigged shovel on board that dragged along and dug into river bottoms in search of phosphate. Five men on the dredge bore the responsibility of mucking through the dredged-up flotsam for phosphate and then moving the rocks onto the washer.[91] There, fourteen men rinsed the rocks, which they moved to lighters, transportation boats that ferried the rock to a drying shed on shore. Two-man crews on board each flat trailed behind the dredge, one man poling the flat along and the other wading, swimming, or diving to salvage rocks that the dredge failed to collect.

River mining was a year-round enterprise, and so the men on the river had to prepare themselves for frigid water in the wintertime or for the possibility of capsizing and drowning during a storm. The pay, once again, ameliorated these fears enough to lure in African Americans eager for a decent wage and with few other options. Engineers, foremen, and tug captains at Coosaw could make $3.71 per day. Land foremen and engineers received $2.41, presumably lower because of the reduced danger of work onshore. Day laborers, like the men who sorted, washed, and collected the rock, could earn $1 a day.[92]

Most river-mining employees worked onshore in processing plants, which used the same kinds of machinery and refining methods as a land-mining company would. Sprawling, multistory factories with wharfs and railways crowded Charleston Neck or loomed up from the otherwise undeveloped riverfronts along the Ashley, Cooper, and Coosaw Rivers. Makeshift quarters erected by the workers themselves huddled near the factory, providing both temporary and permanent housing for the factory workers and their families. In the processing factories, the rocks were dried, crushed, and mixed with sulfuric acid in lead-lined cast iron tubs.[93] The rock and acid mass would come out ready to be packed into 167-pound sacks, which were then sewn shut by young boys, or in a semifluid state that had to dry for two weeks to two months and then be crushed into a fine powder.[94] Work in the factories was filthy with dust and noxious from the fumes of sulfuric acid as well as the arsenic and lead that leaked from the storage tubs, all of which almost certainly had a negative impact on the health and safety of the laborers. Phosphate mining remade the Lowcountry environment, too, infecting the soil with poisons for the next century.[95]

The toxicity of phosphate mining was not the only drawback. Even as the industry reached its peak production of half a million long tons and employment levels totaling over 5,000 in 1889, cracks almost immediately began to appear, undermining the industry just as it seemed to be a success. One was competition, from newly discovered phosphate beds in Florida whose rock averaged 78 percent lime phosphate to the paltry 25 percent of South Carolina rock and a state government that charged a royalty of only fifty cents a ton to South Carolina's dollar.[96] The other was Benjamin Ryan Tillman. Once elected to the governorship in 1890 on a wave of Populist support and white supremacist anger, the rabble-rouser from a wealthy Edgefield family campaigned to increase the phosphate royalty to two dollars per ton explicitly because of his dislike of the coastal elite and his overweening desire to exert control over the "greedy old city of Charleston."[97] It may be that Tillman, who made race-baiting a feature of his campaign, also resented the river-mining industry's employment of African Americans. Driven by antipathy to the excesses of late Gilded Age plutocrats in the North, Tillman made numerous efforts throughout the course of his political career to call corporations to heel and to make them answerable to the authority of the state. This was part of Tillman's quest to make the state "an agent of reform" for white South Carolinians.[98]

In 1890, Tillman seized an opening. The Coosaw Mining Company had taken over operations from the Marine Company, assuming the latter's state-granted monopoly over river mining. Tillman could not undo the general assembly's legislation that gave them that monopoly. But he could establish an executive Phosphate Commission, wait for the company's twenty-one-year license to expire in 1891, force the company to reapply for a license with the commission, and then break the company's sole rights to river mining.[99] And that was just what he did. Coosaw fought back, taking the matter to court and suspending their operations while the litigation unfolded. This essentially froze river mining in South Carolina until 1892. Coosaw played this powerful card because the royalty on river phosphate had previously brought massive amounts of money to the state government. Coosaw had, since its inception, paid $1.5 million in royalties to South Carolina.[100] This was a rich bounty for a state that often struggled to fill its coffers, and the river-mining companies withheld it until they could win a favorable outcome against Tillman. And that was just what *they* did, industry muscling regulatory power back from the state. In return, the government took its due. From September 1892 to September 1893, the state of South Carolina collected $958,955 in taxes total, 25 percent of which came from the phosphate royalty.[101]

African Americans' work in phosphate thereby subsidized state expenditures, which by the 1890s infrequently benefitted them—an old pattern in South Carolina, whose white elite relied on Black labor yet yielded little in return. Phosphate mining gave African Americans an opportunity to engage in flexible, relatively well-paid labor, but at a price, as the toxic legacies of phosphate mining suggest. The dangers of phosphate mining may not have seemed much worse than the life-threatening morass of rice fields, and many African Americans were not in a financial position to forgo a chance at a wage twice what they might receive for fieldwork. Furthermore, as racist oppression gained institutional heft across the South in the years after Reconstruction, African Americans likely understood that they could not squander economic opportunities. Few could afford to shrink from phosphate mining. Though the economy that African Americans on the coast had built was flexible, it was not always forgiving, and it never afforded them easy options.

Coastal African Americans had adapted their economy to a remarkable degree to outmaneuver white mastery of Black labor, and yet it was still a tenuous compromise. Growing kitchen gardens, corn, and sweet potatoes and hunting and fishing fed their families. Working in the phosphate mines, selling small amounts of sea island cotton to local merchants, and hiring out their labor to landowners provided vital cash. But this economy was not without its limits. While the Black majority in the region kept the proportion of white storeowners to Black residents lower in the Lowcountry than elsewhere, white merchants increasingly entrapped Black farmers in debt for cotton seed, farming implements, or groceries in a cycle that would only worsen in years to come.[102] The labor necessary to sustain their livelihoods was difficult and dirty, and they were increasingly hemmed in by white landowners and politicians who resented their relative autonomy, the land that they lived on, and their sheer numbers. Exhaustion, the specter of debt, and the rising tide of post-Reconstruction racial oppression constantly threatened their economy's viability. African Americans were aware of these limits, but it was the best option available to them for the time being.

Pinckney had falsely identified self-determining African American communities as a blight upon the sea islands. But the islands were not rotting from within. The true blight crept toward the coast, menacing Black autonomy because it did not willingly concede profit to white-owned coffers. Legal title to the land and deep knowledge of place were formidable barriers to the blight in the Lowcountry, though it had already eroded them in Charleston and on white-owned farms pocking the countryside. Jim Crow stalked the South, and it awaited a storm to carry it to the Lowcountry.

Chapter 2

The Great Sea Island Storm

African Americans in the Lowcountry watched two horizons in late August 1893. On one loomed the specter of white supremacy threatening to engulf the region, and on the other stirred wind and water that could scour the region of their crops and homes. Both forces were also cyclical: white supremacy mustered a backlash to every step forward, and hurricanes churned across the Atlantic every fall. These patterns of nature and politics intertwined in the Lowcountry for hundreds of years. Some ruptures had rearranged these relations. Emancipation and the subsequent reclamation of thousands of acres of land by people who were formerly enslaved had altered the social and political landscape. African Americans strove to cultivate lasting traditions that preserved their communities, protected their labor from exploitation, and strengthened bonds of kinship that could help them weather the coming storms.

Most residents could expect to experience a significant hurricane once each generation. A sweltering heat wave in 1752 preceded a mid-September hurricane with a storm surge that "came rolling in with great impetuosity," flooding Charles Town in ten feet of water.[1] The deadliest storms rarely incurred their toll within Charleston's city limits, however. In 1804, a hurricane swept across the Caribbean and drowned hundreds, mostly enslaved rice workers, in Georgia and South Carolina.[2] In 1822, a hurricane brushed past Charleston at low tide and did little damage there but wrecked the Santee River delta rice lands and drowned dozens, if not hundreds, of enslaved laborers. After, enslaved workers built a handful of cylindrical "storm towers" rising above the rice marshes to weather future storms.[3] A hurricane in 1854 closely followed the path of its 1804 predecessor, so memorably destructive that it was known as the Great Carolina Hurricane.[4] And in the early months of 1893, many Lowcountry inhabitants likely expected that the hurricane of September 1874, which tore apart the Charleston Battery, or the gale of August 1885 might be the worst storms of the second half of the nineteenth century.[5]

Inhabitants of the region knew that the Lowcountry environment was mercurial. The sea islands grew and shrank according to the pull of tides and the push of sand. As the pulses of the tide swelled and retreated, creeks and rivers slung their curves along new channels. The seasonal arrival of hurricanes eroded beaches, flattened dunes, and deposited those tons of sand to build new shores. These elements constituted an environment that was difficult to control and whose changes were difficult to anticipate with precise accuracy, and they coalesced with changing circumstances to challenge the livelihoods of Black farmers and laborers. Those political and social circumstances in the region were on increasingly unsteady ground. Anyone in the Lowcountry who kept a weather eye on the ocean or paid attention to the political upheavals ravaging Black communities across the South could not deny that something ominous was brewing.

In mid-August 1893, dry northerly winds blew down from the desert plateaus of northern Africa to meet the winds and water off Cape Verde.[6] They collided with southeastern winds gusting from the equator, forming low pressure fronts that drew moisture from the summer-warmed Atlantic Ocean. The humid air condensed into storm clouds, and the trade winds cycled more and more air into the lens of low pressure hovering over the water. That air in turn rose to join the growing storm clouds. The layers of disturbance from the atmosphere to the ocean, of clouds, winds, and sea, took cyclonic shape. Fed by summer heat, ocean moisture, and air pressure, a hurricane began to swirl west. It amassed speed, force, and size as it charged across the Atlantic, dragging along ocean and atmosphere.

Outer arms of the gale swung across the ocean and over the late summer Lowcountry, saturated in heat. Low, heavy-bellied clouds mustered on the eastern horizon. In hot gusts, the wind bullied bands of clouds and goaded long swells of ocean toward the shore. Inhabitants of the Lowcountry turned toward the east as these early warning signs washed over the coast. Alice Louisa Fripp, an eighteen-year-old white woman living many miles from the ocean, noted the gale's approach in the thick cloud cover and sudden showers. She wrote in her diary on Saturday, August 26, "Cloudy and east wind look like a storm is on the Atlantic coast."[7] The weather the next day—"cloudy, misty," with a "high north east wind"—compelled her prediction that "there is certainly a gale coming." Anne Simons Dea, the white woman landowner who inherited her family's rice lands, also wrote of the "sorry-looking" weather. "The clouds are flying rapidly from the east," she noted, "while light hurried showers fall occasionally."[8] These two women, separated in age by decades but joined by their class and race, both relied on Black labor to tend their rice fields. In late August, workers would have been surveying the wide fields of rice recently drained of their "harvest flow," which supported the heavy golden sprays of rice as they ripened. African Americans used the same powers of observation in divining a storm's approach. Walking to the fields along the banks and picking their way through the sucking mud, Black rice workers might have also felt the rising winds, persevered in their work through the periodic showers, and worried what the next few days would bring.

For thousands of years, Indigenous peoples of the Caribbean and the Gulf and Atlantic coasts had come to know the telltale signs of a hurricane's approach—a shrouded horizon at sunset, humid breezes thick with salt, shorebirds finding safe roost, shifts in gusting winds from east to west.[9] In 1893, the method of reading the horizon had not changed much. What continued to elude inhabitants of hurricane-prone coasts was forewarning of intensity. Familiarity with weather patterns made identifying the imminent arrival of a hurricane possible, but Black and white coastal residents alike found it tricky to divine the severity of the coming gale. Black sea islanders attempted to predict the ferocity of storms: every decade, they expected a significant hurricane, or if not, every generation.[10] Some African Americans made predictions based on their livelihood. Fishermen of Charleston's Mosquito Fleet, so called because of their boats' fluttering brown sails and small size, zipping across the harbor, looked to the behavior of their prey.[11] During the late summer, if fish were elusive, that might signal a brewing storm. Atlantic croaker, amberjack, black sea bass, sheepshead—the most prized fish for Charleston fishermen—swim into the still darkness of the ocean depths once they sense subtle shifts in barometric pressure and water temperature, to protect themselves from the turmoil of the storm on the surface.[12]

Other signs from nature, more mystical, could not only foretell the coming of a dangerous hurricane—they could also be interpreted as the judgment of God on human behavior. Just before the hurricane, a Black worker tending the banks of rice fields upstream of Charleston told his white landlord that he had witnessed something remarkable.[13] At each point of the compass, a full moon shone brightly, casting silvery light across the golden-green crop of rice. A sight so beautiful and remarkable, however, did not portend any good. "Dey gwine be strange thing happen," he warned the landowner. The landowner did not believe him. "If that is so," he chuckled, "I think I'll come out on the banks, to-night, and see these things myself." The Black rice worker told him that wouldn't be possible, for God would only "show His signs and wonders to the pure of heart." God had already judged white landowners in the Lowcountry, the descendants of enslavers, and had found them wanting. Their economic wealth, gained at the cost of Black lives, had corrupted their hearts so irrevocably that God would not unveil his portents to them. The hierarchy of heavenly morality was, apparently, inverse to the earthly social and racial hierarchy of South Carolina in 1893.

At moments like this, African Americans' ways of knowing the weather revealed a holistic understanding of the Lowcountry environment, in which the omnipresent legacies of history bound the physical world to a spiritual world. As descendants of different West African peoples, many of whom had specialized knowledge of rice cultivation, Black rice workers in the late nineteenth century understood how their technologies, labor, and knowledge had long been exploited by white South Carolinians in and out of slavery.[14] Some worlds were thus closed to whites because of their abuse.

While residents of the distant sea islands and isolated rice lands were left to their own devices, both prosaic and numinous, cities in the path of hurricanes benefitted from weather warning systems. By the mid-nineteenth century, advances in telegraphic communication and in meteorologists' understanding of Atlantic hurricanes as immense storms that rotate counterclockwise made these new warning systems possible.[15] Since the Civil War, the federal government demonstrated a renewed interest in regimenting weather warning systems to facilitate safer shipping routes and thus greater profit. In 1870, the U.S. Signal Service, a branch of the military, established weather stations along the coast. Charleston received its first signal service officer in 1871, Sergeant J. E. Evans.[16] His office was on the third floor of the Carolina Savings Bank building at the corner of Broad Street and East Bay, and he placed his thermometers, rain and wind gauges, and barometer on its rooftop. Weather monitoring was not a new pastime in Charleston. Daily weather records for Charleston stretch back to at least 1738. But now, Charleston was looped into a nationally orchestrated weather monitoring system that included the use of telegraphs to convey alerts about approaching gales.

By 1890, the U.S. military had wearied of tying up officers in civil service.[17] In response, the federal government created the Weather Bureau as a part of the Department of Agriculture and mandated that the bureau take over weather monitoring.[18] The Weather Bureau dispatched a civilian agent to Charleston, Lewis Jesunofsky. Weather Bureau agents like Jesunofsky recorded hourly temperatures, barometric pressure, wind speeds, and observations about cloud cover, providing detailed meteorological records. A telegraph system connected the observers and allowed them to relay information about their records, to alert each other and the U.S. Weather Bureau headquarters in Washington, D.C., about changes in the weather. The telegraph was undoubtedly a major innovation in hurricane monitoring, but its effectiveness was only made possible through improved federal organization of a weather reporting system.

Charleston's position as a bustling shipping port relied on national connections and benefitted from Jesunofsky's presence. As a federal agent, Jesunofsky played a vital role in protecting the economic interests of wealthy Charleston landowners and merchants. Yet he was also an independent thinker in his own right—a characteristic that would have a meaningful impact on the city's preparation for the August 1893 hurricane. Around the same

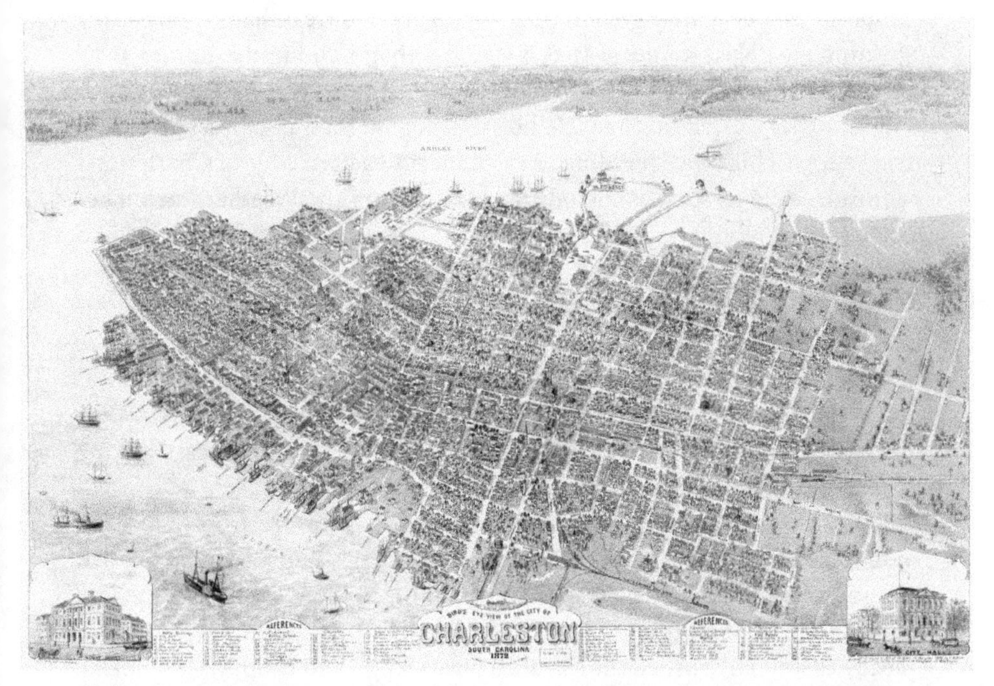

Bird's-eye map of Charleston. This bird's-eye view of Charleston, from 1872, depicts the city before the western edge (here at the top of the map) was filled in to make more land for development, as the bustling port that it was. Note the West Point Rice Mill belching smoke just past Calhoun Street and the bustling wharves off East Bay Street on the east side of the peninsula. C. N. Drie, *Bird's Eye View of the City of Charleston, South Carolina* (1872), Library of Congress Geography and Map Division (LCCN 75696567).

time that Anne Simons Dea, Alice Fripp, Black farmworkers, and others along the coast began to suspect that a hurricane approached, Jesunofsky received alarming news from the Weather Bureau headquarters in Washington, D.C. On August 25, the bureau had received reports via telegraphs from island officials of a storm traveling offshore of Nassau, in the Bahamas.[19] The storm had lurked at sea until that day, when it strayed too close to Nassau to continue eluding detection by those on land.

The news of a hurricane was no great surprise to Jesunofsky. For two weeks, he had observed high seas and foul skies (though these were in fact the result of another hurricane that stayed deep in the Atlantic and did not make landfall).[20] Weather forecasting, after all, remained an inexact science in 1893, more speculation than certainty. The barometers and other instruments then available provided maybe a day's notice of a hurricane, and simple weather observation gave roughly the same. Alas, these reports, transmitted through telegraph lines from the Caribbean to Washington, D.C., and then to agents, composed the best alert system that the Weather Bureau had. A hurricane brushing past Nassau meant that it was less than two hundred miles away from Florida, shockingly close to the mainland United States.

Jesunofsky wasted no time. He hoisted a signal for a northeast gale from the rooftop of his office building, located on the eastern edge of Charleston's peninsula. And as soon as the official reports from the Weather Bureau began to spread, weather-beaten steamers and ragged sailing craft straggled into ports on the eastern seaboard with tales of enormous seas and ferocious winds.[21] As the hurricane and three others simultaneously battered the Atlantic, a catastrophe undocumented until that year, Jesunofsky began to brace for landfall. On Saturday the twenty-sixth, the bureau telegraphed Jesunofsky to alert him that the gale seemed to have swollen in size. However, D.C. officials also predicted that the storm would spin off into the ocean, leaving the southern seaboard of the United States unscathed.

Jesunofsky, Cassandra-like, disagreed. A low-pressure front had been skulking to the northwest of Charleston for days. Jesunofsky suspected that the storm would be drawn up to South Carolina, turn toward the front, and pass inland. As Saturday wore on, Jesunofsky's prescience proved out. A "steady east wind" rustled the sea until it was "high and choppy," Jesunofsky observed, so much moisture saturating the air that "a low haze covers the land and sea."[22] Soon after, intermittent squall lines marched through the Lowcountry from the northeast to the southwest. The barometer continued to fall to frightening lows. That afternoon of Saturday the twenty-sixth, Jesunofsky spread the word of the gale by posting notice of an incoming storm on the doors of the Charleston newspaper the *News and Courier*.[23] He hoped that at least the residents of Charleston could take precautions for the imminent storm. But whether all Charlestonians, or even a substantial percentage of them, noted his warning is not clear. Certainly sea islanders and inhabitants of the rural hinterlands around and between Charleston and Beaufort received no advance notice.

That night on the sea island coast, the wind shifted to the northeast, and a thick canopy of clouds hung at the horizon.[24] The morning of Sunday the twenty-seventh dawned blustery and unsettled in the Lowcountry as the hurricane's fringes grazed the coast of Florida. Dawn brought torrential rain to Jacksonville and winds up to thirty-eight miles per hour to St. Augustine.[25] Those winds, not even at the level of a tropical storm, were far less alarming than the water pouring over the old city's seawall. The hurricane carried a high, angry sea behind it, and that boded ill as it lumbered ashore. As the fringes of the hurricane flooded St. Augustine, Lowcountry residents passed their morning uneasily in feeling but as normal in action. They were unwilling to panic or to discard their usual routine, so they continued about their day. Churchgoers even attended their Sunday services. Susan Hazel Rice, a young white woman in Beaufort, felt the pressure changing in the form of a headache. "I fear we may get the gale yet," she wrote, after a "very stormy" Saturday night.[26] The ill-favored weather did not dissuade her, or others, from church. In Beaufort, congregants sat through service but then "hurried home" as the rain and wind picked up.[27] Perhaps in sea island communities, the praise houses were full, and the congregations that were on a pastor's rotation that week were crowded into churches, like the newly constructed African Methodist Episcopal Queen Chapel on Hilton Head.[28]

But by the afternoon, from Savannah to Georgetown, it grew more difficult to dismiss the severity of the storm. Tybee Island in northeastern Georgia, which juts out into the ocean, was one of the first sites on the coast to feel the full brunt of the storm. At Tybee, a popular vacation spot for white tourists, a few remaining late-season visitors may have looked out at the rising storm and wondered if they should have taken the Central Georgia Railroad back to Savannah, as three hundred other tourists had, instead of lingering over the weekend.[29] Fishermen and boaters who had taken their craft out along the creeks near Beaufort hurried back to safe harbor in the early afternoon.[30] Anne Simons Dea shuddered as "the rain fell in torrents" by three that afternoon.[31] Alice Louisa looked out at the yard around her house, already strewn with branches, as a tree crashed to the ground.[32] By then, Jesunofsky watched "violent gales" whip rain around the Carolina Savings building and climbed to the rooftop to find the barometer "oscillating and pumping and falling rapidly" and the winds reaching eighty-four miles per hour.[33]

In Charleston, city operations continued apace into the late afternoon. Long accustomed to flooding and hurricanes in their water-bound city, Charlestonians yet remained sanguine. In the early afternoon, rain began to fall, and floodwater lapped at the fine marble steps of mansions on the Battery and the wooden steps of tenements on the peninsula.[34] The surf inundated the dozens of wharves that jutted out of Charleston's eastern waterfront, precipitating an emergency.[35] Three Black dockworkers launched a boat from Chisholm's Causeway to a flat anchored nearby to save a friend stranded there by the storm—not a moment too soon, for the flat sank as they rowed away.[36] The ferry to Sullivan's Island left from Charleston at three as scheduled, despite the choppy surf in the harbor. It arrived at Sullivan's safely but did not attempt the return journey to Charleston that evening.[37]

By four o'clock in the afternoon, nearly every street south of Calhoun Street and west of Coming was under water as the salt of the harbor mingled with rainwater. But residents in that western quadrant of the peninsula stayed calm, as that same area had been flooded "half a dozen times during the summer when the drains [were] congested by the unusually heavy rains."[38] The horse-pulled streetcars, which had run along artery roads in Charleston since 1866, continued service along the peninsula until seven o'clock. At the south-facing Battery, the horses bent their heads to fight the lashing rain.[39] Around the same hour, St. Michael's bells rang out not to call white Charlestonians to the Sunday evening service but to comfort citizens in earshot with its "glorious old hymns."[40] That evensong would soon be stifled by the shriek of high winds and the crash of waves across the city. As the hurricane crescendoed through the early evening, Charlestonians must have realized that it would intensify right as the tide hit its high point, at 8:07 P.M.[41] As high tide drew near, Charlestonians could no longer ignore the misfortune that would soon wash over them.

The true calamity of the Great Sea Island Storm was not the wind alone but the timing and height of the storm surge. The hurricane would not reach its peak winds in the Lowcountry until midnight, but by then the storm surge had already rolled in and drowned hundreds, if not thousands. The waves were unsparing, submerging entire sea islands. Kiawah Island, Sullivan's Island, Lady's Island, and many of the small islands near Beaufort were entirely subsumed by anywhere from four to twenty feet of water. A storm surge of this magnitude had never been recorded in the Lowcountry.[42] And while some hurricanes may have had impressive surges near Georgetown or Beaufort, none brought this hurricane's level of power to such a long strip of the coast. That night, the sea grew so monstrously because the hurricane, high tide, and a full moon crested at once. High tides during the full moon, called "spring" tides, could be a foot and a half higher than the neap tides of the quarter moon. Though layers of storm clouds veiled the bright moon, its gravitational pull could not be slackened. The hurricane held the spring tide in place and heaved the ocean over it, swamping barrier islands and marshy estuaries.

Indeed, the storm surge was so violent in its approach that night that many observers, from Beaufort to Charleston, described it as a "tidal wave." Some took it as a fact: "A rise of the sea so sudden and so great," the journalist Ambrose Gonzales wrote, "cannot be classified under any other name."[43] Rachel Mather, a New England schoolteacher who had settled in Beaufort County in 1867, wrote of the "encroaching tidal wave," which was "driven in by the fierce hurricane."[44] This tidal wave, she recounted, struck the death blow for thousands of Black sea islanders. Others tempered their description of the storm surge. C. Mabel Burn, a well-to-do white woman living in a fine house on the Point neighborhood in Beaufort, watched the storm surge as it took its toll on her neighbors. "Tide was due to be high at 5 p.m.," she said, but "the wind increased in velocity and the tide was held up and could not fall, so the next tide piled on top of the first."[45] State senator W. J. Verdier, also of Beaufort, equivocated: "It was more like a tidal wave than anything we have ever seen hereabouts."[46] The South Carolina naturalist James Henry Rice outright scoffed at an acquaintance's insistence that the hurricane brought a tidal wave that night. "There was no tidal wave and nothing to cause one," he confided in a friend, though "I have never attempted to set him right, because of his passionate feeling on the subject."[47] No, Rice said, the hurricane had "piled up three tides in the creeks" and "sent the ocean over" the coast. "That was all," Rice asserted.

There was some truth to both the tidal wave evangelists and the doubters. It is easy to understand why survivors would recall the sudden rise in the waters or the high waves that engulfed their homes and fields as akin to a tidal wave. That night loomed large and dark in their minds' eyes, its most terrifying hours distorting their memories. For journalists and others recording the hurricane for posterity, a tidal wave—rather than simply a storm surge—as an oceanic phenomenon surely inspired shock and fascination. Whether through the trick of memory or the cynical exaggeration of tragedy for a hungry audience, the hurricane took on chimerical dimensions.

However, the memory of a tidal wave is not without some scientific grounding. Many environmental and meteorological factors influence a storm surge's intensity. The wind speed of a hurricane and the circumference of the storm have the greatest impact on how much water the hurricane pushes along with it—and this was an enormous gale, with winds reaching 120 miles per hour. The wide, shallow continental shelf off the coast of Georgia and South Carolina gives hurricanes more time to build a storm surge's velocity. The porous brim of the Lowcountry, with its estuaries, marshes, and low islands, welcomes the tidal flow of water and does not hinder storm surges with the same friction that a rocky coast would. A hurricane, then, can pile up billows of water and drag that swell of ocean with it, which is what the Great Sea Island Storm did. When the hurricane rolled into the low, sandy coast, which eases into the shallow continental shelf, so too did its storm surge. That storm surge did not strike in a single wall of water but in an unfurling flood that rushed into sea islanders' homes.

The hurricane's unprecedented storm surge would have been deadly enough, coinciding as it did with a high tide. What compounded its harm was that the hurricane raged all night. Many people had difficulty reaching high ground or making it to sturdier homes and safer havens in the midnight darkness before the storm surge swept the islands. Outside the city, away from the townhouses and infrastructure that gave a safety net to coastal residents in Charleston and Beaufort, the hurricane was even deadlier. The distance between homes in rural areas made finding shelter less likely. The whipping winds drowned out cries for help, and the roaring waters overtook thousands of sea islanders who had no other choice but to shelter in their homes. "As night drew on and darkness obscured their vision of the appalling scenes," the schoolteacher Mather recalled, “the horror of the poor islanders increased.”[48] As the water rose on the islands, “men, women and children, knowing that the sea surrounded them on all sides, groped vainly for higher ground, and many perished in the attempt.” The storm surge rolling in over a spring tide; the unusually long duration of the hurricane, from early afternoon on Sunday till mid-morning Monday; the strong, circling winds—all these natural elements diminished the chance for survival.

The hurricane was an exceptional weather event, but the built environment, community bonds, and generations of knowledge shaped the human response to the storm's fury. That night was the deadliest fourteen hours in Lowcountry history. Residents of the coast, the vast majority of whom were African American, looked to their loved ones for strength, engaged in acts of astonishing bravery, and drew upon the vernacular landscape of the Lowcountry to endure. Though the hurricane disoriented the senses and swept away homes, trees, and other familiar landmarks, African Americans nonetheless found features of the landscape to guide them to safety. They turned to the tools that they used every day in fishing, boating, and farming, to their traditions and their knowledge of the land, to carry their families through, much as they had drawn upon those to survive centuries of white supremacy.

The magnitude of the storm surge was frighteningly sudden, and many families had no choice but to decide between staying in their houses and fleeing for higher ground. Margaret Weary, one of only a few white families living on St. Helena Island, was cooking supper when she looked out the window and “saw the sea all around the house” with “the waves rushing up to the door.”[49] Her mother seized her little sister Grace and ran to a neighbor's house on a hill, and in the confusion, Margaret and her brother lost them. They plunged into the waters to make their way to a house but found it already destroyed and had to face the storm again to reach another. Maggie Waring, also on St. Helena, told a similar story. Her family threw themselves against windows and doors to keep the hurricane out until her mother sunk to the floor in despair.[50] When they opened the door to abandon their house, “the tide met us, roaring and filling the house instantly,” Maggie recounted, and as they all “plunged into the water . . . it floated off.” Her family had to flee to two different houses that night, one so crowded they could barely squeeze in. The residents of nearby Parris Island, exposed to the ocean through the wide mouth of Port Royal Sound, also weathered a frightening night. Charlotte Edwards of Parris Island stayed in her house, which filled with three feet of water.[51] Dozens of white sea islanders struggled through waist-deep water to take shelter within its walls. At midnight, the winds shook the house fearfully. Though the walls held, the interior stairs were no match for the disturbance, and collapsed on a neighbor's neck, killing her.

Lewis Grimball, a white landowner, wrote about the storm surge from the Pon Pon River in cruel, dismissive terms. "Does it not seem remarkable," he mused to his brother, "that so many of the negroes down here, were stupidly asleep, and opened their eyes merely to close them again in death?"[52] It was cheap to speak ill of the dead, from the vantage of a large house built on high ground by enslaved workers. The waters came in too rapidly for flight, or, out on exposed, low-lying land, there was nowhere else for them to flee to. And in truth, African Americans fought all night to protect their loved ones. Their relationships—through family and community—constituted the basis of survival. One African American man on St. Helena threw his weight against the wall of his house, which had jammed against a fig tree as it started to sweep suddenly out to sea, as his wife and two children fled to the upstairs loft. Part of the wall fell and cracked open his skull, but, as his wife Essie Roberts recounted, "half-crazed as he was," he kept the house from floating away.[53] One father, also on St. Helena, buttoned his baby into his jacket so that his hands could be free to carry his wife and two other children.[54] Though he was twice knocked over by mammoth waves, the infant and the rest of his family lived through the night. On the same island, two Black men spent the entire night "wading swimming" through the floodwaters for people in danger, saving at least thirty.[55] Another African American man, who had just delivered his family to a neighbor's house, "heard the cries of a woman" and "plunged again into the water, to save a widow and child some distance off."[56]

For sea islanders surrounded by the storm, trees often became places of refuge. They favored live oaks, huge trees easy to find in the landscape and perhaps closer than the next neighbor's house. Families clambered up into trees for the night, sometimes lashing children to a sturdy trunk until the storm passed.[57] With a storm surge of up to twenty feet on the low-lying areas of the coast, this would have been the only viable option for riding out the night. Alice Louisa recalled, "We got tied off the oak trees and held on most all night," with "the water waist deep all around."[58] One African American woman spent the night holding tight to her three children until daybreak, without the reassurance of a cord.[59]

Escape into the trees could still be a deadly enterprise. A friend of the woman with three children was not so lucky. Though she somehow made it into a tree with her five children, the branch supporting three broke and plunged them "into the whirling flood."[60] One woman, holding onto her baby by its clothes with her teeth, fell from a snapped tree limb. Both drowned.[61] Onlookers, clinging to the rough limbs of live oaks nearby, would have hoped that theirs would hold firm. The sea islanders preferred live oak trees to serve as their rooted lifeboats for good reasons: they are easy to climb, with their broad trunks and low-hanging branches, and their ample size and reaching branches emit an air of tranquility. Though wealthy enslavers often chose live oaks for their elegant alleys leading to their mansions, the live oaks took on new meaning for Black sea islanders enduring hurricanes, holding memories of slavery and of fortitude alike in their branches.

Other emblems of Lowcountry life took on new valence in the hurricane. Another common method was to ride out hurricanes aboard boats. Some would moor the family boat to a tree trunk, pile everyone in, and weather the hurricane aboard. On Daufuskie and Pine Islands, over which floodwaters roiled all night, hundreds of African Americans took to their vessels, roped to trees.[62] The craft that allowed them to fish and travel here became a tool of survival. On Sullivan's Island, then a favorite vacation spot near Charleston, which the storm surge consumed, two men saved thirty-two people—some of whom had to grip a rope leading from the boat to the windows of houses nearly battered apart by the flood.[63] Absent a strong tree to secure the boat to, one man put his family into it and, "guiding it into a thicket," kept them safe till morning.[64]

Survival strategies were many, and sea islanders had to decide quickly which from the motley assortment of methods they had heard about from older family members or that they may have earlier used themselves would make the most sense under the circumstances. To prevent the storm surge from lifting a house off its bricks and smashing it to pieces, one African American man "took his axe and cut a hole in the floor" which "let the water in to weight down the house."[65] The twenty people who had taken shelter in his home lived through the night because of his fast action. Another man, also on St. Helena, took ox chains and "put them out at one window, round a tree stump, in at another window, so saving his family and house."[66] Two brothers, living nearby, dragged out their two-wheeled cart, a traditional conveyance on the sea islands, "brought out their old mother, put her in the cart, piled the children in with her, and took the shafts in their own hands."[67] "The boys helped, the women pushed," and they "lifted their precious charge to higher ground." African American sea islanders transformed the everyday items necessary for their livelihood—axes, ropes, carts, ox chains, boats—into vital objects that could mean the difference between life and death.

Some sea islanders had to make horrifying choices. The hurricane cornered one father, living on St. Helena with his wife and two children.[68] As the sea rose to the door of his small cabin, his wife, delirious with an illness, lay in their bed. He tucked his two children into the cabin's loft, picked up his wife, and battled the breakers to carry her to safety. He turned back to the home to rescue his children, to find that the water had risen too high. But when he came back to his cabin the next day, the two children were sleeping snugly in the loft. This miraculous ending, though, is contradicted by another telling of the same story.[69] The account began much the same way, also with an African American man, his sick wife, and two children dreading the sea's approach. He nestled his two young children into bed and kissed them before swaddling his delirious wife in their thin blankets and wading out into the storm. He heard a thunderous crack of lightning and whirled around to see an oak crash into the cabin. The ocean swallowed his broken home and his children. His wife, when she recovered, was riven with grief and wracked with sobs. "It was a hard thing to choose," said the man.[70]

The array of survival methods could not ensure life. Thousands of African Americans had to bear witness to the death of loved ones that night. A seventeen-year-old boy, Johnson Atkins of St. Helena, watched as his house collapsed on top of his parents, himself, and his four-year-old brother, Buba.[71] They swam through the wreckage to the surface of the water. Johnson held tight to the little boy, who was so exhausted that he begged his big brother to let him go. Johnson refused and swam with him to the gable of their house. From that vantage, Johnson marveled, it "seem like you can see the whole ocean," and "the wind blow so hard and the wave so high" that he nearly gave up himself. Before the stricken eyes of Johnson and Buba, a wave knocked their father unconscious and ripped their mother from his hand. They never saw her again. One African American man lost eight children and his wife in the flood.[72] Another, thirteen, his wife, and his mother-in-law.[73] Not all witnesses lived. Many families perished entirely.[74] On St. Helena Island, "whole settlements are swept away," as one eyewitness recalled, with "the men, women, and children of families . . . all gone."[75]

While survival depended upon many factors, community connections were vital. The bonds of kinship sustained many through the night. While community alone could not secure life, to be solitary was to risk death without witnesses or saviors. One young white woman, who lived on the nearly uninhabited, ocean-facing Dewee's Island, found herself in a position that demonstrated the vulnerability of those who were isolated.[76] Her husband, John Roberts, had gone terrapin fishing in Bull's Bay when the hurricane struck. She had given birth only two days earlier and had in her care the new infant and four other children. She heaved the children into their small house's loft after the wind blew off two sides of the house and then swam with one child at a time to a pinpoint of land. There, they crouched in quilts without food for two days until her husband, who had miraculously made it through, returned. Other women found themselves in similar positions, responsible for the care of a child and with their partner away for work. Laura Hamilton, an African American rice worker on the Pon Pon River, ran from her flooding house and tried to swim "holding her child above water with her teeth." Though "she struggled bravely for some time," she "was finally exhausted, and drowned with the child."[77] On the surface, these two women faced a similar situation, and the hurricane did not spare them any danger. However, it would be facile to argue that the hurricane did not discriminate: indeed, where Laura Hamilton lived and worked were inextricably tied to her race and her labor. While white coastal residents did indeed suffer terrible losses that night, African Americans did so in much greater numbers and often by design of the working environments in which they labored and lived.

This was an exceptional hurricane, a confluence of wind and water rarely matched along the southeastern coast. But as always in the Lowcountry, a region defined by the saga of struggle between the hubris of human control and the volatility of environmental change, histories of slavery, emancipation, and control over labor and resources wound their way into the circulation of the storm. Rice fields had to be built along low-lying rivers that ran to the ocean. Sea island cotton grew best on barrier islands, when wafted with humid ocean breezes. Phosphate works, a more recent economic driver in the region, similarly lay on the banks of tidal rivers and at the edges of salt marshes. Black workers died by the hundreds while encamped near their work sites, laboring in rice or phosphate, grinding, filthy work that forced that they accept vulnerability to disaster in exchange for a meager wage. Even a place of employment that African Americans considered a worthy addition to the area, the U.S. Naval Station on Port Royal Island, was exposed to the Atlantic at the mouth of the deep-water Port Royal Sound.

Just as enslavers had compelled enslaved African Americans into rice fields amid hurricanes to shore up embankments, white employers also deliberately endangered the lives of Black workers to protect property.[78] That outright cruelty, though, was less common than the pervasive and insidious practice of compelling African American labor on the sinking edges of the coast. Exploitation of labor in the Lowcountry, whether enslaved or waged, was firmly yoked with increased vulnerability to environmental harm. It was all by design.

First, as Africans kidnapped and trafficked to the Lowcountry; next, as enslaved laborers who died in droves from overwork and sickness; then as newly emancipated people striving to establish a homeland and determined to do what must be done to support themselves and their families—the reasons why African Americans lived on the sea island coast were rooted deep. Of course, they knew that sea islands bent and marshes shifted unpredictably, that rice lands flooded, that sea island cotton fields lay open to storm surges, that phosphate camps were miserable places to live, and that hurricanes could, any given autumn, destroy everything they had built. But where else could they go? And why would they wish to leave behind their communities and their

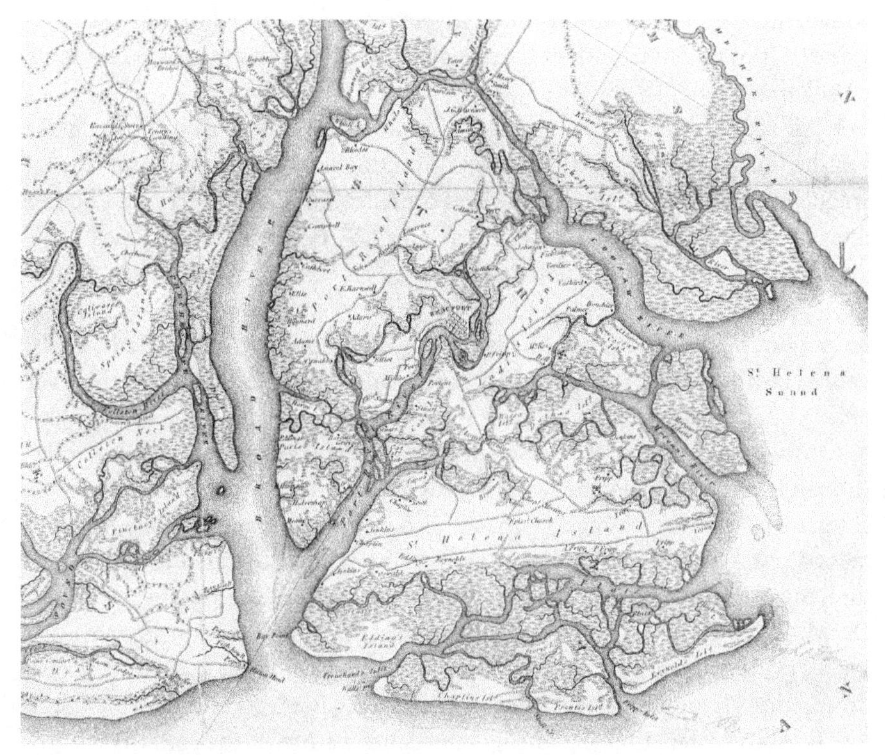

Beaufort and its environs. This is a detail of the Robert Mills map of Beaufort County from 1825, cropped to show Beaufort in the center of the map, and the sea islands around it that comprised a cradle of Black landownership and political participation in the late nineteenth century. It also reveals the vulnerability of the islands to the ocean. Robert Mills, Charles Blacker Vignoles, Henry Schenck Tanner, and Henry Ravenel, *Beaufort District, South Carolina* (Baltimore: F. Lucas Jr. for Mill's atlas, 1825), Library of Congress Geography and Map Division (LCCN 2012590212).

land, to abandon the traditions and knowledge that imbued the sea island landscape?

In contrast, the white landowners who refused to forsake rice, despite the lack of enthusiasm for commercial-level cultivation of the crop among Black workers, preferred to grasp at the fraying threads of their power. Phosphate bosses saw themselves as forward-looking capitalists bringing much-needed industry to a backwater region. Farms and phosphate mines could not be located anywhere else but where they were. Employers viewed the mass death of workers as an unfortunate byproduct of doing business, not admitting that the systems of labor predicated on racial hierarchy from which they benefitted were rotten to their core. Though the original choice had not been theirs, African Americans stayed in the Lowcountry because it was now their homeland. White employers not only exploited their labor but also demanded it in an environment that heightened the exposure of workers to its most lethal elements.

Hundreds of African Americans who rented land and worked on rice farms perished in the storm, sacrifices to white landowners' dreams of resurrecting antebellum wealth. Black rice workers on Alice Louisa property huddled in their flimsy cabins, caught "like rats in a trap."[79] Henry White, a Black boatman and rice worker at Bischoff Place on the Pon Pon River, inland of Charleston, spent the night watching his fellow laborers' cabins wash away.[80] As he floated on the roof of his house, his friends Richmond Branham, Bell Smalls, Laura Hamilton and her baby, Bain White, and Bowles, drowned as the river broke the rice fields' banks. Seventy rice workers drowned in the swollen Coosaw River, winding north of Beaufort to the farms upstream.[81] On the Savannah River rice farms, the storm surge charged over the banks and drowned at least 150.[82] Cotton plantations, positioned near the sea, also became scenes of mass death. On Edisto Island, an ocean-facing island dedicated to cotton culture, thirty-two African Americans died as their homes were swept away.[83] On Dr. M. Jenkins's farm, every single house vanished into the storm's maw. On Eustis Place, a sprawling cotton farm on Lady's Island, at least fifty people drowned.[84] Though the days of enslavers ordering Black men and women back into fields to protect the crop during hurricanes were over, the patterns of habitation and labor that they had established were yet embedded in the landscape.

Phosphate works, emblems of the New South rather than the old, were also sites of mass death. The new world of wage labor and industrial production was no less deadly and demanded no less of its employees than the old. Phosphate workers lived in shacks provided by the companies, clustered around the mines and exposed to storm surges. At the Pacific Works on Chisolm Island, a tiny patch of land barely protruding above the marsh, seventy-nine workers and their family members drowned.[85] At Coosaw Mines, a handful of workers lost their lives in the storm.[86] Some phosphate companies issued a mandate that workers stay with the expensive dredges, the backbone of the river-mining operations—with mortal consequences. Bosses of the Beaufort Phosphate Company ordered workers back onto the $300,000 phosphate dredge *Kennedy*, the flagship of their operation, during the hurricane.[87] Arthur Wilson, a twelve-year-old deckhand, was aboard the *Kennedy*, moored on the Beaufort River, when the hurricane struck.[88] Engineer Hand, a white man, and his African American crew did their best to protect the boy, strapping him into a life preserver. The gale sent the dredge spinning into the river, which had catapulted to twelve to eighteen feet above its spring tide mark.[89] A gigantic wave slammed into the dredge so hard that it flipped over. Hand and the five crewmen were trapped in the engine room and drowned, sacrificed for the preservation of property. Wilson, who was on the deck when the wave hit, floated unconscious many miles through storm-swollen rivers and creeks to Beaufort, where he washed ashore—somehow still alive. That he survived was nothing short of miraculous. But forces of nature did not put him in that position: the forces of capital did.

At the U.S. Naval Station on Parris Island, twenty workers drowned. The naval station represented the hope for federal investment in the region, a commitment to placing infrastructure in the Lowcountry that could bolster the local economy. There, the most famous death of the hurricane occurred. Dr. Gowan Hazel, the station's physician, died when he dove into the rushing floodwaters that covered the island, fifteen to twenty feet deep, to save two African American boys also employed at the station.[90] All three drowned, and their bodies were found clustered together once the waters receded. Given that Dr. Hazel was from a prominent white Beaufort family, his death was widely reported. In these remembrances, his death presented an implicit, conciliatory narrative demonstrating the heroism and beneficence of white paternalism. The descendants of enslavers, Hazel among them, were not evil oppressors interested only in what capital Black sea islanders could produce but instead brave caretakers willing to sacrifice everything for their charges. The gears to reconfigure enslavement and to justify its continued practices of white control over Black lives clicked into place. Eighteen Black men drowned alongside Dr. Hazel at the naval station.[91] Their deaths apparently merited little commemoration, and, like far too many Black survivors and victims of the hurricane, they are unnamed in the white-dominated records of the storm.

Some white landowners leaned on their African American tenants and servants directly, using their power to extract subservience from them amid the crisis. Dantzler, a storeowner in Beaufort, "sent a negro up an oak tree, had the limbs cut off, and then moved up in the tree for the night."[92] May Elliott, a wealthy white woman, held on to a rope tied between two trees. But to protect her valuables, rings that no longer fit on her fingers, she summoned a "little negro boy who helped take care of the chickens" and put the jewels in his pockets. Others in her family followed suit, and then they tied the boy up in the tree to guarantee that, should he drown, at least their valuables would not float away.[93] This white family, led by the matriarch, did not hesitate to use a child likely younger than ten as a glorified purse. He may have been safer up in the tree. But the family only trussed him there once they had turned to him as a human vessel, originally content to let him hold tight to the rope between the trees, suggesting that they were less concerned with his safety than with the protection of their property.

The Lowcountry's history inflected the experiences of African Americans in subtler ways as well. C. Mabel Burn of Beaufort recalled that at 1:00 A.M., a tall African American man rang their doorbell.[94] Dripping wet and desperate but still aware of the racial order that confined him, he asked Burn's father if he could shelter a large group of women and children on their piazza. Until then, this party had crowded into boats that he had scouted, but the seething storm surge endangered their lives. Even so, he did not dare to ask much of a white man. Burn's father told the man that the piazza was liable to collapse and invited them into the house. The man ushered Black men, women, and children into the fine, spacious home and then dashed back out into the night on the largest of the boats. He rescued thirty more Black Beaufortonians that night. The next morning, the people who had taken refuge at the Burn house had to step shudderingly over the drowned bodies of an older African American couple, Thomas and Kate Huger, in a chilling reminder of what their fate might have been.[95] Indeed, the racial disparity in death was so great that the hurricane claimed the life of only one white woman: Mrs. Talbird, of Beaufort, went into "spasms" from "fright," as her husband desperately tried to calm her.[96]

During these catastrophic hours, African Americans chorused in song against the wind. They gave voice to their faith in God, singing hymns to raise their spirits, to comfort their loved ones, and to remind one another of the power of salvation that could be found in God and in each other. On St. Helena, Black sea islanders shouted into the wind "Jesus, lover of my soul, / Let me to Thy bosom fly, / While the nearer waters roll, / While the tempest still is high."[97] As Arthur Tolliday ran to a neighbor's house, "I heard them singing," he remembered, "'Hide me, O my Savior, hide, / Till the storm of life be past.'"[98] The family who had enthroned their mother in an ox cart raised their voices as they pushed, "answering back the roaring wind with their fearless song: 'The Lord's our Rock, in Him we hide, / A shelter in the time of storm.'"[99] In Charleston, African American residents of tenement houses on Gadsden Street hunched together in a prayer meeting at midnight.[100] They met the worst of the hurricane with "song and prayer" and sang "On Jordan's Stormy Banks I Stand" so fervently that "semi-occasional bars" "mingled with the accompaniment of the whistling, shrieking gale." The metaphorical storm of life—the individual struggles to be better and to maintain faith, and the communal trials and tribulations of hardship in the burgeoning Jim Crow South—mingled with the hurricane.[101] For African Americans in the Lowcountry, this was an old storm that had buffeted their shores for centuries. They knew how to meet it.

As the hurricane churned through the Lowcountry, colliding with the detritus of two centuries of toil, it spun itself into a fury. It reached its apex between midnight and 1:00 A.M., east winds wringing the live oaks and pine forests of the islands, flinging itself across wide waters, and ricocheting off the tenements, churches, and storefronts of towns from Savannah to Georgetown.[102] Jesunofsky's anxiety lurked in the jittery crookedness of his usually neat handwriting, his letters' loops loosening as the weather grew more and more dire.[103] At 11:00 P.M., he shouldered through the spray of the storm to the wharves along the east side of the city. There, he saw a vast, chaotic sea where one hundred wharves once stood—boats smashed against each other, waves heaping lumber in tangled piles, a flood ten feet deep billowing over the Battery. At 12:10, the barometer sank to 29.234; and at 12:45, it hit its low at 29.099. At 12:50, the "winds screamed to a twisting 120 mph," he recorded, "and the tides are now twelve feet deep over the Battery and the extreme lower part of Charleston; from Tradd Street to the Battery is three to five feet under water." Ben Oree, an African American man living at Coming and Warren, left his house to batten down his tin roof but was blown to the sidewalk and died from the fall.[104] At the City Hospital at the peak of the storm, a physician remembered that three women, "poor demented creatures," had been quarantined in the pest house.[105] He found them nearly drowned and carried them through chest-deep waves, "holding them in his arms above the water."

The wind, having built to its hour of greatest intensity, demolished warehouses and stores, ripped off roofs, toppled church steeples, snatched the leaves from the trees, carried tugboats, schooners, and steamers miles from their docks, knocked down nearly every telegraph and telephone pole in Charleston, and whipped the storm surge into ferocious whitecaps. As wind and water wrecked town and countryside, the hurricane reconfigured the landscape. The ocean jumped the Beaufort seawall like it was a pebble, surging twenty feet high.[106] In Mount Pleasant, the hurricane swept over the sand hills on the west end and cut into the bluff.[107] The dunes on Sullivan's—some ten, twenty, and twenty-five feet high—were obliterated in the surf.[108] It carried away every bridge on Edisto Island, flattened the causeway to St. Helena, and tore apart the iron-and-steel drawbridge to James Island from Charleston.[109] On Daufuskie Island, the waters pounded so hard against the 700-foot long, seventy-five-foot high bluff that thirty feet of its length slid into the ocean.[110] The hill that protected Alice Louisa's house from inundation crumbled nearly up to the house, sinking the piazza.[111] The flood threw aside railroad tracks on Tybee Island, carrying them 200 to 500 feet away.[112] The storm surge scraped the "fertile soil" from atop fields, seeding them with salt.[113] It even gilled a catfish on the iron fence around the yard of the Methodist Church in Beaufort.[114]

Just before dawn, "in the chill gray light," "the wind turned to the southeast in a tremendous gust."[115] Anne Simons Dea peered out of her house at "a wild and desolate looking scene" and looked on as "the rain was swept in misty sheets before the roaring wind, the yard like a sheet of water." The wind continued to howl at sixty to eighty miles per hour well into the daylight hours.[116]

But away from the ocean, the hurricane's power had lessened. There, the storm brought destruction to crops and tree branches and left peoples' lives alone. "Columbia stood it sturdily," a resident reported. "Our trees went down but not our houses."[117] All throughout the state, corn and cotton had "been laid flat," with "barns, gin-houses, etc." "blown away."[118] Shade trees splintered in half, roads "blockaded" with fallen trees and carpeted with leaves.[119] The night must have been nerve-wracking for everyone from Orangeburg to Greenville, but absent the storm surge, it was not deadly.

The storm spun rapidly northward, its form disintegrating, once over land. By Tuesday morning, it had already reached Oswego, New York.[120] As far north as New York City, the hurricane blew off roofs, uprooted hundreds of trees, toppled chimneys, and flooded the low-lying fringes of the city.[121] It downed the telegraph service along the entire eastern coast of the United States, plunging the region into silence from Boston to north Florida and as far west as Chicago. Once north, the storm was drained of its rain, dropping less than an inch of precipitation. Its final breaths of wind, with gusts up to forty-eight miles per hour, and the waves it dragged ashore at last exhausted its power. It fell apart over the St. Lawrence River by Tuesday night, after a journey over Lake Champlain and across the border to Canada. No deaths because of the hurricane were recorded on land beyond North Carolina, but because of the hurricane's strength and size, it made its presence felt and revealed the vulnerability of the new systems of communication that represented innovation in the late Gilded Age.

Over eight days, the hurricane had cut a path of 3,300 miles.[122] As it dissolved and dried up, disconnected from the humid late-summer heat of the ocean, it left behind thousands of dead and tens of thousands of homeless, grieving survivors. While residents of the sea island coast were no strangers to hurricanes, the hurricane of August 1893 was the largest and deadliest to ever strike the South Carolina coast—before and since. The giant storm slammed into northern Georgia and the South Carolina sea islands, a fist punching into the porous concavity of the sea island coast. The hurricane roared through the region, from Tybee Island in northern Georgia to Murrells Inlet in South Carolina. It pitched the Lowcountry into a stormy, dark chaos and formed an uninterrupted front of unprecedented destruction from about three o'clock Sunday afternoon, August 27, until late Monday morning, August 28. One Sunday morning in August 1893, sea island residents had awakened to a day that seemed like any other. They tended their gardens, chickens, hogs, and cows, had breakfast with their families, and went to church. By the next morning, thousands were dead, and those left behind found themselves suddenly impoverished, now facing a battered, flooded, stripped landscape that reflected the disorienting, uncertain future that now lay ahead.

As it wrecked the coastline, the hurricane also marked memory. "Big storm?" one African American woman responded, when asked about the hurricane forty years later. "Yinnah talk about big storm hang people up on tree?"[123] It battered the sea island coast for a sliver of time, but because of the trauma it caused, those hours left an outsized impact on the survivors' lives. Years before and after may have been compressed, details barely remembered, but the hurricane left a deep impression. For most survivors, it became a waypoint: before the hurricane or after ordered events and marked episodes of change in their lives. Hundreds of years of occupation of the Lowcountry by people of European descent and inhabitation by enslaved Africans and their descendants had incised patterns onto the landscape of labor and exploitation, of trauma and endurance, of knowledge and community. The Great Sea Island Storm of 1893 challenged those histories' legacies, deepening some, wiping out others, and carving fresh channels.

Chapter 3

The Survivors

In the gray dawn, inhabitants of the Lowcountry looked out at an unfamiliar landscape. They gazed upon the silent faces of the dead, their soaked, flattened fields, and their damaged houses, or bare patches of ground where their homes had once stood. The ruin seemed as intractable and insurmountable as perhaps any they had yet faced. The hurricane had rendered their homeland a strange place. Not only had the hurricane scrubbed the Lowcountry of its harvest, its communities, and its August verdancy—it had swallowed families whole and ripped loved ones from the reaching hands of their desperate kin. The blighted scenes made the wrenching sorrow of coastal African Americans tangible, a physical manifestation of their misery.

In the days immediately following the hurricane, before relief efforts reached a coordinated level of organization, Lowcountry residents grappled with the tragedy around them. But, as with any disaster, the hurricane's derangement was not uniform. The extent to which the hurricane had disordered their lives depended on the hierarchy and strictures of New South capitalism and Jim Crow racism. Though some white Lowcountry residents did suffer terrible losses, of loved ones and property, their wounds were not deepened by the pointed, spiteful apathy of white southerners who suggested that the hurricane was a natural outcome for a population poorly fitted for property, citizenship, and humane treatment.

Indeed, rather than the hurricane as a disturbance and its devastating death toll as the sign of a world gone awry, many white South Carolinians came dangerously close to welcoming the storm as righting the wrongs that the Civil War and Reconstruction had set into motion. The hurricane recalibrated the racial order, reducing once-autonomous African Americans to conditions of poverty and dependence and revealing, according to white South Carolinians, their inability to maintain a steady livelihood outside the bonds of white control. The white elite sought to naturalize the penury and despair of African Americans after the hurricane and to make those conditions permanent. As a step further, they used the hurricane's damage to contend that, because so many African Americans perished in the storm and faced starvation after, they were in fact incapable of supporting themselves and deciding their own future—that they were, essentially, unfit for freedom.

The white elite, most of all those living in Charleston, hastened to downplay the hurricane's effects, to urge a return to normalcy, and to leverage their wealth and power over the hungry, grieving African Americans on the coast. But the normalcy that they sought was impossible for those caught in the grip of sorrow and was anathema to Black communities. The white elites' equilibrium required white dominance and Black servility. In their actions, speeches, and writings, they wasted no time in articulating a vision for the Lowcountry that naturalized their power and delegitimized Black self-determination. The hurricane was, for them, not so much a disaster as an opportunity to consolidate power and to assert an ordered regime that prioritized their political and economic ascendancy.

The same hurricane that the white elite greeted as a blessing in disguise appeared as the angel of death to African Americans, a harbinger of troubles to come. Over a night, the hurricane had brushed away homesteads and livelihoods that had taken decades to secure and communities that had struggled to assert their autonomy. And in Beaufort County, which had borne the brunt of the storm's fury, it was difficult even for wealthy whites to pretend that there could be a hasty return to business as usual. The volume of death and destruction weighed upon the sea islands south from Charleston to the Savannah River. The hurricane had damaged every wharf, warehouse, barn, boat, field, and house in the region. Even counting the dead proved an overwhelming challenge. African Americans had to fight off the looming specters of hunger, poverty, and disease, which deepened the trauma of that one stormy night that had so profoundly changed their world.

It was hard to imagine a way forward through the muddled terrain and the dislocation from daily routines, familiar landscapes, and the comfort of kin now gone. Many were too heavyhearted to do more than try to survive, one day at a time. As sickness, starvation, and exposure cast a further pall on the ravaged coast, the hurricane morphed from a singular event into a long, ongoing disaster, which sank its claws into the survivors and did not let go. In the howl of the hurricane, African Americans heard an echo of the lives they had lived before, and which they feared would come roaring back: a return to a white supremacist regime where they lacked meaningful control over their labor, their lives, their communities, and their futures. Disorder or equilibrium, hierarchy or self-determination—the hurricane brought forth competing visions over what was natural in the southern political environment.

On the rain-lashed morning of Monday, August 28, Charlestonians left their houses to find "a city which had been through a great siege of shot and shell and a subsequent deluge."[1] The old city was in total disarray, much to the chagrin of white Charlestonians, who prided themselves on the city's grace and beauty as a reflection of the supposed gentility and paternalism that overlaid an ordered racial hierarchy. Waves wrecked the new iron-and-steel drawbridge over the Ashley River, which connected Charleston to the truck farming districts on James Island and was supposed to represent and facilitate the new vitality of Lowcountry agriculture.[2] Charleston's dozens of wharves, which stretched around the peninsula, and the hundreds of tugs, schooners, barks, canoes, and fishing boats moored to them were pummeled to splinters by the grinding waters. The commercial fringes of the city had been thrashed to pieces by the storm surge and buffeting winds.

The destruction extended well into the heart of the city. Brick-and-mortar sidewalks had crumbled from rushing floodwaters. King Street was cluttered with tin roofs, felled branches, and telegraph, telephone, and trolley wires. The handsome shade trees on Rutledge Avenue had been carried away in the torrent. As one Charleston reporter commented wryly, "lumber, window-shutters & indiscriminate woodwork to be found on Rutledge Avenue and Lynch Street when the tide [went] down would start a comfortable woodyard business."[3] White Point Gardens, which the *News and Courier* declared the "pride of the city," looked disgraceful: the newly planted umbrella trees had toppled, and the live oaks had fallen.[4] The city, the newspaper lamented, was "almost in ruins." The fine houses of the white elite, the beautiful Battery upon which white Charlestonians loved to promenade, and the wharves and warehouses that symbolized and constituted the city's commercial wealth had all undergone serious damage. Glossy displays of the wealth built by Black labor and funded by international capital had been dimmed by the gray spray of the ocean. Gruesomely, dead chickens floated in stagnant pools of saltwater across the city, which could not bode well for the city's water supply.[5]

Yet white Charlestonians did not dwell on the scenes of destruction for long. Instead, they quickly crafted a New South narrative of unity and triumph in which the hierarchies of economy, society, and politics could not be disordered by the ravages of man or nature. No hurricane, not even the most destructive in the city's history, could long cloud their fortunes. An attitude of defeat could not be allowed to take hold. Progress required a steady drumbeat of optimism. Indeed, these delusions of unity surmounted the divides of race, as "white men and negroes, merchants and laborers" were "all working together to clear the city."[6] The *News and Courier* hastily scrapped its catalog of the city's wounds and described Charlestonians' response to the hurricane as evidence of the city's historic resilience. Charleston had overcome a litany of woes: it "arose from its ashes at the end of the late war of secession," "battled with the plague for years and finally conquered it," "emerged from the

The High Battery of Charleston, after the hurricane. The storm surge leaped over the seawall and, as this cyanotype shows, jumbled the massive, newly laid flagstones on the High Battery. "7. East Battery Looking North," Charleston Hurricane, 1893 Vertical File, 30-29-7, South Carolina Historical Society.

wreck of the great cyclone[s] of 1874 . . . and 1883," and, "by indomitable will and unflinching courage and the energy of its sons, aided by the practical sympathy engendered in the bond of brotherhood of man," was rebuilt after the earthquake of 1886.[7] Indeed, the "battalion[s] of stalwart workmen" on the streets hacking at fallen trees and debris only proved that "the people of Charleston have indeed learned the knack of never surrendering." If the force of the federal government could not bow Charleston in 1865, then a hurricane surely could not either.

Despite the many Charlestonians who could not help but observe the hundreds of thousands of dollars' worth of damage done, the white Charlestonian elite crowed that the hurricane had posed no real threat to the city. Ex-mayor George D. Bryan urged a quick return to the city's "march of progress."[8] A former Confederate captain, Henry Schachte, boasted, "We are solid as a gold dollar, and with plenty of money being brought into circulation, good is sure to result." The businessman Sam Israel assured a *News and Courier* editor that "we have no reason to expect anything worse than we would have

Charleston wharves, after the hurricane. The hurricane's waves tore apart the wharves jutting out from the eastern edge of the Charleston peninsula. "6. South Commercial & Southern Wharves," Charleston Hurricane, 1893 Vertical File, 30-29-7, South Carolina Historical Society.

gotten had there been no cyclone." The *News and Courier* soon informed its readers that "there is nothing to cry about, and nobody is crying." As several businessmen and leaders pointed out, the city had weathered the storms of war, epidemic, and natural disaster before, and nothing had yet conquered it.

The statements of these politicians and businessmen read as a defiant stance against any challenges to the Charlestonian status quo that, by 1893, they were successfully defending. These leaders saw themselves as titans of a city that had been the heart of the Confederacy and that had overcome intruding foes, most of them connected to the federal government and national-level events: the U.S. Army, Reconstruction, alterations to currency and fiscal policy, an economic depression. And now, the city had faced down a terrible storm. Charleston could stand strong against them all and persist with a specific vision for economic, political, and social success that extended narrowly to rich white Charlestonians. The white elite were continually under attack and continually victorious, simultaneously.[9] The hurricane gave them another opportunity to pursue this triumphalist narrative, which justified limiting the rights of a broader, mostly African American populace; lining their own pockets with the profits of cotton, rice, phosphate, and lumber; and posturing as victims of a larger system of federal power and global capital, which did not understand or respect their way of life.

None of this was fair or accurate, but that did not matter. The success of this story's propagation, in 1893, became increasingly apparent as federal will to enforce civil rights in the South had waned. A more sympathetic view might construe their hearty proclamations of future success as the necessary optimism of civic leaders cheering a demoralized people. But the broader context in which white Charlestonians time and time again refused to acknowledge defeat instead points to a group of men declining to construct a world less pitiless—and less personally profitable.[10] Before the Civil War, white Charlestonians in the antebellum era found that the world of slavery and white supremacy that they had constructed brought them both massive power and massive insecurity. After the Civil War, their refurbished socioeconomic systems of agrarian and industrial use relied on restrictive laws limiting African American rights and granting themselves, through new corporations, unfettered access to labor and the final word on environment management.[11] Their insecurities nonetheless did not lessen, and the instability of the new system upon which they relied for wealth and power required a vigilant and vigorous campaign of words, money, policies, and racialized violence to bolster it. Harnessing the hurricane into their accepted narrative, then, was important in the days immediately after the storm.

The city's government marshalled resources to restore order. Charleston's Street Department, under acting superintendent W. S. Ogilvie, joined with a civilian force to clear the streets of rubble and debris.[12] The elegant magnolias on Rutledge Avenue were no more, same as the flowering, poisonous oleanders and young poplars on Broad Street.[13] King Street was a tangle of wires, tin roofs, and branches, though King, Calhoun, and Meeting Streets were more passable than tree-lined Rutledge and other shaded thoroughfares.[14] The city's Street Department employees labored alongside anyone who was willing to swing an ax, shoulder trunks off of the road, or, more dangerously, remove live electrical wires from the streets. One hundred thirty-eight employees manning fifty-one carts undertook the task of digging out the city. They cleared 7,528 loads of debris, 589 stumps of downed trees on city property, and 449 drowned animals, costing the city $8,000. The Street Department also spread on the streets and distributed chloride of lime, slacked lime, and "Copperas Solution" to residents as disinfectant that they hoped would ward off waterborne illnesses or others related to poor sanitation conditions.[15]

Some of the labor was not voluntary, casting doubt on the *News and Courier*'s cheerful claim that "the people are fairly bubbling over with pluck and

Charleston street scene. In this photograph, a scan of a postcard, a group of African American men on Meeting Street may be gathering to clean debris, since a few seem to be holding tools, while a group of little boys tag along and women watch from second-floor windows. "Meeting Street Road, near Power House," Charleston Hurricane, 1893 Vertical File, 30-29-7, South Carolina Historical Society.

patriotism and energy."[16] The city of Charleston's convict force, under Superintendent W. H. Halsall, sent ten chain gangs onto the streets to assist citizens and municipal employees of the Street Department in their work.[17] The city's convict force was an invaluable, and quite new, pool of labor for the Charleston Street Department. The city of Charleston began to use people convicted of misdemeanors by the municipal courts in 1892 for free, forced labor and would continue to mine that source for decades.[18] Convict laborers were a common sight around the state (though since the mid-1880s the South Carolina General Assembly had passed bills that made it prohibitively expensive for private businesses to contract imprisoned laborers from the state penitentiary).[19] The state itself put convicts to work instead, and counties and cities around the state replicated this exploitation of forced imprisoned laborers to improve infrastructure.

Charleston followed the state's example. In 1893, 258 convicts from the city jail sweated on the streets of Charleston, digging ditches and grading roads, at the cost to the city of fifty-five cents per convict's room and board, per day.[20] The Charleston city police department recorded that in 1893, 442 Black men were sentenced to hard labor, as opposed to only 96 white men.[21] Disorderly conduct, drunk and disorderly charges, and larceny were the three misdemeanors with the highest arrest rates, most of all for African Americans. The use of forced labor might not seem to mesh with reporters' and politicians' optimistic characterization of Charleston as a metropolis of progress, determination, and resilience or a rosy vision of postdisaster unity, in which the hurricane served as a leveler joining people of different races and classes to save their beleaguered city. However, the sight of forced laborers moving debris was commonplace to Charlestonians. Indeed, to white Charlestonians, convicts' presence in the streets might have been a sign of the city's system functioning as normal. The exploitation of Black labor undergirded the hierarchy that made Charleston run for its white elite. White Charlestonians set about immediately reifying their dominion over Black Charlestonians, eager to prove that a great derangement of nature could not diminish their power.

The white elite did not limit themselves to making cheery statements to reporters. In fact, they began to formulate their own solution to the hurricane's destruction, the economic future of the Lowcountry, and what they saw as the problem of autonomous, politically active Black sea island communities. Harry Hammond proposed a plan that would circumvent the need for state aid and permanently drain the coast of its African American population. As one of the most prominent agricultural policy makers in South Carolina after the Civil War, Hammond's voice mattered. It mattered, too, because of his father, James Henry Hammond, who had once declared, "Cotton is king": despite a public reputation as a serial sexual abuser, the former governor and U.S. senator from 1857 until South Carolina's secession in 1860 remained a powerful man until his death in 1864.[22] Harry benefitted from this connection, surely deriving legitimacy in the eyes of the white elite of South Carolina, for whom pedigree counted foremost among inherent virtues. Harry served briefly as a professor of natural science at the University of Georgia before fighting for the Confederacy during the Civil War.[23] In his postwar career, he maintained an occupation as the owner of a cotton farm but held significant positions in state government, including supervisor of the 1880 census in South Carolina.[24]

Hammond was thus a well-known figure in the state for his agricultural expertise and family name, which he matched with a vision that advocated for careful land management and promoted control of Black labor. Hammond inflected his racism with scientific principles, economic thrift, and frequent appeals to what he saw as plain common sense but was in fact a logic of white supremacy. If his father represented the nastiest, worst excesses of the United States' slave empire, so the younger Hammond embodied a newer, if no less racist, form of governance in which a seemingly mild, benign interest in improving southern agricultural production coalesced with a desire to control

Black labor at any cost. He was, in short, a planner of the New South, which would harness Black labor for efficient agricultural production untethered from the interferences and redundancies of the federal government.[25]

After the storm, Hammond saw a new opportunity to implement his vision. In an editorial published in *The State* soon after the hurricane, he suggested that agricultural organizations like the Farmers' Alliance should give Black sea islanders train tickets to the cotton, pea, and cornfields of the upcountry South, where "all may engage" in cheap labor, including "women and children from eight years to extreme old age."[26] Hammond argued that the "peasant proprietory [*sic*] of the Sea Islands had reached the end of their career," having misused the "genial climate," "abundant supplies of fish and fruit," and "fertile soil" of the Lowcountry and taken advantage of lands that were "almost a free gift to them," despite having been "aided by government largesses [*sic*] and magnificent charities," "the building of a new railroad," and "the development of the extensive phosphate works," all while remaining "protected from any unfriendly outside influence."

Hammond described the natural beauty and abundance of the Lowcountry as a present to Black sea islanders who had only recently thrown off the shackles of slavery. He constructed a convenient but false historical narrative that omitted the fact that their labor over hundreds of years had constructed a Lowcountry environment that yielded profit to white enslavers. The Lowcountry, a bounteous landscape only with an immense amount of hard work and constant maintenance, was no gift to African Americans. They had earned it through centuries of abuse and forced labor.

Hammond saw it differently. In his view, the beneficence of nature and government were for naught. "The result?" he demanded: "One thousand perish in a catastrophe that destroys only three of their white neighbors! What promise is there that any outside help can enable them to build up their waste places[?]" In Hammond's view, coastal African Americans had squandered their opportunities since emancipation, and they could rely upon their relative isolation and white charity no longer. They were fundamentally unfit to live in the region, to grow their own crops, and to determine their own future. They had to be cleared out for their own benefit, and Hammond envisioned a hasty exodus that would solve the problem of a Black-majority Lowcountry. Hammond's benevolence would allow Black men, women, and children to work in land not their own, uprooted a second time from their homeland and "spread out among the white population." This mass dispossession would clear the sea islands of Black landowners—and Black voters. It would solve two problems: the problem of the hurricane's degradation of Lowcountry agriculture and the problem of the region's Black majority.

Hammond's proposal did not come to pass, but it reflected the animosity that wealthy white South Carolinians held toward Lowcountry African Americans. Like Hammond and his Charleston counterparts, other white landowners in the region sought to downplay the grievances of the hurricane and to deride African Americans as deserving of their misery. Lewis Grimball, in his typical haughty fashion, wrote to his sister Elizabeth about the crisis's impact on him and his rice plantation.[27] Lewis knew of other losses, but in the same letter, he both confirmed the terrible calamity and waved it away as exaggerated or as an opportunity for abuse of charity:

> The loss however has fallen heaviest on those who had anything to lose—it will turn out badly in the case of the negro—they are now making a row and howl all over the country about their desperate condition, when really there are only those down near Beaufort and the Islands lying immediately on the ocean who need looking after, and that only for a little while. Around here I have not heard of a case of utter destitution—The result will be to make the negro lazy and sassy and less inclined to work than he was before, and what is intended as a great charity and blessing, may turn out quite the reverse in many instances.

Grimball was not actually troubled by the potential for Black indolence. Indeed, any sign of laziness among African American workers would be taken by white landowners as evidence of white superiority and Black inferiority. Grimball believed that African Americans could not be trusted to wield control over relief efforts, and Grimball implicitly contended for the necessity of controlling African American access to power. Grimball invoked a white supremacist vision of Reconstruction, suggesting that any formal structures that assisted African Americans were inept and useless at best, corrupt and harmful at worst.

As Hammond and the white elite saw it, the hurricane could be an opportunity for white South Carolinians to empty the region of the African Americans who had long held onto the land and used the leverage of a high proportion of African Americans and their land ownership to maintain the franchise. Unseating Black sea islanders would be a final blow against the legacies of Reconstruction, forcing them to leave behind their land, their political power, and their culture, with results as devastating as those of the hurricane.

Many of the people living in Beaufort and the surrounding countryside must have seen through white Charlestonians' narrative of progress and perseverance as a false front. Once the floodwaters began to recede in the late morning, the sun poked through the clouds to shine weakly on the slippery, black pluff mud of the salt marshes. A thin film of evaporated salt and dried sediment stretched over the bare mud as the hurricane's storm surge drew back. Spartina, the cordgrass that dominates the expansive flats of intertidal marshes, began to weep salt from its spiky leaves. In the warmth and stillness of an August low tide, salt crystalized on the bright green leaves. The saline, sulfuric funk of pluff mud—emitting from the bacteria that thrive on the decayed biomass of spartina—permeated the humid air around Beaufort.

But by the afternoon, a stench not of the usual bouquet of sulfur, salt, and sea breeze began to rise from the marsh. George A. Reed, the Beaufort County sheriff and one of the few (if not the only) Black sheriffs in the country, looked out over the tidal flats around Beaufort. He had survived the night because his house, sixty yards off a beach, had two stories: the bottom floor was inundated with two and a half feet of water.[28] The hurricane had damaged his home, and he spent the night "occasionally hear[ing] shrieks and screams" that pierced through the gale, but his family had made it through, for which he gave thanks to a "protecting Providence."[29] Now that the storm had passed, he rushed into town to learn of the damage. The destruction assaulted him on every front: the sickly smell of death; the sight of the wrecked wharves, warehouses, and homes; the cries from distraught sea islanders rending the air. Reed suspected the extent of the harm was as complete as any he had seen.

Desperate, grieving survivors paddled a paltry few canoes from the nearby islands, the only watercraft on the sea islands that could be salvaged from the storm. They brought with them powerful hunger and thirst and tragic tales of families swept away and entire islands flattened of trees, crops, and homes. Reed heard early reports of three hundred dead on St. Helena Island and of soil salted by the ocean so that the crops were already rotting in the fields. Bodies crammed the roads, the retreating storm surge having deposited them there. Through his telescope, Reed could even see some of the sea island sufferers just across the river from Beaufort, but without a means of crossing. Not that Beaufort could offer much relief: its wells were brimming with seawater, dozens of homes and businesses were in ruin, and most of the town's boats were gone. Reed knew that "the wreck in Beaufort" was "the worst that the oldest inhabitant has ever seen in his life," with "everything seriously injured."[30] And yet he also recognized that Beaufort would be the point of convergence for relief efforts and a beacon of hope for Black sea islanders in need of food and clothing.

In Beaufort, as in Charleston, the hurricane took a serious toll on personal property, affecting African Americans and wealthy whites. The storm damaged, for example, the grand home of state senator William J. Verdier, a moderate Democrat who worked alongside Black Republicans in Beaufort County in 1888 to form a compromise "People's Ticket," a pragmatic political bargain that staved off the rise of Tillmanite politics in the county. The loss of his library and study were lamented in the newspapers as a tragedy on par with human death, based on the amount of coverage it received.[31] Several blocks away from Verdier lived Robert Smalls in an elegant antebellum house that had once belonged to his enslaver (who was also his father). In 1893, Smalls was three years into a thirteen-year term as the port of Beaufort's customs collector. Within the more racially integrated context of Beaufort, Smalls largely maintained his position as a respected town leader. His house, an old, solid building, emerged from the storm apparently unscathed.[32]

The storm incapacitated the Beaufort business district too. Niels Christensen emerged from his house the morning after the hurricane and looked at his once-prosperous lumberyard with dismay.[33] He was a Dutchman who had migrated to the United States during the Civil War and served in the U.S. Army, eventually as captain of the Forty-Fourth U.S. Colored Infantry; he settled in Beaufort to start a family and a business. The waves crested the seawall surrounding his garden, lush with fig trees and rose bushes, filled his house and his store, and scattered his lumber.[34] His cow and his $200 horse drowned when the winds blew away his barn. The tone of his letters to his wife, Abbie, then living in Massachusetts, became more and more despairing as he narrowed in on an accurate tally of his losses.[35] Other Beaufort businessmen also suffered heavy financial losses. F. W. Scheper, a ship chandler who owned a grocery store on Beaufort's waterfront Bay Street, lost property up to $20,000. George Waterhouse, one of Niels's closest friends and associates, ran a cotton warehouse and wharf on Bay Street. The cotton warehouse, which in late August was stuffed with the year's harvest, was an utter wreck, cotton either soaked to ruination or swept away, and the wharf was simply gone. The cotton-ginning machinery lay in the muck at the bottom of the Beaufort River.[36]

Indeed, as in Charleston, every wharf and dock that jutted out from Beaufort's waterfront was wrecked by the hurricane's surge. The delicate threads of connection from the sandy sea island soil, to the long, silky strands of sea island cotton, to Black labor and harvest, to sale and distribution by white businessmen, had been severed by the hurricane. The hurricane, in erasing the mercantile life of Beaufort, irrevocably broke some of the town's merchants, too. Waterhouse, a "quiet New Englander" who dedicated himself to the Baptist Church of Beaufort, died of a heart attack in 1894, and his family repeatedly blamed the hurricane's annihilation of his business as "the proximate cause of his death."[37] The hurricane killed long after it disintegrated over the St. Lawrence River, its destruction amplified by the human exigencies of capitalism.

"By noon," *The State* newspaper declared the day after the hurricane, "every one on the chain of islands had become a grave-digger."[38] All capable inhabitants emerged from their battered homes or their friends' or neighbors' homes, crawled down from the branches of a sturdy tree, or waded out of floodwaters to begin the hard, heart-wrenching work of recovery. This was recovery in the most immediate sense: finding loved ones, burying bodies, clearing streets and houses. Without functioning telegraph wires or trains, and with few craft remaining seaworthy, coastal residents found themselves disconnected from other islands, towns, cities, and locations farther inland. The coast had been, for many hours, a "country transformed into an almost unbroken swamp by the cloudburst."[39]

The ocean fully receded by the early afternoon, but it left fields a flooded mess of vegetation logged with saltwater and cisterns filled with brackish, unhealthy water. It had ripped houses down and pushed the corpses of people and animals alike onto roads and into the marshes.[40] Essie Roberts of St. Helena recalled shudderingly that the "blood of animals looked like it was all over the place when the tide went down."[41] The winds had died down, but the trees left standing had been stripped of their foliage, reducing the deep, glossy green of August to a strange winter desolation. The hurricane rendered the sea island coast unfamiliar to its own residents, and the visual dissonance compounded the confusion and pain that many felt.

The enormity of the situation was paralyzing: "I simply didn't know what to do," Sheriff Reed said. Still, Reed fought through his exhaustion after the wild night and his certainty that worse things were to come. He helped clear streets, listen to the stories of survivors who were "only saved by climbing to the tops of trees and lashing themselves there," and bury bodies "in revolting condition."[42] The survivors, as they searched, had to desensitize themselves to what surrounded them. Reed observed that "unless two or three or perhaps five or six bodies are found at one time the survivors display absolutely no feeling."[43] The effort to shove down emotions that could overwhelm them and prevent them from assisting in the search must have been immense. Rather than collapse into sorrow, the searchers would "simply say, 'there's another,' and dump [the body] in" to the shallow graves that they dug along the marshes. Just as Reed weathered the night listening to the roar of the hurricane, pierced by the shouts of the storm's victims, he spent the two days after hearing "screams, shrieks, and cries all over the town" as families learned of the deaths of loved ones on the islands.[44] "It was horrible," Reed recounted a week after the storm struck, "and has been pretty well kept up right along ever since."

Sea islanders discovered hundreds of bodies lodged in the marshes, floating along intertidal rivers and creeks, and huddled in their ruined houses. White landowners counted the dead Black workers and sharecroppers who lived on their land: Julian Mitchell found eleven; E. M. Whaley, nineteen; and at the Clay Hill property on the Combahee River, eighty drowned.[45] Along the Savannah River, 150 rice workers, none recorded as white, drowned.[46] People in Beaufort pulled seventy bodies of Black rice workers from the Coosaw River, many "so far advanced in putrefaction that the sight was sickening and revolting."[47] The bodies of twenty-one sailors washed up on Rockville in South Carolina, buried on the shore where residents found them.[48] Two African American men who rowed their small boat to Beaufort from St. Helena Island reported that their whole families had perished. One had lost a wife and eight children; the other, a wife, thirteen children, and his mother-in-law.[49] Eustis Place, a former cotton labor camp on Lady's Island, was the burial ground of fifty-four Black sea islanders, fifty-one of whom were placed in the same grave just two feet deep.[50] The dire circumstances forced Black sea islanders into foreign burial practices—no church services, no mass gathering of their loved ones, and nobody laid to rest deep in the woods in the communal graveyards of the Gullah, where their spirits could dwell with those of their ancestors.[51]

The State Phosphate Inspector A. W. Jones, who monitored the coast's phosphate industry and facilitated relations between the state government and the industry, estimated that no fewer than 1,500 had drowned in the hurricane.[52] The day after the hurricane, he rode out with Sheriff Reed on the *St. Cecilia*, one of the only larger craft that the storm spared, to assess the extensive damage to the area phosphate mines. He judged that the Beaufort-area companies had lost a total of $350,000. At Cain's Neck, Coosaw Island, and the Pacific Phosphate Works, he tallied forty-two coffins and ninety-three dead.[53] In one house on the Johnson River, fifteen bodies were recovered in one house. That, however, was not "even a small proportion of the bodies."[54] Many had been "swept far up in the marshes where man cannot go," Jones noted, and often the corpses of "deer, cows, hogs, snakes and all kinds of wild animals piled up together" rendered certain creeks, marshes, shorelines, or fields "so offensive that no one can go there." He speculated that gathering precise data about the dead would likely be impossible.

That insurmountable task fell to the Beaufort County coroner, who struggled to collect and identify the deceased. He appointed six new "coroners" in an attempt to gain a handle on the number of dead, but this was a piecemeal effort in a system overwhelmed by the sheer volume of death.[55] He and his deputies eventually compiled a list of over 800 dead, but even this was surely but a fraction of the total.[56] Death overwhelmed the sea islands, from the personal calamities of one's whole family gone in a single night to the communal trauma of finding a kinship group diminished, swallowed by the ocean's maw. The hurricane had destroyed too many boats to launch a proper body recovery effort. Heat, scavengers, and saltwater disfigured many bodies beyond recognition. Many of the African Americans who combed the marshes for their dead buried the bodies on the spot, perhaps without reporting numbers to the local coroner. Gathered over the course of several days by the Beaufort coroner and Charleston city officials just up the coast and trumpeted in newspaper headlines, the tallies of the dead were incomplete, and many sea islanders were too exhausted and heartsick to keep up the count.

Weathering the hurricane itself was no guarantee of survival either. The specter of death haunted the living, many of whom were homeless, their crops gone, their subsistence gardens ripped out at the roots, and their belongings swept away. This was true particularly for African Americans who lived on the rice farms upriver from Charleston and Beaufort and on the sea islands, the areas most vulnerable to the storm surge. The newspapers cataloged the destruction of rice fields, and implicit in those descriptions is the toll those same floodwaters took on Black rice laborers: along the Edisto River, 2,500 to 3,000 acres of rice flooded; along the Combahee, 7,500 acres; the Santee and the Cooper Rivers, 3,000 acres.[57] Rivers along the coast spilled their banks, leaving land far inland underwater, and most of the laborers' homes were swept away.[58]

The sea islands were in a similar state. On Eustis Place on Lady's Island, where those fifty-four African Americans died, only three of fifty houses were left standing. On Coosaw, just twelve of seventy-five.[59] A survey of houses on St. Helena, Coosaw, and Wassa Islands, and Eustis Place found that of 1,078 houses, 217 had been washed away, 254 were wrecked, and 220 had suffered damage, whether a lost roof or splintered walls.[60] Rebuilding homes was impossible without lumber and tools. While sea islanders could potentially scavenge wood for their homes, tools were in short supply. Many hastily constructed rudimentary shelters—tents, or "bush camps," grouped in "whole settlements" along the islands, including a refugee camp of hundreds on the grounds of the Penn School on St. Helena.[61]

African Americans also feared that hunger and thirst would combine with exposure to weaken them. Food supplies had already been ruined. Riotous green potato vines rotted into black lines of slime scoring the earth, and most potatoes had dissolved into a salty mush where they grew. The potatoes that Black sea islanders did dig from the ground were "salty . . . speckled and small," and "far from being as nutritious as they usually are."[62] Former Democratic congressman William Elliott reported to newspapers that "starvation is sure and imminent."[63] One journalist wrote that on Lady's Island, "desolation and ruin stand out conspicuously from a background of starvation and terror."[64] Black sea islanders rowed their canoes to Beaufort to ask for food the day after the hurricane, wearing only undershirts and torn trousers. Boats that arrived on sea islands were met by hungry people hoping that they had brought provisions with them. One observer reported seeing three hundred waiting at a landing on St. Helena a week after the hurricane.[65] Diana Brown, an African American woman living on Edisto Island, had weathered the storm in her boarded-up house but watched days later as her neighbors "come home naked" after "de tide carry dem out."[66] They had slaked their intense hunger, they told her, by catching and eating birds raw.

Thousands of African American sea islanders tried to obtain employment as soon as possible. In the few days after the hurricane, Moses E. Lopez, known as the "king of Coosaw," turned away hundreds of men daily who came looking for work at his now-defunct phosphate operation.[67] Unfortunately, though, "there [was] little for them to do" after they had found and buried the dead and cleared the roads.[68] The crops were ruined, stored food was sodden, and the phosphate fields were not hiring. A sense of helplessness settled over the islands.

With drinking water and earth contaminated with salt, Lowcountry residents also began to fear sickness. Not only did the bad water, lack of food, exposure, and isolation suggest that widespread disease was forthcoming—so too did the pervasive stink in the air.[69] Though by the early 1890s, the miasma theory had largely been discarded by medical professionals, it still informed how many sea islanders understood their environment. Rachel Mather, the northern schoolteacher who moved to Beaufort County during Reconstruction, posited that the very air was poisonous to health. "How," she asked, "shall all this debris with decaying animal and vegetable matter be buried or thrust into the sea—debris whose foul exhalation in that hot, moist climate would quickly poison the air and breed fatal disease or perhaps extensive pestilence?"[70]

Lowcountry residents fixated on the smells around them, certain that the dying animals, dead vegetation, and human bodies emitted "effusive fumes."[71] One Beaufortonian recounted that "the smell of decayed animals and dead bodies scattered in every direction, which even the buzzards from far and near cannot clean up, and the decaying leaves" "make the very atmosphere fetid."[72] He feared that "unless the cruel hand of further death by pestilence is not stayed, it would look as if this whole people must be swept out of existence by means slower but no less surer than the quick and sudden sweep of the hurricane." While the theory of disease that many sea islanders still ascribed to was not precisely correct, they did grasp the dire need for action and assistance and the dangerous effect of decaying organic material in drinking water.

Physicians debated the chances of an epidemic. In both towns and rural areas, physicians knew that dysentery could result from poor water quality and that diarrhea, dehydrating as it was, could prove dangerous among a weakened population without sufficient access to food, shelter, and medical care. After all, nearly every well and cistern on the coast was now filled with brackish, unhealthy water. Mosquito-borne illnesses, common and feared occurrences along the coast, also threatened sea islanders' lives. That fall, as a yellow fever epidemic raged in New Brunswick, Georgia, Charleston kept a strict and successful quarantine to prevent its spread. As doctors observed the large area devastated by the hurricane, they were doubtful that quarantine would be achievable. Dr. Daniel T. Pope of St. Helena Island demonstrated adherence to the idea that the entire countryside was unhealthful: "The whole country smells so bad that I hate to ride the roads," he wrote, and he feared that the decay would "produce a great deal of sickness."[73] "Pestilence," he declared, "seems imminent." Dr. James Wood Babcock of Columbia, sent to the coast as an emissary of Governor Tillman, ordered that five pounds of quinine, five gallons of castor oil, 500 compound cathartic pills (a laxative), two gallons of paregoric (a tincture of opium that worked as an antidiarrheal), one gallon of laudanum, and two pounds of calomel (a compound of mercury taken as a laxative or disinfectant) be sent to Beaufort for distribution to cleanse the city and its residents of pestilence.[74]

In the days after the storm, physicians began to report numerous cases of fever and bowel illnesses across the sea islands. Dr. Peters of St. Helena diagnosed one hundred people suffering from diarrhea, and one hundred more exhibited symptoms of "malarial fever."[75] Along with illness, he treated seventy injuries that required surgery. Three weeks after the hurricane, a doctor on Edisto Island noted that cases of fever on the island spiked as the temperature rose, and on Young's Island, six cases of "malarial fever" appeared on one plantation over the course of two days.[76] More and more sea islanders sickened over the weeks, in desperate need of medical care.[77] Rachel Mather described disease as common on St. Helena, where "in most of the houses one or more is prostrate with fever now prevalent on the islands, and has been ever since the cyclone. A scanty diet and exposure aggravate the disease."[78] As physicians saw it and as sea islanders experienced it, the hurricane had exaggerated the worst natural features of the Lowcountry—the humidity, the quickness of decay, the low ground's tendency to flood—and left behind a region ripe for epidemic. Disease, hunger, thirst, and exposure were a potent mix within a landscape that medical authorities characterized as unhealthful to begin with.

A less material element clouded survivors' futures: heartache. Susan Hazel Rice, the teenaged sister of Dr. Hazel, who had perished on Parris Island, wrote of her brother's "poor discolored face" once his body was brought home and lamented that "sorrow has overwhelmed me & I have no judgement left."[79] A few days later, she wrote, "The smells around are fearful.... The loss of life & suffering is harrowing. Aid is coming in for the sufferers. But what can help me?"[80] Where she had once taken comfort in her belief in God, she found little solace: "I am too sad to live," she wrote. "God help me to feel his love."[81] Her anguish persisted in a months-long depression, with no relief but time. G. J. Wilkins, a Beaufort businessman and cotton broker, several months after the storm "was very much concerned in regard to the effect the night of the storm" had on his wife and admitted that "he stills feels the effect on his own system. It was some time after that night before he felt he could speak to any one in a natural way."[82] Wilkins's anxiety emerged, it seemed, both from the trauma of surviving the storm and the long-term damages it incurred on his livelihood, fearing that his "firm was ruined."[83]

Berkeley Grimball, owner of a rice farm, saw the vast majority of the rice crop drowned, his rice mill wrecked, and six draft animals killed. Berkeley, whom his brother Lewis described as a man who "seems to meet misfortune when success seemed in his grasp," saw himself as perpetually aggrieved.[84] Sure enough, in a letter to his sister Elizabeth, Berkeley wrote that "it seems impossible to become accustomed to calamities, although I have been through so many." A year later, Lewis wrote to Elizabeth that Berkeley was taking strychnine pills, then a cure for depression, because he had not yet recovered from the storm's blow.[85] Lewis wrote to Berkeley in June 1899:

> Your treatment lies largely with yourself, cheer up and don't look on the dark side of life. I really don't see any reason why you cannot be sound and well. My advice to you is not to go to any quiet retreat. What you want is to get away from yourself, don't think of yourself and keep your mind and thoughts occupied, you should go to some place where in addition to a bracing climate you should be amused.... Go about, Berkeley, occupy your mind.... Keep up your tonics and again I beg you not to worry over yourself, your mental condition is more than any bodily ailment.[86]

Lewis, with his usual sensitivity, urged Berkeley to simply will his way out of his depression, even as he recognized the dire weight Berkeley's "mental condition" had on his physical health. Berkeley never did recover. He died only a month after Lewis's letter, after he underwent an ill-fated operation intended to cure his melancholy at a sanatorium in North Carolina.[87]

Berkeley's mental illness, sparked by the hurricane's degradation of his rice farm, raises questions about similar suffering among coastal African Americans. Sources that give details about the state of mind of the tens of thousands of African Americans who found themselves in desperate straits after the hurricane are even more rare than those on whites. Nonetheless, a couple windows into the nature of grief and the long-term effects of the hurricane on African Americans' psyche are cracked open. Joseph Elkinton, a Quaker minister who toured the sea islands in January 1894, visited a Black meeting house on St. Helena. There, he talked with many African Americans who expressed a sense of lingering distress both physical and mental. "One intelligent man seemed to have been overtaxed and had paralysis," Elkinton recorded, "and not a few others showed by their countenances that care and anxiety had furrowed and wrinkled them."[88] Diana Murray, an African American woman and laundress who lived on Little Edisto Island, lost her husband, two children, and baby "in the deep water" of the storm surge, which flooded the entire island.[89] She survived the night in a tree and starved for days after, chased away from a barrel of potatoes by "an alligator holed up in it." "She was never the same again," an acquaintance recounted. "It left her fierce." Years later, she rejected an offer to harvest oysters and asserted that "My foot ain't touch water since the '93 storm." The hurricane had even alienated her from the lifeblood waterways of the Lowcountry.

The long-term affliction of grief and depression, exacerbated by their physical surroundings and the poor outlook for the future, was an undeniable feature of the months after the hurricane. Rachel Mather's house became a waypoint for many Black sea islanders seeking food and assistance, and she bore witness to their sorrow. One afternoon, a man trudged to her door, and when she asked him his troubles, he responded, "Too much trouble, Missis, ever since the storm, that night my house falls in, my wife she die of fright and fever, nex' day my father who live with me follows right on and dies the next day. In less den a week, one chile die, and now I leave three chillun down sick wid de fever and notin' tall to give um for eat."[90]

In such hopeless circumstances, with troubles compounding daily, some survivors could not see a way forward. One starving woman visited Mather as a last resort, telling her, "If you fail us, deres' nuffin more to hope for, we must just lay down an die."[91] Mather recounted a chilling story that, even if hyperbolic, reflects the mood and the crisis: "I remembered how a mother not long since traveled from Combahee seeking food, and finding none, returned in despair and fell down dead on the threshold of her own door."[92] Johnson Atkins, who had saved his brother Buba on the night of the hurricane but lost his mother, had to be the one to care for his family because his father was "most broken down all together and 'stracted" with guilt and distress after his wife's death.[93] Juliana, a nurse on St. Helena, lamented that "I wish I been dead, for all los' something. . . . So much sorrow an' misery I never want to see again."[94] Abram, a resident of St. Helena, reported years later to the island storekeeper Mr. Macdonald that "things ain't go well since that storm. It took my wife and took my house, and now it seem like I can't make no crop like I used to, nor earn no money."[95] The hurricane left thousands of traumatized and grieving African Americans in its wake, and the survivors' scars lasted for a lifetime.

Joel Chandler Harris, the folklorist who recorded derivative versions of the Brer Rabbit tales, journeyed to the sea islands a couple of months after the hurricane to write about the storm for *Scribner's Magazine*. In the first of two features, "The Devastation," he wrote about an African American woman who was resolutely "standing apart by the water's edge."[96] The woman had placed her five children high in a tree during the "fury and confusion of the storm," but the winds snapped the trunk and sent the woman and her children spinning into the raging floodwaters. Three were swept out to sea. A neighbor of hers told Harris (who recorded it in his approximation of the Gullah dialect), "She los' dem chillun, suh. She have trouble." Harris turned her long sorrow into simile: she was "as lonely as the never-ending marshes." Harris dwelt upon the marshes as a desolate landscape. In metaphor, they were not rich biological communities teeming with life but vast, empty expanses of marsh grass carved through with saline creeks, inhospitable and barren with the baking sun overhead and viscous pluff mud underfoot. Empty in Harris's eyes, the marshes reflected the endless distress that the hurricane brought to the Lowcountry.

The effects of the storm, both material and emotional, were long-lived and mutually reinforcing. The hurricane wrecked homes, farms, and businesses up and down the coast, and it forced different lives upon the tens of thousands of African Americans left to cope with the deaths of their loved ones. Numerous white South Carolinians, especially the Charlestonian elite, fomented a neat, progressive narrative of quick recovery that ignored the grim reality of the hurricane's damage. Far too many viewed African American deaths as a justification for their own bigoted ambitions. They pathologized the vulnerability of African American communities to environmental disaster as a signal of a dying race unsuited to independence rather than considering their own complicity in the manufacture of that vulnerability. Many African Americans knew that recovery would not be straightforward. Dealing with the material devastation of the storm would take months, if not years; and memories of the hurricane and lost loved ones would linger much longer than that.

Sam Polite, almost fifty years after the hurricane struck, calmly recounted his life story as part of the region's history as he wove a shrimp net on a sunny day on St. Helena Island.[97] The history that he had lived through and the milestones of his own life were one and the same. When asked his age, he said he did not know, but he'd "been doing man's work when first guns shoot" in 1861. "I was born a slave," Polite relayed in a steady, measured voice. "I remember gettin' flogged till I can't stand it no more." A change in his fortune—emancipation—was soon met with more hardships. "I see de big wind come, and de tidal wave, an' knock over de house and drown de people," he said, placing the Great Sea Island Storm immediately after the Civil War in his memory. The hurricane unleashed "a heap of trouble": "I member when de fever come and de smallpox and de people fall down dead in de fields. Den come de drought and dem kill all de crops. I see folks so hung'y and weak they can't walk. And de well so low, nothin' but mud water for drink."

The hurricane was not a discrete calamity but part of a cascade. Speaking over the decades, and with the benefit of that long view, Polite knew this to be true. He ended not with distinct historical event, because he understood there could be no clear end to his story, but with testimony about endurance: "We ain't got much dollars an' cents, but we harbor before de Lord. Everybody is free." Lowcountry African Americans gradually fit the hurricane into a larger narrative of loss and survival.[98] For some, the hurricane fit into a vision that linked the power of God, the power of nature, and the power of human resilience. The hurricane was at once a tragedy of incalculable dimensions and a sign to place your salvation into the hands of an awe-inspiring God.[99] Many were dubious that anyone else would arrive to assist them: for the time being, they had God, and they had whoever had survived along with them.

PART II

Aftermath

Chapter 4

Relief for Sea Island Sufferers

"How shall these famishing thousands stripped of everything be fed till relief comes? Where shall food be obtained and how transported hither? How shall the salt water be removed from wells and springs so men and animals shall quench their thirst?"[1] Rachel Mather, whose house and yard on St. Helena Island were thronged with Black sea islanders searching for food, medical care, and clothing, grasped the imminent peril of the moment. She asked questions laden with the weight of experience, because she had lived in South Carolina through Reconstruction and was aware that white dominance over Black South Carolina depended upon Black deprivation. She doubted white beneficence and feared mass death. "Can we let them now perish in the rice swamps and cotton fields where they and their fathers passed so many years of unrequited toil?" she asked—and she warned that, if white South Carolinians failed to act, "justice is slow-footed but sure and retributive."[2] The debts that the white elite of South Carolina owed to African Americans were as yet unpaid, and should they fail to act after the hurricane, they would only increase the amount that they owed.

The questions that Mather asked, that Black sea islanders discussed among themselves, and that many white South Carolinians contested dominated public discourse in South Carolina for the five weeks between the hurricane and the arrival of the American Red Cross. South Carolinians fought over who should take responsibility for the "sea island sufferers," as they came to be called in newspaper headlines and in published appeals for aid. The plight of the sea island sufferers, the vast majority of whom were Black, was swallowed by the politics of 1893. Never simply about helping one's fellow man, the question of responsibility for the sufferers became enmeshed in debates over the legacy of Reconstruction, the future of white supremacy, and the role of state governance in a burgeoning Jim Crow South.[3] The weeks after the hurricane demonstrate a transitional period in the history of disaster recovery: in the late nineteenth and early twentieth centuries, the federal government had no standardized practice for providing aid after a calamitous event and frequently left the choice of whether to fund and assist in a recovery effort to state and local governments. That inaction folded responsibility for disaster relief into debates over democracy on the coast of South Carolina.

These weeks encompassed a dangerous time. Not only did starvation, disease, and exposure threaten Black sea islanders' lives; for white South Carolinians, recovery offered an opportunity to advance the consolidation of white supremacy over what they saw as the fractious African Americans and white allies of the southern coast. In fact, perhaps the preeminent question on the minds of white Democrats during those weeks was how they could manipulate the relief effort to dislodge the Black community in Beaufort County. The hurricane's devastating effects on Black sea islanders, who had thus far eluded the control of the white supremacist state, could be an entrée for white Democrats to crush their political autonomy. Weakened, grieving, and vulnerable, Black sea islanders might not be able to defend themselves against a final assault.

Black sea islanders, disproportionately affected by the material losses of the hurricane, instead worked toward a recovery that would rebuild their communities and their political power. Through the integrated, Beaufort-based Sea Island Relief Committee (and later the American Red Cross), they worked to make that a reality. They sought to define recovery for themselves, just as they had worked to lay claim to their own vision for liberation during the Civil War. This was not always easy, for what could a robust recovery look like for African Americans standing hip deep in the runoff of slavery, amid the rising tide of Jim Crow?

Disputes over the state's political future hamstrung the debates over disaster recovery and relief. This was a test period for contentious visions of recovery between white Democrats, on the one hand, and Black sea islanders and their few (and at times unreliable) white allies, on the other. These groups vied to implement at-odds plans for recovery that would determine not only the future of the sea islands but the future of the white supremacist state itself. Heated debates among Black sea islanders, white Democrats, and white Republicans about relief, who guided it, and who received it never existed within a purely humanitarian sphere. The humble materiality of rations and donated clothing were imbued with greater meaning, and they were central to Black South Carolinians' long journey toward racial and economic justice.[4] Black sea islanders knew they had to act quickly. The degree of the calamity forced the white elite to entertain the discussion. Something had to be done: the lingering rot of the harvest, the demands of the stomach, the sobs echoing across the marsh flats, and the ominous buzzing of the *Anopheles* mosquito grabbed attention.

Assessing the damage took on politicized dimensions. White politicians, landowners, and others stoked racist fears over aid distribution to Black storm sufferers both in the Lowcountry and across South Carolina. Though the accounts splashed across the state's newspapers seemed indisputable, some white South Carolinians doubted that the situation was indeed so dire. Furthermore, they were loath to dispense aid to African Americans because doing so dredged up memories of Reconstruction. White South Carolina Democrats had a simple relationship to state solutions, in that they supported state solutions until the state no longer curried their favor. After Reconstruction, government solutions at the *state* level enjoyed Democratic favor, as long as that party could control governance.[5] White South Carolinians developed a mythology of Reconstruction that they used to deny African Americans civil rights—and those same people were often pivotal to local and state relief efforts.

White South Carolinians believed that aid—most of all, government-sanctioned aid—would herald a rebirth of Reconstruction-era redistributive policies, which in reality had been practically nonexistent, that would undermine white power and wealth. Dr. Daniel T. Pope, a Lowcountry physician, wrote in an op-ed that "if the negroes are fed they will not work any more; so that if there not be a very judicious distribution of the charity, more harm may be done than good. Many are looking to the government to come to their rescue. They have been the 'government's wards' so long that they look upon it as a certainty, and are not disposed to go after work to better their condition."[6] He presented no evidence for this claim, but most white readers of the paper would need none, as it was confirmation of their own biases.

White-owned newspapers fed this narrative. The *News and Courier* relayed the story of an African American man, R. W. Wright, who came to their office to opine on the recovery effort: he wanted a "colored committee" to be appointed. However, the editors of the paper did not bestow upon him any noble virtues; nor did they consider why a Black South Carolinian might not trust white oversight: "He was well-dressed and apparently well-fed. The point of what Wright had to say was that Wright would like to take a hand in the distribution of the supplies."[7] This description was a parody drawn from the racist imaginations of white South Carolinians. A portly Black man, dressed in fine clothes, slyly implying embezzlement from public funds: the paper had painted a familiar portrait to white Democrats of Reconstruction-era corruption. Federal assistance to southern African Americans, a concept that white southern Democrats despised, did not bode well for white supremacist dreams of power and control.

However, white South Carolinians' aversion to aid distribution for Black sea islanders did not preclude rebuilding their own houses and businesses, and Black sea islanders and a few white allies worked to facilitate a substantive recovery. These efforts incidentally and occasionally ran along parallel lines in form if not goal. Coastal residents developed committees to raise money and to distribute food and clothing. These organizations worked toward material recovery but frequently embodied competing visions. In the week after the hurricane, businessmen, politicians, landowners, and prominent citizens formed two ad hoc organizations: one in Beaufort, the Sea Island Relief Committee; and one in Charleston, the Charleston Relief Committee. The committees tried to accomplish what state and federal governments were unsure they could, or should, do.

The engagement of private citizens in charitable endeavors was less controversial than the state government of South Carolina distributing money and material aid to a predominantly African American population, which many white South Carolinians compared to their ballyhooed bogeyman of Reconstruction-style governance. However, the two committees represented very different interests. Beaufort's Sea Island Relief Committee, an integrated body, had a greater burden than the all-white Charleston Relief Committee because of the concentration of death and destruction around Beaufort. White landowners, politicians, merchants, and bankers alone staffed the Charleston Relief Committee, despite the city's populous Black professional and working classes. Due to less loss of life in the Charleston area in comparison to Beaufort County, the Charleston committee could afford to be more restrained in its distribution and mission, which suited the politics of the committee members and most white Charlestonians.

The extent to which the committees were unable to escape some degree of state control demonstrates the fraught nature of state regulatory power at the time. While many white southerners resented federal control, they were also amid a transition as they began to consolidate white supremacist power within the South through focusing on state governance. In South Carolina, where African Americans in the Lowcountry had maintained autonomy in part through working to maintain their voting rights within a small corner of the state, white Democrats were leery of allowing the exercise of local autonomy in the region. Governor Benjamin Ryan Tillman, bent on enforcing his authority over the Lowcountry for that very reason, insisted on oversight of the committees. The white leadership in Charleston despised Tillman for his vulgarity, his frequent taunts of the white Lowcountry elite, and his mimed populism. Tillman hated them back as obstructionist, backward-thinking snobs, famously deriding the Citadel, the city's venerable military college, as a "dude factory" that produced soft-handed dandies.[8] And of course, because Tillman saw himself as a champion of white supremacy in South Carolina, the Black residents of the Lowcountry felt no allegiance or trust toward him. They were not among his voting constituency, and neither were many of the white

Democrats of Charleston who constituted the Charleston Relief Committee. There was little trust or goodwill to go around, and with good reason.

Reluctant though Tillman was to run to the Lowcountry's rescue, the staggering loss of life, infrastructure, and capital at the coast necessitated some sort of response. Tillman paced his office as he pondered the challenge.[9] He had just returned from a two-week visit to the Chicago World's Fair.[10] The storm, which blew shade trees down and roofs away in the state's capital of Columbia, had cut rail and telegraph lines from Beaufort and Charleston to Columbia, and only on August 31 did a telegraph reach Tillman from the coast.[11] The telegraph, from J. H. Averill, the receiver of the Port Royal and Augusta Railroad, read with breathless urgency: "The loss of life by the recent cyclone on the islands adjacent to Beaufort and Port Royal will number not less than 600 people. There are 7,000 people on the islands entirely destitute of provisions, all they had being washed away and their crops entirely lost. Great destitution will prevail among them unless they have speedy relief."[12]

Tillman's first public statement on the hurricane was, by his standards, quite benevolent, and he seemed to be impressed with the size of the storm. Tillman urged "the people throughout the length and breadth of the State to come to the aid of their suffering fellow-citizens."[13] He suggested that the state owed a special paternalism toward African Americans, saying that areas hit outside of the sea islands "can take care of themselves," in contrast to the "poor colored farmers whose homes have been ruined and crops destroyed" along the coast.[14] He requested that "contributions . . . be made in money, food, clothing, and other necessaries of life sufficient to meet the present emergency" from "all classes of people, both white and colored." It was an unusual statement of unity from a man who was ordinarily deliberately divisive.

But Tillman knew that words were cheap, and he was an expansive orator. Before he would lay out any state resources for the ailing coast, Tillman required confirmation of Averill's account and so appointed a representative to travel to the Lowcountry. Tillman called on James Woods Babcock, superintendent of the South Carolina State Lunatic Asylum, to be his emissary to "secure accurate information of the losses of life and property and the exact condition of things along the coast."[15] Tillman needed to know if people needed tents, if there was danger of an epidemic, and if "sanitary arrangements" should be made to prevent the spread of illness. And in an aside that would eventually transform the recovery effort, Tillman wondered whether "it will be of any use to get the Red Cross Society to send its representatives there," since "this society does this kind of work wherever there is great loss of life by war or otherwise."

After this appropriately expressed concern, the more familiar, abrasive Tillman returned: "I do not want any abuse of charity. . . . The people have the fish of the sea there to prevent them from starving. . . . They doubtless have their potatoes left in the ground. I hope, too, that some one will make them go to work at once and plant turnips on the islands."[16] He also alluded to Harry Hammond's suggestion that the sufferers be transported to other parts of the state, saying that instead "they can live where they are cheaper and better for the present." Tillman had struck up the refrain that would be repeated by countless white South Carolinians: Black storm sufferers deserved assistance only if they proved their utility as laborers, and starvation would be the fault of their own laziness. Tillman also announced his intention to appoint a committee to guide relief and to manage donations, pending Averill's suggestion of seven potential members from Beaufort and Port Royal, two of whom Tillman requested be African American. Averill worked quickly.[17] He wired Tillman with the committee appointees: Beaufort mayor George Holmes, E. F. Courvoysier, W. H. Lockwood, Niels Christensen, S. E. Danner, T. F. Welch, Robert Smalls, and T. T. Ford (the last two were African American, as per Tillman's instructions).[18] The committee was a firmly well-to-do and Republican enterprise, consisting of local politicians and businessmen both Black and white.

However, neither Tillman nor Averill had mobilized the committee so speedily: the people of Beaufort had already taken steps to organize before Tillman's orders, and Averill was simply passing along names of the Beaufortonians who had already agreed to serve on the committee. On September 1, the people of Beaufort had held a citizens' meeting during which they appointed their own committee to "hurry the work of cleaning up the town" and another, led by Smalls, to "issue a public appeal to the country."[19] That same day, papers across the country published an appeal that Smalls wrote with the support of the committee in which he declared an early, confirmed death toll of 600. He described Black sea islanders as "proverbial for thrift and enterprise," living on islands among those of their own race.[20] "By thrift and industry," he said, they "have made themselves homes, with none to molest or make them afraid"—but "in one night all have been swept away." On behalf of the committee, Smalls asked for "aid in feeding and clothing the hungry and naked."

Smalls was a canny, pragmatic politician.[21] He knew that impressing upon readers across the country the self-reliant diligence of beleaguered Black sea islanders would incline them to open their wallets wider, if they believed that Black sea islanders were hardworking and thus deserving of assistance in times of trial. He reassured readers that these African Americans did not conform to popular stereotypes, such as those that gave rise to the "lazy and sassy" African Americans of the white imagination; they were, instead, respectable, industrious, and thus worthy of assistance in their hour of need.[22] Smalls himself had come to embody a Black, middle-class respectability, with a heroic backstory, that appealed to white northerners as much as it rankled white southerners, who pursued segregationist measures in part because of the rise of a Black middle class in the South.[23] Given his audience, then, Smalls generalized Black sea islanders as possessing diligent, modest characteristics. Furthermore, he snuck in a reference that would surely pique the self-righteousness of white northerners, many of whom reveled in a self-righteous excoriation of the vicious racism of white southerners. Black sea islanders, he insinuated, were surrounded by unfriendly whites who would not hesitate to do them harm: they lived in a protected environment in which they were shielded from outside forces that might "molest" them or "make them afraid."[24] Donations could help maintain the safety and autonomy of these threatened Black communities.

Smalls may have also played up this image of Black sea islanders as hardworking yeoman farmers because of rumors of violence. An incendiary report in the *New York Times* claimed that while "the white people . . . are doing all in their power to relieve the suffering," African Americans were growing restive and that "several" were "killed in a fight for provisions."[25] Niels Christensen, the Beaufort businessman, wrote about these reports to his wife and son in Massachusetts. His wife, Abbie, then had the letter published in various newspapers in the Northeast to calm the flames. He refuted those rumors as "*sheer nonsense*." He elaborated that while Black sea islanders "*are* flocking to Beaufort" as a potential center for ration distribution or employment, they "behave with perfect propriety. There has not been a fight or threatened demonstration, or personal threats of any kind."[26]

The weeks immediately after the hurricane were not entirely void of flares of anger and action, though such episodes were rare. In perhaps the only recorded case of a murder stemming from the hurricane, a group of African American men set fire to the house of Edward Thompson, the night watchman for the Ashley Phosphate Works, outside Charleston in the early morning hours of September 7.[27] The only existing account of this incident recounts that the "desperadoes" shot at the night watchman's wife, Mrs. Thompson, when she ran out of the house to extinguish the flames. The men then fired at and killed her son, George. All five perpetrators swiftly fled the scene. *The State*, which reported on the murder, speculated that their goal was to "outrage a daughter" of the family. But it may have been more likely, if speculative, that the men were hungry survivors of the hurricane, perhaps even former employees of the phosphate works, who set the fire as an act of anger against the shuttered company. Much more common was the everyday violence of white supremacy, which demurred over distribution of food and aid to the Black storm sufferers.

In any case, Lowcountry residents, especially those in Beaufort, were aware of public scrutiny and defensive about implications of misconduct within their community. But those accusations existed within a larger context, for white Democrats abhorred Beaufort County for reasons other than its significant Black population's political mobilization. Indeed, Beaufort's Sea Island Relief Committee was, in many ways, the fulfillment of white Democrats' greatest fears: an integrated coalition of locally mobilized, sufficiently well-to-do businessmen and politicians. Smalls was one of a group of Black professionals, lawyers, and politicians living and working in Beaufort in the late nineteenth century, a modestly prosperous Black middle class fairly well integrated into the social fabric of the town.[28] Black Beaufortonians did not yet see any reason to bow to political pressure from Columbia and Charleston.

In Charleston, white citizens had formed their own relief committee by September 3, with Joseph Barnwell—a Confederate veteran, white landowner, and Democratic state congressman—as the chairman.[29] Other prominent merchants and businessmen joined Barnwell, including Mayor John F. Ficken, George D. Bryan (a veteran of the Confederate navy and himself a former mayor of Charleston), George I. Cunningham (the last Republican mayor of Charleston), G. W. Dingle, George A. Wagener (a Confederate veteran, grocer, railroad booster, and phosphate industrialist), Markley Lee, and Reverend Daniel J. Jenkins (a Baptist minister and founder of an African American orphanage in 1891).[30] A final member, Morris Israel—a German-born Jew, president of the Beth Elohim Synagogue in Charleston, and a bank director—had prior experience in the field of disaster relief. After the 1886 earthquake that racked Charleston, he had served as chair of the Executive Relief Committee's subcommittee on immediate relief.[31] The Charleston committee was, like the Sea Island Relief Committee, representative of the city's leadership and a reflection of how white Charlestonians wished to see their city managed: with Confederate veterans and descendants of enslavers working alongside railroad magnates, financiers, and phosphate industrialists. Many of them straddled what they preferred to describe as the old and New South in their backgrounds and current occupations.

Cavalier expressions of racism fit within this paradigm by design. The committee's chairman Barnwell, at least, did not bother to hide his contempt for Black South Carolinians. A Quaker minister from Pennsylvania, Joseph S. Elkinton, toured the sea islands in January 1894 and met with many of the leaders of the recovery efforts, including Barnwell. Elkinton, who made a great show of abhorring southern racism, "found [Barnwell] a man apparently not much in sympathy with the negro" and noted that Barnwell thought that "accounts [of hurricane destruction and suffering] were exaggerated."[32] Barnwell also emphasized to Elkinton the difference between the Charleston and Beaufort areas: the islands around Charleston were "chiefly owned by the whites, whilst those roundabout Beaufort were chiefly owned by colored people." This suggested to Elkinton that Barnwell resented the flow of assistance to Beaufort-area islands because of their legacy of African American autonomy and his own disdain for African American governance.

While Tillman rarely curried favor with Charleston, he now saw the city's white elite as his allies against Beaufort. Accordingly, Tillman saw no need to appoint the Charleston committee, by either special envoy or his own choice. He did, however, announce on September 10 that the Charleston Relief Committee would take charge of all others along the coast and would be renamed as the General Relief Committee. Mayor Ficken, another Confederate veteran and a state Democratic representative from 1877 to 1891, traveled to Columbia to meet with Tillman to discuss this centralization of the committees. The intention, as they summarized it in a public report, was to ensure that a "committee of very intelligent and earnest men" should oversee the collection and distribution of aid both monetary and material and to "perfect . . . a systematic plan of distributing the contributions from Charleston as the central point."[33] The new centralized committee structure would ensure accountability through a chain of command. Other committees would receive contributions and money from the General Relief Committee and report to it about the needs of their districts. Tillman claimed that this would enable aid to be sent as quickly as possible to any area with a sudden shortage.

Above all, Tillman and Ficken wanted to "guard against any demoralization of labor" and requested that "all parties on the coast who can supply work" should send a description of the number of workers and the amount of pay needed to the Charleston-based committee. The General Relief Committee would employ a bevy of inspectors to check on each subordinate committee's adherence to the system of employment, and Tillman himself wanted to "keep a general supervision of this feature of the relief work" and would "write to the various county commissioners in all the coast counties to give these people all of the public work they can."[34] If any county was found wanting, Tillman would perfunctorily cease all aid.

Tillman wrote a letter of instruction to the Sea Island Relief Committee to spell out the new hierarchy and the committee's responsibilities to it. He demanded a detailed "statement of every contribution that has been or will be received" and for the committee to complete a survey of the land south of the Ashepoo River and to determine "distributing agents," who would make weekly reports about need for help and desire for employment.[35] "I think," he continued, "the policy of the committee should in all cases be to give work when possible, and require work where needed before giving any assistance. Help the destitute to help themselves, and it will be the greatest charity." Work "of a public nature," such as "clearing roads, repairing bridges, etc.," or of a private nature, such as rebuilding houses, should be paid for in rations. To obtain rations and any other form of aid, the Sea Island Relief Committee would "make requisition" of the Charleston committee. Tillman also, without consulting the Beaufortonians, added Dr. Allen Stuart and John J. Dale, a Beaufort cotton merchant originally from Vermont, to the committee, perhaps to serve as informants on the committee's doings.

It is unclear how the Beaufortonians on the committee privately reacted to Tillman's subjugation of the Sea Island Relief Committee to the General Relief Committee. It seems clearer, though, that Tillman deliberately promoted Charleston over Beaufort because Charleston was, though not a city terribly loyal to Tillman, a stronghold of stalwart, white supremacist Democrats. Thus, the General Relief Committee could be counted on to keep an eye out for any subversive machinations by African American and Republican Beaufortonians and would prevent the distribution of food and clothing to any African American who did not meet their labor requirements.

The Sea Island Relief Committee followed Tillman's instructions—for the most part. The committee members assembled a concise explication of the committee's responsibilities.[36] The committee divided the region into nine districts overseen by subcommittees, with very specific borders: St. Helena, Pollywanny, and Dataw Islands; Parris Island; Lady's Island; Corn, Coosaw, Morgan, Hutchinson, Fenwick, Beet, Hog, and Warren Islands and Bennett's Point; Eustice Plantation and Warsaw; Port Royal Island north of the Saltwater Bridge; Port Royal south of the Saltwater Bridge; the mainland; Pacific Phosphate Works; and Hilton Head Island. Totaling twenty-two, the members of each subcommittee were an integrated group of reverends, the phosphate industrialist Moses Lopez, landowners, storeowners, businessmen, and farmers. On the subcommittees, former enslavers and Confederate veterans were in short supply. African Americans were not. They constituted at least half of the subcommittee members, appointed by the Sea Island Relief Committee. Tillman may have wanted to smother any yet-smoldering fires of the Freedmen's Bureau, Reconstruction, and the Port Royal Experiment, but the Sea Island Relief Committee's inclusion of so many freedmen suggests that Black sea islanders did not.

The Sea Island Relief Committee delineated the rules in ten sections, following the contours of Tillman's instructions: that the subcommittees must answer to the Sea Island Relief Committee and then to the General Relief Committee; that each subcommittee must "canvass each of their respective districts and ascertain by personal examination" the district's losses and determine their needs therefrom; that the names of the sufferers and the likely duration of their affliction would be recorded to the General Relief Committee; that medicine and loaned tents would be distributed as needed; that subcommittees could not contract debts on behalf of the General Relief Committee; that sufferers could only receive aid from within their district; and that each subcommittee was to hold themselves to a strict, standard, prudent, and economical distribution.[37] The rules did not contain any mention of the labor requirements to receive rations for which Tillman had so emphatically pressed.

And while Dr. Stuart was appointed as a member of the Parris Island subcommittee, he was not on the concluding list of the Sea Island Relief Committee's executive members—and neither was J. J. Dale. That the committee refused to seat Tillman's handpicked appointee may have been a deliberate act of resistance to the governor's attempts at curbing the Sea Island Relief Committee's power and a reminder to the governor that he did not command power in Beaufort County. The absence of any labor requirements to receive rations must have been a deliberate act of resistance, given Tillman's explicit insistence on such provisions.

Local relief committees of this kind were common in the aftermath of hurricanes within the western Atlantic in the nineteenth century, with municipal authorities, politicians, and reformers in cities like Havana, Cuba, and San Juan, Puerto Rico, forming "*juntas*" of "*beneficiencia*," or assistance.[38] These committees afforded provisions and an organizational structure to which sufferers could appeal for recourse in a transitional era when governments were timid in their approach to disaster relief and recovery. Authorities within the hurricane zone had largely moved beyond envisioning these storms as a divine punishment and had embraced scientific understandings of hurricanes as meteorological phenomena.[39] However, in the late nineteenth-century United States, the government had not yet fully accepted responsibility for the victims of such calamities. This liminal stage in the evolution of federal and state governance meant that survivors of hurricanes, earthquakes, forest fires, floods, or other hazards might hope and vie for federal aid, but government assistance was far from reliable. Though the federal government did frequently dispense relief through the Senate's Committee on Appropriations after events like hurricanes and floods, it did not yet have a built-in bureaucracy to which citizens could appeal directly for aid, and no such provisions had been made for the sea island coast. Absent any accessible government structure, local organizations could respond more nimbly and more effectively. In the Lowcountry, the relief committees thus busied themselves with collecting and distributing funds, food, and other goods to compensate for the lack of federal or state assistance.

A bourgeois, progressive spirit of reform and collective organization inspired many politicians, businessmen, and citizens around the country to help the sea island sufferers. Relief organizations, like those on the South Carolina coast, were simultaneously private and public, untethered from government bodies but, through elected boards, they at least nominally derived power from the citizenry. They were less threatening to race and class hierarchies than a complete transformation of the federal government into the administrator of a welfare state. Often central to the mission of relief organizations was the dispensation of not only material assistance but the dispersion of values that should be impressed upon the beneficiaries.[40] The rights that many relief organizations promoted were not revolutionary: the right to survive an unexpected calamity, like a hurricane or an epidemic, and to have access to resources that could guarantee that survival. But these organizations nonetheless provided indispensable services to victims, and the Black storm sufferers turned that assistance toward their own goals of self-determination and mutual aid.

The primary goal of the South Carolina committees was to provide immediate sustenance and support for starving people living in the open. Rebuilding infrastructure, homes, and fields was of concern, but none of the committees had the bureaucratic structure or the funds necessary to execute those projects. The committees were nonetheless capable of raising significant sums of money, though perhaps not as generous as the amount might have been before the economic downturn of 1893. A network of smaller, local committees spun out from the coast to serve the Charleston-based General Relief Committee. Residents of towns across the state, from Abbeville to Rock Hill, formed committees to collect money and goods to send to Charleston and Beaufort. These efforts were not restricted to white South Carolinians, for newspapers reported that "colored people . . . also respond . . . to the call as liberally as their means will allow."[41] The General Relief Committee, though, had the fullest coffers. The committee raised $1,500 at their first meeting, and until the arrival of the Red Cross, they accumulated $8,834.93 in cash donations alone.[42] That sum included money raised by several different committees and sent to Charleston as per Tillman's reshuffling of the committee structure. Charlestonians donated $3,348.03 of that total, and $700 of that portion was from Black Charlestonians.

The committees received support from private citizens, churches, charitable organizations, and businesses alike. Charity, in the reformist spirit of the day, was intended to introduce the disadvantaged and impoverished to proper modes of moral citizenship and control their behavior. The coast's demography thus may have increased donations because the afflicted population was constituted of rural African Americans, whom benevolent whites saw as prime subjects to be guided into proper modes of behavior.[43] (This was ironic, given that mutual aid societies to help impoverished or hurting community members had flourished among Black sea islanders since the Civil War.[44]) Organizations and individuals sent money, groceries, and supplies of all sorts. In Columbia, a variety of local businesses and fraternal organizations were designated recipients for money and goods: the Carolina National Bank, the Loan and Exchange Bank, the Young Men's Christian Association, Mrs. A. C. Squier's, Dr. E. E. Jackon's drugstore, and the three daily newspaper offices.[45] Squier had been chosen from the Columbia chapter of the King's Daughters, a nationwide interdenominational Christian service organization with roots in New York, to receive, pack, and ship bedding and clothing. Squier's charity work also exemplified an acceptable public sphere for middle- and upper-middle-class white women, who organized efforts to collect, pack, and ship provisions and supplies to the coast.

The relief committees relied on the beneficence of businessmen to donate goods, money, and even transportation. The results were sufficiently gratifying. Indeed, every railroad and delivery company in the Charleston area agreed to transport provisions for the sufferers without charge. Agent Leath of the Southern Express Company announced that the company's superintendent O. M. Saddler had decided to forward, gratis, all relief packages to Beaufort—with the caveat, "Don't make the packages too large."[46] The Southern Express Company sent its vans out to gather clothes and bedding from Charleston. Charlestonians donated goods along with money. H. O. Heiser gave 1,000 loaves of bread to hungry sea islanders; Dr. T. Grange Simmons, twenty-five pounds of corned beef; O. G. Margenhoff, ten boxes of soda crackers and 500 loaves of bread a day for five days; Alexander Melchers, 1,000 loaves of bread; J. H. Devereux, ten barrels of potatoes; and F. W. Wagener & Co., fifty barrels of grist and five barrels of molasses.[47] Food was more practical than money, as long as the committees stayed focused on relief rather than recovery. In the immediate days and weeks after the hurricane, sea islanders needed food more than anything else.

The fundraising superseded regional boundaries. Newspapers up and down the East Coast raised the alarm on behalf of the sea island sufferers. Besides South Carolina papers, the *New York Evening Post*, the *New York Tribune*, the *Philadelphia Press*, the *Washington Star*, the *Boston Herald*, the *Cincinnati Tribune*, and the Providence, Rhode Island, *Journal* published appeals for citizens to send money and provisions to the sea island sufferers.[48] J. J. Dale traveled to New York City to raise money and came home with $1,649 for the General Relief Committee.[49] Another relief fund from New York City held nearly $5,000 by September 10.[50] Some actions suggest that a degree of economic self-interest in the cotton-producing South drove the donations: the

Loading rations for St. Helena and Lady's Islands. This photograph, taken at Massey's Ferry on Lady's Island with Beaufort across the river, shows a busy scene as rations are distributed for St. Helena and Lady's Islands. Note the vernacular watercraft in the background, reminiscent of the old periaguas of Indigenous and West African origins but with schooner rigs, and the numerous two-wheeled carts for transporting rations across the islands. Barton, *Red Cross*, 236.

New York Commercial Exchanges assisted with this fundraising, with the Produce and Cotton Exchanges posting an appeal on the floor for donations.

One of the most esteemed men to call for donations was Frederick Douglass. In the *Brooklyn Eagle* on September 9, 1893, the elder statesman requested that contributions be sent specifically to Robert Smalls for distribution.[51] He vouched for Smalls, asserting that "into no hands could they be placed with greater certainty to being promptly and faithfully distributed to the needy. In a case of this kind no lengthy appeal need be made. The calamity is trumpet tongued and speaks for itself." The plight of Black sea islanders may have particularly moved Douglass, as he was likely aware of their precarious position in a state hurtling toward Jim Crow. Smalls's reach and reputation still meant a great deal, and he marshalled powerful allies to support the cause of hurricane relief.

Individual donations were listed almost daily in *The State*, as the editors hoped that the recognition might encourage others to give too. The paper noted twenty-eight donations from across the country on September 12 of substantial sums.[52] Mrs. Frances G. Shaw of New York gave $500; W. P. Fife, "The Drummer Evangelist" of Charlotte, North Carolina, donated $25. Three churches appeared on the list. New York, Washington, D.C., Maryland, Virginia, South Carolina, Mississippi, Pennsylvania, and Delaware were all represented, too. On that same day, the Gaffney Manufacturing Company in the upstate donated 5,000 yards of shirting and sheeting to the General Relief Committee for bedding and clothes, much needed to replace clothing lost in the storm or tattered by a night of hurricane conditions.[53] A nine-year-old girl, Ethel, and her ten-year-old friends Belle and Alice wrote Tillman from Ridgewood, New York, to inform him that they had sewed items for a fair and raised $5.38 to donate to the relief effort. The newspaper could not resist printing their letter, with the sentimental title "Even the Children Work."[54] Charity could be both women's work and girls' work, especially if it improved feminine skills such as needlework.

Two donations stand out because of their origins: in the realization of citizens that the government, at the federal or state level, was generally inefficient in its response to the effects of disaster events and that the victims of those events should have institutional recourse. The Citizens' Permanent Relief Committee of Philadelphia and Jacksonville, Florida's, semipermanent relief committee offered substantive funds to Tillman, demonstrating their acquisition of capital and ease of mobilization. The mayor of Philadelphia relayed a message to Tillman from the Citizens' Permanent Relief Committee, voting to appropriate $5,000 to help the sea island sufferers.[55] This committee, founded in 1885 by businessmen and philanthropists, had already participated in relief during the Johnstown Flood of 1889 and had even provided aid for international causes, most recently the Russian famine of 1892–93. The apparently formidable resources at their disposal and their willingness to provide aid at a time when the federal government shied from relief efforts did not seem to impress Tillman. He responded curtly, "If money send to me. If supplies, ship to relief committee, Charleston, S.C. Thanks."[56]

Mayor Fletcher of Jacksonville, Florida, likewise called a "public mass meeting of the board of trade" to discuss ways of assisting sea islanders. Jacksonville's relief committee of the "colored Auxiliary Sanitary Association" shipped twelve barrels of flour, 250 pounds of bacon, ten barrels of grits, and fifty suits of clothing to Charleston.[57] Residents of Jacksonville had founded the Auxiliary Sanitary Association in response to the city's 1888 yellow fever epidemic. To improve response to public crises, the city's citizens felt that "it was necessary for the citizens to organize with the authorities and meet the crisis as a unit."[58] These two organizations emerged from the activism of private citizens and crossed into the public sphere to aid the general citizenry, thus blurring the lines between private enterprise and public service.

Churches in Columbia, Charleston, and Beaufort also served as key nexuses of charity. On the Sundays after the storm, numerous reverends chose the hurricane as their topic and many congregations collected money and goods for the relief effort. Rev. P. D. Hay of Beaufort's St. Helena Church mused on the storm's origins. "Whose hand," he asked, "guided that great storm up from the tropics over thousands of miles of sea and land to smite us in our repose and helplessness? And for what reason?"[59] While he stopped shy of declaring the hurricane an explicit punishment from God and did not name any specific crimes that would have brought God's wrath down upon them, he urged his congregants to consider their transgressions and engage in continual repentance, actions that would make them better Christians. Charity toward one's fellow man was one aspect of the process of self-improvement.

The congregants of Charleston's Mount Zion AME, one of the oldest Black congregations in the South, accumulated forty dollars in donations during their service on September 3—even though the hurricane had caved in the roof and gable of their brand-new church building, compelling them to hold services in the damp basement until they could raise their own funds for repair.[60] At the Morris Street Baptist Church in Charleston on September 8, A. C. Kaufman and the Reverend Dr. J. A. Clifton organized "a colored mass meeting" with "prominent colored ministers and gentlemen" "in the interest of the sufferers on the sea islands."[61] Other Black churches across Charleston planned for meetings two nights in a row to raise a proper relief fund for sufferers further down the coast.[62] Black pastors organized "a mass meeting" at the Bethel AME church in Columbia, too, and called a meeting the morning of September 2, during which they agreed that they would urge their congregations to donate to relief efforts.[63] Black churches rarely had deep pockets, but their members organized rapidly and mobilized in a "united course of action" to assist their ailing compatriots on the coast.[64]

The General Relief Committee and its subsidiaries—particularly the Sea Island Relief Committee—proved fairly effective distributors of aid. But donations could only give temporary relief, not permanent recovery. They could not revitalize communities or provide employment for the thousands of Black farmers, phosphate workers, and agricultural laborers who clamored for those opportunities. Distrust infused the process. White South Carolinians, whether Tillman supporters or wealthy white Charlestonians, fretted that Black sea islanders should be provided employment, not just temporary aid. Racist stereotypes about African Americans drove their proclamations about what Tillman called the "demoralization of labor" and the "laziness and idleness" of the African American populace.[65] They were so sure that Black sea islanders sought to shirk work and live off donated supplies that they ignored the fact that Black sea islanders did indeed want to rebuild their livelihoods.

Waiting for work in Beaufort, South Carolina. Taken near the American Red Cross headquarters in Beaufort, which was likely at present-day 901 Bay Street, this photograph represents a common scene: African Americans arriving in Beaufort in the hopes of finding work so that they could feed themselves and their loved ones. They traveled from distant islands or inland farms, a difficult journey under the best of conditions, starting the day after the hurricane and for many months after. While this photograph is from sometime in the winter of 1893–94, after the Red Cross began the recovery effort, it is demonstrative of what would have been a daily occurrence in Beaufort from August 28, 1893, until spring 1894. *Waiting for Work—Headquarters, Sea Island Red Cross Relief, 1893–94*, Clara Barton Collection, Library of Congress Prints and Photographs Division (LCCN 98519362).

Conversely, Black sea islanders had no reason to trust assistance from Governor Tillman, the Democratic state government, or the all-white relief committee in Charleston. Black sea islanders had only salted earth, rotten potatoes, wrecked boats, and limited fishing supplies. While Black sea islanders gladly accepted food and clothing to keep their families and loved ones from starvation, survival supplies were a stopgap, not a long-term solution to the hurricane's deprivation. Rebuilding Black sea island communities, spaces that had been able to operate semiautonomously; to nurture Black churches, schools, and property rights; and to protect a degree of Black political power stood out to African Americans as the most desirable set of goals, all of which were focused on preserving autonomy and well-being. If that did not succeed, Black sea islanders might find that they had few other options than to move away from their homes, family, and sea islands to places that might better foster the political, educational, and employment opportunities that they had long considered foundational to their freedom.

Sensing the potential long-term inadequacies of the local relief committees, both white and Black South Carolinians sought out money from the federal government. Senator Matthew Calbraith Butler was the first South Carolina elected official to attempt to make headway in Washington, D.C. Senator Butler had held the office of U.S. senator since his avid participation in the Hamburg Massacre and the violent overthrow of Reconstruction in South Carolina in 1876. Furthermore, his service as a major general in the Confederate army and his loss of a leg at the Battle of Brandy Station helped establish his bona fides as a protector of white supremacy (though he clashed with Tillman, who was not an easy man to ally with).[66] Butler called on the secretary of war to request a supply of tents and rations for the homeless and hungry. Butler doubted that the U.S. Congress would approve any legislative appropriations for aid because Congress had "always, in the past, declined to appropriate for this sort of relief, holding that that was the duty of the State."[67] But Butler nonetheless considered returning the question of responsibility to the federal government. Butler's attempts to bring federal aid to South Carolina, however, were thwarted in part by the state government's resentment of federal oversight, which generated little sympathy among many Republican politicians in D.C. for the state's distress in the wake of the hurricane.

A coterie of representatives who also felt a responsibility to seek federal aid joined Senator Butler. Representative William Brawley, a Confederate veteran and later a lawyer who practiced in Charleston, hailed from upstate but had served as a Democratic congressman for South Carolina's First Congressional District since 1891.[68] Brawley met with Rufus Ezekiel Lester—a Confederate veteran, former mayor of Savannah, and a Democratic representative of the district since 1889—to strategize.[69] The two men asked Representative Joseph Draper Sayers (a Democrat who had served Texas since 1885 and an officer in the Confederate army), chairman of the Committee on Appropriations, to "ascertain whether it is possible to secure an appropriation for the relief of the sufferers."[70] Though Butler was leery that an appropriation was possible, the entire federal delegation of South Carolina representatives and senators agreed on September 5 to request relief from the U.S. Congress.

Representative George Washington Murray, at the time the only Black representative in the U.S. House, volunteered to draft a joint resolution on the subject.[71] Murray, born into slavery in Sumter County, began his political career as a leader of the Colored Farmers Alliance in the late 1880s.[72] He gained further prominence when Republican president Benjamin Harrison appointed him as the federal inspector of customs of the port of Charleston in 1890. Murray then rode the fusionist wave of white Democrat–Black Republican alliance in South Carolina, which split the Democratic vote between conservative and Tillmanite Democrats, to win a seat in the U.S. House in 1892. Murray had to fight tooth and nail to retain his seat as Democratic challengers, fraudulent election practices, accusations of corruption, and vigilante violence undercut his position. He deliberately wrote the joint resolution for relief in such a way as to appeal for help specifically for African Americans, cognizant that northern congressmen might bristle at a southern state pleading for federal assistance for hostile whites. In Murray's resolution, he asked for $200,000 specifically for "the colored people" of the state.[73] If the South Carolina delegation could gather support from other southern states, *The State* speculated that "it may pass, though there is no precedent in such cases and some of the members of the House think it dangerous to establish one."[74]

Though the federal government might periodically choose to bestow a congressional appropriation on the victims of a sudden disaster, that coverage was far from guaranteed. Each new disaster required that the citizenry of the afflicted area mount their own campaign, send their own delegates, and advocate for themselves. Congress had, on some occasions, made provision for some aid. In 1890, floods inundated the Mississippi River's Tensas Basin in Louisiana and the Yazoo Basin in Mississippi.[75] In part because these were sugar districts that enriched and empowered white landowners, federal support came in two forms: some rations for the people living in flooded districts and an elimination of a proviso stating that "levees could be built only to protect navigation."[76] The yellow fever epidemic in Brunswick, Georgia, earlier in 1893 also suggested that federal recourse might be made available to the storm sufferers: the surgeon general had been legally authorized to dispense funds to Brunswick to prevent the spread of the disease.[77] However, Congress had not created a new fund specifically designated for disaster relief but had instead drawn money from "the general funds to prevent epidemics," treating the epidemic as a public health crisis.[78] The congressional process limited access to federal resources from communities that were too impoverished and too Black to muster the political muscle necessary to win an appropriation. Localities were instead expected to provide for their own.

Several white Charlestonians and the General Relief Committee asked former Democratic congressman and Confederate colonel William Elliott from Beaufort, son of a wealthy enslaver, to travel to Washington to help with the movement for federal aid.[79] Elliott was a foe of George Washington Murray and Thomas E. Miller, another Black Republican in the Lowcountry. Miller had contested Elliott's 1887 election to the U.S. House, which declared the election fraudulent and gave the seat to Miller in 1890. Elliott nonetheless won another election against Miller in 1891 and served until March 1893. At the time of the hurricane, then, Elliott was between congressional campaigns.[80] That the General Relief Committee called on Elliott to head up the effort, even though he held no public office, reveals the committee's bias against Murray, who presumably was perfectly capable of doing his job as a congressman from South Carolina. The committee charged Elliott to "interest himself on behalf of the negroes," citing the support of Rep. Brawley, a strange turn of events given that the committee had chosen Elliott over an actual African American congressman.[81]

Some members of the Sea Island Relief Committee traveled to Charleston for an official meeting to appoint a delegate to Washington, D.C., to work on behalf of the storm sufferers. Niels Christensen, eager for any assistance, moved that Elliott go to Washington on behalf of the General Relief Committee.[82] Robert Smalls, hoping to find an angle that might result in federal aid and drawing on the example of Brunswick, suggested that the South Carolina delegation should request appropriations from a medical relief fund rather than a separate fund exclusively for use of the sea island sufferers. The committee decided, then, that Elliott should begin with a meeting with the U.S. surgeon general of the Marine Hospital Service. The strategy was for Elliott to inform the surgeon general that, following hurricanes in South Carolina in 1817, 1854, and 1871, malaria epidemics had taken hold in the storm-stricken areas. The committee hoped that conveying fears of an epidemic would convince the surgeon general to send a few physicians to the sea islands, especially since emphasizing the dangers of epidemic disease had resulted in assistance for yellow fever victims in Brunswick.[83] The fear of epidemic was sufficient for South Carolina's advocates in Congress to lean on that anxiety as a pretext to obtain funds, and they envisioned a public health crisis that they argued the federal government was obligated to ameliorate.

Rep. Murray, too, formulated a vision of a posthurricane world in which deprivation rather than the storm itself would cost lives: in his appeals to Congress, he argued that "more will die from privation and starvation than were immediately taken off by the flood and the storm."[84] The necessity for manipulation of the federal system was a reminder to South Carolinians that the federal government had no real interest yet in instituting a formal set of funding structures to relieve the victims of calamitous events. Though South Carolina congressmen were willing to mobilize because of the severity of the crisis, none seemed optimistic that they would be able to single-handedly convince the U.S. Congress that the federal government should have a unique and standing obligation to disaster survivors. The South Carolina delegates realized that they probably could not compel the federal government to feel responsible for regionalized catastrophes, but they could construct an argument about the duty of the federal government to its citizens when inefficient distribution of food, water, or medicine threatened the lives of Americans. The issue, then, was not with the hurricane but with the failures of distribution after the storm, a logistical problem that the federal government could help solve.

Members of the delegation pursued multiple avenues to work out some sort of aid package. Murray engaged in some grassroots organizing. At the Colored Young Men's Christian Association in Washington, D.C., on September 8, Murray addressed the organization's members and "recit[ed] the unfortunate condition of the negroes there."[85] The CYMCA members, moved by Murray's description of the plight of Black sea islanders, called for a second meeting at which they could adopt a series of measures for relief, which included encouraging Black pastors in D.C. to collectively take a plea for relief to their congregations. Murray also weighed the joint resolution that he had drafted and, with the consensus of the South Carolina delegation, decided to drop the bill. He and the other delegates agreed that Congress would not grant them their own fund for the sea island sufferers. Instead, he wrote a new bill requesting 150,000 rations to be distributed "in the cyclone district" by the War Department—a bill that would ultimately be unsuccessful.[86]

Senator Butler agreed with this strategy of asking for rations or medical aid rather than money. Butler, with the cooperation of the secretary of agriculture, requisitioned 5,000 packages of turnip seed for the sea island sufferers upon hearing Butler's plea.[87] While not the most popular subsistence crop among sea islanders, who preferred the sweet potato as their staple root vegetable, the secretary reasoned that turnips were the only crop that could be planted in September and harvested before midwinter. Elliott cleared other avenues for relief. He presented the supervising surgeon general of the Marine Hospital Service a letter from Dr. Peters of Beaufort, who reported 700 cases of malarial and other fevers and a near epidemic of diarrhea and bowel trouble because of poor water conditions.[88] President Grover Cleveland even granted Elliott an audience and then issued an order to the U.S. surgeon general to take to Beaufort a force of officers and a revenue cutter, a small boat usually used to enforce customs regulation but here to enable travel among the sea islands, with a shipment of disinfectants, medicine, and rations for the sick.[89]

With these piecemeal efforts, South Carolinians cobbled together some federal assistance. Through direct appeal to federal agencies and cabinet departments rather than congressional appropriation, Rep. Murray, Sen. Butler, and Elliott negotiated and wheedled for federal resources. They depicted the crisis on the sea islands not as the result of an uncontrollable act of nature but the product of flawed human systems of distribution. That required a less morally complex way of looking at the problem—at least, once ensconced in Washington, D.C., removed from the vicious political debates in which white Democrat South Carolinians mired themselves. Once in D.C., men like Butler and Elliott, both hostile to Black South Carolinians, felt sometimes obligated to mask their more explicit racism for the sake of statesmanship and to maintain more equitable relationships with more centrist Democrats and even Republicans in Congress. Sometimes political expediency required the appearance of moderation.

In this case, the South Carolina delegates found it necessary to change both the question at hand and to move its location. No longer was the question the same as in South Carolina: Do Black South Carolinians deserve direct aid? Instead, they asked: Can we treat this as an issue of mitigating harm and staving off epidemic disease and starvation through distribution? No longer was it located on the floor of either the U.S. House or Senate. The debate shifted to private offices and meeting rooms in D.C. The South Carolina delegation did not have the time to pressure Congress to shift its perspective on disaster relief and instead altered the terms of the debate to suit the delegation's needs and the needs of the sea island sufferers.

During the weeks immediately after the hurricane, South Carolinians developed an array of strategies for dealing with the storm's destruction at local, state, and federal levels, and they contended the meaning of relief and recovery along lines of race and class. These debates occurred at a crucial moment in South Carolina history. Governor Tillman had seized his office for the banner of white supremacy. His quick alliance with the Charleston Relief Committee suggested that his sparring with the white Lowcountry elite mattered less than defeating interracial alliance and Black political power in the region. But the 1868 Reconstruction constitution still stood, and Jim Crow had not yet been fully institutionalized. Black sea islanders yet clung to the voting rights and political power they had maintained since the Civil War.

After the hurricane, South Carolinians fought over every aspect of the recovery. White South Carolinians sniped over the extent of damage and harm done to Black sea islanders. Black sea islanders and their white Republican allies resisted the imposition of Tillman's control over their recovery. South Carolinians in Washington, D.C., jostled over the question of whether a member of the white elite, such as Elliott, should be given precedence over an elected Black representative like Murray. White and Black South Carolinians looped themselves into national-level debates over what support the federal government ought to supply to its citizens, most of all those in danger from the growing power of white supremacist state governments in the South. These conflicts were anything but petty. The hurricane laid bare the dire situation Black South Carolinians were in, and the fight over relief and recovery was, in fact, a fight over the state's political future.

A lack of institutionalized federal support necessitated citizen-driven organization. White and Black South Carolinians responded through local relief organizations, newspaper editorials and appeals for donations, and connections with political figures and organizations across the country. Black sea islanders already had a long tradition of mutual aid and communal care. The reformist interest in progressive improvement drove donations and organizational vigor among northern donors. This widespread mobilization in the wake of the storm facilitated donations, but it also demonstrated that processes of recovery were highly political, especially when so many of the victims and survivors were African Americans. Black sea islanders were not naïve. Better than anyone, they knew how white supremacy threatened their lives as much as any hurricane, and their organizing after the storm sought to counterbalance the assaults they sustained from white Democrats in the state. Neither racism nor hurricanes were new phenomena to Black sea islanders. Both had dealt them blows for centuries. But in the weeks after the hurricane, the struggle took on urgent dimensions as both Black and white South Carolinians recognized that they were gearing up for a pitched battle between Black political power and white supremacy.

Chapter 5

Red Cross Recovery

Clara Barton, president of the American Red Cross, watched the crisis in the Lowcountry unfold in the newspapers. She read with interest when Tillman wondered aloud "if the Red Cross could be asked to aid."[1] Shortly thereafter, she approached Senator Matthew Calbraith Butler to see if he thought it was appropriate for the Red Cross to assess the coast, with an eye to administering a large-scale recovery effort on the sea islands. Butler agreed to serve as an intermediary to Tillman, and he communicated to the governor that the Red Cross would like to tour the afflicted region and determine whether their skills could be put to good use.[2] The American Red Cross, which Barton founded in 1881 after doing volunteer work for the International Committee of the Red Cross, had engaged in a few significant recovery efforts previously. None were nearly so challenging as that which the Lowcountry posed.

But Tillman perfunctorily refused. He then received a flurry of letters from Clara Barton and two of her closest associates, Dr. Joseph Gardner and George H. Pullman. Barton addressed Tillman's misconceptions as to the scope and nature of the Red Cross's mission.[3] Her letter forcefully laid out the case for the Red Cross to a man who had fundamentally misunderstood what she offered: "Be assured, Governor, that no one knows better how to sympathize with you in the great and trying charge that has fallen to your lot." Tillman's flat denial may also have irritated Barton, especially since he dismissed Barton's offer to help as merely the assistance of "a few nurses [who] would come to care for the sick, homeless, hungry and naked."[4] Dr. Gardner and Pullman echoed Barton's sentiments, emphasizing the Red Cross's ethos of hard work, self-reliance, and community empowerment—all solid, industrious values that they would instill in Black sea islanders.

The letters worked. Tillman accepted the Red Cross's offer to visit South Carolina and, once there, to discuss further the possibility that they would assist with the work of relief. He also extended an olive branch after his initial, terse refusal: "Permit me in closing to express my appreciation and admiration of the grand association of which you are the head, and believe me, with the highest respect, very truly yours, B. R. Tillman, Governor S.C."[5] The Red Cross was coming to visit the South Carolina coast, with Tillman's blessing.

At the invitation of Tillman, Clara Barton boarded the steamer *Catherine* to tour the storm-stricken sea islands at 3:00 A.M. on September 18, 1893.[6] The steamer journeyed north from Beaufort to Charleston, winding through the Inner Passage of intertidal waterways, passing the hurricane-wrecked dredge the *Wimbee*, navigating through treacherous waters into the Stono River, and landing at a wharf on the Cooper River. Along the way, the boat stopped at several sea islands. An unofficial emissary delivered news of each island's travails. Barton listened to these reports, dispensed food stored in the hull of the steamer, and considered the scope of the disaster. What Barton saw and heard during that steamer trip convinced her that the Red Cross was obligated to lead the recovery effort. On September 29, she wrote in her diary that she had "decided to take charge of the relief of the Sea Islands" after a nighttime meeting with her closest colleagues.[7] Misgivings plagued her. She did not write of the nobility of charity work but instead commented on what a "great undertaking" it would be to "feed, clothe . . . doctor & nurse 30,000 human beings for eight months, and do it all upon charity gathered as one goes along."

The Lowcountry effort was much larger than any previous field that the Red Cross had handled to that point and indeed may have been the largest coordinated, centralized recovery effort from a disaster event in the nation's history.[8] Despite these logistical difficulties, the Red Cross served the storm-stricken coast from October 1, 1893, until June 1894, taking the reins from the local relief committees. Throughout the recovery process, the Red Cross attempted to nurture the long-term self-sufficiency of Black sea islanders through short-term aid. Barton and her fellow volunteers had a definitive vision for what recovery meant for Black sea islanders: to nurture modest, self-reliant prosperity among the devastated population.[9] The Red Cross's recovery effort was extraordinary in scale, and in some significant ways, the organization acted in solidarity with the goals of Black sea islanders themselves.

For Black sea islanders, maintaining their autonomy after the hurricane through the acceptance of short-term relief anchored their fight against the white elite, who resisted the presence of the Red Cross. The history of the South Carolina Lowcountry affected Black sea islanders' expectations for and their engagement with the Red Cross's recovery effort, and they brought to bear their own familiarity with labor organization and land ownership. Black sea islanders were no strangers to outside intervention or to molding it toward their own ends. Indeed, the Red Cross's recovery mapped onto the Port Royal Experiment. Amid the Port Royal Experiment, white northerners worked to transform enslaved African Americans into obedient wage laborers. The federal government took advantage of "critical shortages of the basic material necessities of life" to deliver relief only to Black sea islanders who worked, even if that meant returning to white-owned land—an unwitting precursor of the labor-based relief structure that the Red Cross implemented.[10]

Thirty years after the Port Royal Experiment, Black sea islanders would, with the Red Cross as a valuable buffer against hostile white South Carolinians, use the recovery effort to protect their community. The position of Black sea islanders may have been better than elsewhere in the South, but they still lived and labored precariously, surrounded by a hostile state government that had, through fraud and violence, ascended almost twenty years earlier.[11] Black sea islanders, capable of managing outside intervention and maintaining their own autonomy, knew how to leverage their labor and landownership to their advantage. Aware of the danger that white Democrats posed to them generally and after a hurricane particularly, African Americans accepted Barton and the Red Cross because the organization helped them to survive starvation and destitution after the storm and supported their desire for autonomy.

Barton's steamer trip along the South Carolina coast was not her first visit to the state. During the Civil War, Barton worked as a U.S. Army nurse. In 1863, she was stationed at Hilton Head Island. From a tent on the wide sandy beach, she watched as the Fifty-Fourth Massachusetts Volunteer Infantry Regiment, comprising African American soldiers, launched their assault against the Confederate-held Fort Wagner. With bullets still flying overhead and shells landing on the beach, Barton dashed to injured Black soldiers. The experience of caring for Black soldiers during that failed attack was transformative for Barton. While she had grown up in a reform-minded, antislavery Massachusetts household that attended the local Universalist Church, she was not a radical abolitionist—until that moment on the beach.[12] The sight of dying and wounded Black soldiers giving their lives for freedom changed her mind and made her support immediate emancipation and citizenship rights for African Americans. Barton set a high price indeed.

Barton's tenure on Hilton Head Island made a lasting impression on those whom she tended. Soon after Barton's return to the sea islands in 1893, three Black veterans arrived at her door. Oliver "Middeten" (likely Barton misspelled or misheard Middleton, a common Lowcountry surname derived from a wealthy white family of enslavers), Abram Fuller, and Dago Haywood came to see the woman whom they remembered from the battlefield.[13] They had served in the Second South Carolina Volunteer Infantry Regiment, and, at Wagner, all three had been injured.[14] Barton had bandaged Middleton's head, Fuller's arm, and Haywood's leg. Fuller lifted his sleeve to reveal a scar that sliced from his elbow to his shoulder and said to Barton, "You dressed it for me that night, when I crawled down the beach—'cause my leg was broke too. But we all got to you, Miss Clare. And now you've got to us. We've talked about you a heap of times, but we never expected to see you. We never forgot it."[15]

Perhaps the veterans wished simply to thank Barton for binding their wounds, because they had been too ill to do so during the Civil War. Perhaps they were trying to remind Barton that they had served in the Civil War to battle for their emancipation and citizenship, which were increasingly tenuous. Back in South Carolina, Barton ought to remember whose side she had once been on and whose side she should be on now. The scars were their struggles made manifest, and they bared them for her. Normally so confident, Barton did not know how to respond. She reflected later that "the memory, dark and sad, stood out before us all. It was a moment not to be forgotten."[16] She gleaned a mutual bond and a sense of responsibility, but she pushed it away in favor of a more facile interpretation of Black sea islanders' characteristics. She introduced this moment in her recollections as an example of a "tender memory of the childlike confidence and obedience of this ebony-faced population."[17]

This rhetoric of Barton's was pervasive throughout the organization and the recovery effort. The Red Cross shared similar objectives with Black sea islanders and accomplished a great deal of material good, but their attitude toward Black sea islanders was laced with a paternalism that doubted the capabilities of Black sea islanders, demonstrated both in their language and in the repeated need to teach Black sea islanders "how to live."[18] The Red Cross hoped to deploy lessons in moral behavior and modern farming practices alongside donated grits and pork to Black sea islanders, whom they perceived as requiring instruction. As Barton declared when she laid out her vision for the recovery effort, "They must not eat the bread of idleness. We must not leave a race of beggars, but teach them the manliness of self-support, and methods of self-dependence."[19]

When she began the effort on October 1, 1893, Barton had with her a small crew of trusted, longtime Red Cross volunteers to manage the organization and administration of the recovery effort from a headquarters and warehouse in Beaufort.[20] Barton also brought on the Beaufort-based Sea Island Relief Committee as her advisory board. This motley assortment of medical professionals, area leaders, businessmen, and seasoned Red Cross volunteers reported to Barton, whose authority over the effort was total.[21] This hierarchical structure was sometimes a source of frustration for the whites in Beaufort who assisted Barton as local fixers.[22] However, the seventy-one-year-old Barton's willingness to toil tirelessly, frequently for fourteen or sixteen hours a day—as "the first to enter upon any work and the last to leave it" as one associate remarked—stifled most criticism from the people working with the Red

A Beaufort street scene, winter 1893–94. Bay Street in Beaufort would have been a crowded place for the ten months after the hurricane struck. Here, illustrator Daniel Smith has placed himself on the south side of Bay Street, looking down the street to the west. A line of African American men looking for work and rations stretches down the block from the Red Cross headquarters past William J. Verdier's mansion, with its double-tiered portico. Two men, perhaps committeemen carrying supplies back to their communities, load goods onto a two-wheeled cart in the foreground. Daniel Smith, *In Beaufort, S.C.*, in Harris, "Sea Island Hurricane," 280.

Cross.[23] Her coterie of advisers and colleagues, both longtime and newly arrived, felt deep loyalty toward her and the Red Cross. Such bonds of trust would be necessary to face the field of recovery. The Red Cross had few precedents from which to choose. None of their previous efforts matched this one: after forest fires in Michigan in 1881, the 1882 Mississippi River flood, the 1884 Ohio River freshet, an 1886 drought in Texas and the Charleston earthquake, an Illinois tornado in 1887, the 1888 yellow fever epidemic in Jacksonville, and the 1889 Johnstown disaster.[24]

Late nineteenth-century ideas of practical benevolence and scientific charity, which would come to inform the Progressive movement, shaped Barton's ethos. Dissatisfied with the rote brutality of industrialized labor and the severity of urban poverty and distressed by the sectional chauvinism of their younger years during and immediately following the Civil War, middle- and upper-middle-class white Americans like Barton sought new institutions and ethics to change what they saw as a fractured world in crisis. While other reformers tackled the plight of impoverished immigrants in northern cities, fought for white women's suffrage, or attempted to curtail business abuses without capitulating to radical labor organizers, Barton rushed to sites of calamity.[25] In those moments of crisis, Barton found a venue for experimentation. As Barton saw it, the sea island coast was rich ground for a project, and so were its residents, whom she believed she could shape into a model population.[26] The Red Cross intended to guide them away from "habits of begging and conditions of pauperism; to teach them self-dependence, economy, thrift; [and] how to provide for themselves and against future want."[27] The supposed introduction of these values and morals to Black sea islanders, Barton believed, would be an antidote to the degradation of African American farmers across the region fomented by the hurricane's destruction.

Like many other educated and well-to-do white northerners with an interest in benevolent aid, Barton believed that African Americans, particularly freedpeople and their descendants, were not biologically inferior to whites—but they *were* inferior because of the social and economic conditions that debased their condition. White northerners like Barton conceived of slavery as a sin not just because the hereditary bondage of a race of people was inherently immoral and exploitative but because it had provided little incentive for self-sufficiency and hard work. Barton assumed that the legacy of those lessons was still among Black southerners, and she sought to teach freedmen new ones. Barton wrote that the Black sea islanders were only "twenty-five years out of serfdom with all the habits of dependence and the memory of a Freedman's Bureau still with them."[28] Barton's dim view of the Freedmen's Bureau provided a counterexample to what she hoped to inculcate. The Red Cross would require labor for assistance, and that strict exchange would school Black sea islanders in lessons of industry that Barton assumed were necessary. She ignored the insidious, creeping Jim Crow regime and favored a moralistic judgment of Black sea islanders' perceived work habits, eager to address the latter and unable to perceive the former.

White South Carolinians warned Barton and the Red Cross that the legacy of slavery had rendered Black sea islanders "demoralized, lazy, and worthless, and probably uncontrollable" if "provided for," and that "being unused to them [the Red Cross] could not manage them."[29] Nonetheless, Barton was optimistic that the organization could coax Black sea islanders toward habits that would uplift their condition through reform of their behaviors, homesteads, and farming practices. Perhaps, too, she wished to prove white South

Clara Barton, 1904. While this photograph is of Barton a decade after she led the Red Cross recovery effort in the Lowcountry, it is a striking portrait of Barton and one of few from her later years that shows her facing the camera. James E. Purdy, 1904, Library of Congress Prints and Photographs Division.

Carolinians, no friends of hers, wrong. Barton and the Red Cross seemed to believe that, remade into economical, thrifty, modestly prosperous farmers, the plight of Black sea islanders would end—without acknowledging that Black sea islanders' plight, beyond the hurricane, was caused by the combination of racism and economic exploitation.

The response of local relief committees may have been insufficient to overcome the ravages of the hurricane, but the Red Cross's vision for recovery, so focused on improving African American habits and behaviors, had profound deficiencies. Barton and the Red Cross wished to leave Black sea islanders better than they were before the storm, but this was a difficult task, constrained as the organization was by its refusal to face racism as one of the forces bringing South Carolina to its late nineteenth-century crisis. What redeemed Barton and the Red Cross's recovery effort was the vision of Black sea islanders, who leveraged those values toward something more radical: a liberatory model for Black southerners that centered Black landownership and collective action.

Though it may have appeared to be a humanitarian project, this was in fact a citizenship-building endeavor. Just as Tillman had warned against the "abuse of charity," Barton and Red Cross volunteers forged a system of relief linked to work. This was also new for the Red Cross. They had not required labor for rations in either of their two previous biggest recovery efforts after the 1884 Ohio River floods or the horrific Johnstown Flood in 1889, whose survivors were almost all white. The Red Cross's decision to do so may have aimed to ward off criticisms from the white South Carolinians who already begrudged their presence in the state, but the Red Cross and Barton emphasized this narrative of labor for relief in a way that indicates the importance of the narrative to their ethos. Barton even questioned the condition of Black citizenship among the sea islanders, declaring the Red Cross's mission was "to fit them for the citizenship which, wisely or unwisely, we had endowed them with."[30] She had no sense of the hard-won fight of enslaved African Americans in seizing their own emancipation and bending the federal government to acknowledge their fundamental right to citizenship. For the Red Cross, Black sea islanders were still pupils to the white Americans who had denied the rights of citizenship and freedom to African Americans generations ago.

Barton and the Red Cross divided up the issues facing the storm sufferers into categories, each of which would require its own structure: distributing rations, mending garments and cloth goods, providing medical care, raising money to support the effort, rebuilding infrastructure, and preparing for the next planting season.[31] Each element was an integral part of a recovery that would fill the stomachs, clothe the bodies, busy the hands, and heal the wounds of sea islanders.[32] But first Barton had the challenge of dispersing the crowds of sea island refugees who had gathered in Beaufort. Barton arrived in Beaufort, the small town whose population hovered around 5,000, and found it bursting at the seams. Fifteen to twenty thousand sea islanders had thronged there, first to collect goods from Beaufort's Sea Island Relief Committee and later in anticipation of Barton's arrival.[33]

The crowds of hungry, ragged sea islanders underscored the insufficiencies of the local relief committees' efforts and highlighted the challenges that the Red Cross would face. The organization did not have more money at its disposal than either the Beaufort or Charleston relief committees. Instead, the Red Cross had expertise. Before even setting up a headquarters, Barton decided that the thousands of sea islanders must return to their homes. She announced to the crowds that rations would no longer be distributed from Beaufort but would be brought to them.[34] Once informed of this decision, the crowds melted away, to make the journey home across the Beaufort River or on roads inland, still cluttered with detritus. Barton and her volunteers could now settle into Beaufort, divide the field, and decide on the rest of their methods for relief.

Barton determined that a system of community-administrated relief distribution would serve the needs of the sea islanders best.[35] She began by carving the sea island coast into districts. She then assigned to each district an administrator who would canvass the district to determine the needs of the people and provide general oversight for relief operations. Each administrator would answer to Barton. On October 3, a warm, rainy day, Barton met with eleven delegations of African Americans from the various islands to speak with them about their needs and about how this plan would work.[36] Barton asked them to return to their communities and elect from among themselves "somebody in whom they had confidence."[37] They apparently "went home satisfied" with turnip seed from Sen. Butler and the Department of Agriculture in tow. The elected "committeeman" subsequently traveled once a week or month by foot, by boat, or by wagon to Beaufort to deliver a stamp card to a Red Cross volunteer, who filled their ration order. Their responsibilities went beyond ration distribution. The committeemen were also representatives of their constituency and "the director of various kinds of work that should be carried on among his people."[38] The Red Cross chose democratic election of committeemen, according to Dr. E. Winfield Egan, to "avert criticism . . . because [white South Carolinians] could not in that case accuse us of having any motives in selecting the persons that we did."[39]

Barton administered the relief effort from downtown Beaufort, within an empty building likely at the northwest corner of Bay and West Streets.[40] There, she and her assistants "camp[ed] out indoors."[41] Initially, her desk was a dry goods box, and her bed was a cot. The Red Cross dedicated each section of the warehouse to a different aspect of recovery: living quarters, Barton's office, a storage space, and a medical clinic. Barton designated the physician, Dr. Egan, as the manager of the valuable rations and donated goods, though he found only a few sacks of grits, a barrel or two of lard, a barrel of syrup, and a few assorted items "all mixed up together."[42] Indeed, Egan was perturbed by the disorder that greeted the Red Cross in Beaufort. Beyond the thousands of refugees, Egan grumbled that there was "no system" and that "everybody was at the head" of the relief effort before the Red Cross's arrival. He could not discern a system of ration distribution either. "Anybody who wished," he claimed, "could come there and say they had four motherless children, and that settled it; they could get anything they had on hand. The thing they could reach the quickest, went out; there was no system."[43] Barton wrote in her diary that she was "not content to let it go on in the old way."[44]

The headquarters of the American Red Cross in Beaufort. This small watercolor sketch by Smith is a side view of the building at the northwest corner of Bay and West that was a key node of activity in Beaufort. Daniel Smith, *In Beaufort, S.C.*, Harris, "Sea Island Hurricane," 274.

The charity of the relief committees was, at least in the eyes of the Red Cross, chaotic and inefficient, anathema to the Red Cross's penchant for orderliness. Egan and the Red Cross may have been ignoring the power of local knowledge, in which perhaps members of the relief committee were familiar with the people who came to them requesting aid; and they may have also been playing into stereotypes of deceptive African Americans eager to leech off a system. Nonetheless, it was true that distributing aid to 30,000 people over the course of many months required a system.

In October and November, the Red Cross focused on immediate relief. Thirty thousand sea islanders received rations of one peck of grits and one pound of bacon for each family of seven on a weekly basis, collected and redistributed by committeemen across the sea islands; 40,000 South Carolinians on the mainland obtained rations too, either by a special envoy or a committeeman.[45] Barton knew that was not enough and fretted privately that those meager provisions would have to be supplemented by fish, oysters, and crabs and "now and then a day's work" for the Red Cross.[46] The Red Cross did provide boats and nets to families as they obtained them, and it was certainly the case that Black sea islanders had long enriched their diets with creatures dwelling in the twisting waterways of the sea islands.

But Barton was aware that the grits and pork were only "insurance against starvation."[47] The Red Cross had to formulate other ways of helping the sea islanders beyond feeding them. While rebuilding homes, fencing gardens, and

African American boatmen loading rations. One of Smith's spare, elegant watercolors from his illustrations of the sea islands; here, African American men are loading supplies onto a boat with a schooner rig, from a floating dock. Daniel Smith, *Rations for Lady's Island*, in Harris, 268.

clearing and digging drainage ditches to improve agriculture would not begin in earnest until early 1894, Barton ordered 500,000 feet of yellow pine lumber at four dollars a foot in mid-October from Mr. Cummings, "the lumber man of Beaufort," who was "quite solid and honest" despite being, Barton remarked suspiciously, "a Carolinian."[48] Floated down the Combahee River to Beaufort, the pine would be carried in wagons and on boats around the sea islands to frame the walls of new houses and to protect fresh green seedlings in the gardens from hogs and chickens.

In the meantime, Barton distributed turnip seeds and put groups of African American women to work cutting Early Rose potatoes for planting, as winter crops that could tide the sufferers over. However, Black sea islanders had to be convinced that white, or "Irish," potatoes were a viable subsistence crop. Egan found that white merchants in the region claimed that "the soil would not raise them; the negro would not take care of them; they did not know what they were, and if they did raise them they would not eat them. Inquiry showed them to cost $5.00 a barrel, and was it any wonder they did not eat them?"[49] Indeed, Egan found that white merchants in the region generally perpetuated the myth that seeds could not be collected from the crops that Black sea islanders grew; that, instead, new seed had to be purchased every year and that none could be saved. Local merchants, including J. J. Dale, also discouraged Black sea islanders from growing more subsistence crops than cotton, as Egan found, noting that "the prejudice" against the Red Cross was very strong among whites on St. Helena, "simply because we were advising them to plant less cotton and more garden stuff and food for their families."[50]

This fallacy ensured Black sea islanders as loyal customers each spring for new seed and continued because, while Black sea islanders were autonomous in many ways, no African American–owned stores seem to have existed on the sea islands. Though it is not clear that every Black sea islander believed the myth that seed collected from their crops was barren, given that they were themselves highly experienced gardeners and farmers, enough must have for the racket to continue. The simple action of sowing a seed into the warm sea island earth was political in South Carolina. A well-known and respected Black farmer and committeeman on Port Royal Island, Jack Owen, was upset after learning of the deception and immediately returned to his home to establish a seed exchange with his neighbors.[51] He, too, knew that his ability to control his harvest was of utmost importance. In South Carolina, the fight between white supremacy and Black liberation took place not only on the floor of the Columbia statehouse but in the fields of the sea islands. Each seed counted.

Once the Red Cross arranged the system for ration distribution, Barton tackled another dilemma. Hundreds of barrels of donated clothing clogged the Beaufort warehouse. However, much of the clothing was poorly suited to the needs of sea islanders. Barton's solution, which she brainstormed with Betty Middleton, May Chaplin, and Mrs. Brown, all Black women, was as community minded as the ration distribution system.[52] She traveled to various sea islands, called meetings at local churches, and solicited female volunteers who, in groups of six, took one-week shifts in a sewing tent to repair and remake garments to distribute among their community.[53] Her first trip, to the settlement near the Coosaw Works, a phosphate mine upriver from Beaufort, had the appearance of a triumphal parade. Her boat was greeted by multiple teams of drivers, wagons, and beasts of burden. The dirt road to the church was lined with people waiting to see the head of the Red Cross. The church was packed, and a ceremony of singing and praying preceded Barton's appeal. Barton explained what the purpose and needs were and what a useful structure might look like for a sewing circle, but she left the organization of it up to the sea islanders. Similar scenes occurred on other sea islands. After one such meeting in November, the women in the audience were "delighted, and all rush[ed] up to shake hands and thank" Barton for her work.[54]

Black women on the coast mobilized their communities in the wake of the hurricane. They ran and operated the sewing circles. Long at the core of an exploitative slave economy in the rice swamps, sea island cotton fields, and kitchens and houses of South Carolina, these same women had played a key role in demanding and shaping their own emancipation.[55] While in other parts of the South, some freedwomen retreated, when economically possible, from agricultural labor as a means of asserting a new domestic identity that had so often been denied them during slavery, freedwomen in the Lowcountry forged a new identity that embraced the value of their fieldwork as well as domestic tasks.[56] The organizational skills of Black women had bolstered Black independence on the sea islands, and they used those skills for a recovery that the Red Cross subsidized.[57] Red Cross volunteers, however, were not always sensible to their history of community care. Mrs. Gardner, a white woman and Red Cross regular who Barton placed in charge of the sewing circles, claimed that the sewing circles, "perhaps the most important feature of the field" after food distribution, had taught Black women "not alone the lesson of self-help, but of mutual help, which they had never known before."[58]

Not all the Black women who ran or participated in the sewing circles were agricultural laborers. In fact, they came from a variety of occupations and backgrounds. Some were, according to Mrs. Gardner, "strong, matronly women, whose childhood and youth had been passed in the service" of white enslavers; "others were sewing girls, some of whom had partially learned trades," and "a few were teachers."[59] One was Robert Smalls's second wife, Annie, a "lovable and accomplished" woman from an established, relatively well-to-do African American family in Charleston, where she was a schoolteacher until her marriage to Smalls in 1890.[60] Others were the wives of committeemen, like Mrs. Sam Green (referred to in Red Cross documents only by her husband's name), whose husband was a skilled boatman and farmer on Lady's Island who loaned his vessel and his services to the Red Cross regularly. Another, Mary Jenkins, was a committeewoman in her own right and a prominent religious and cultural leader on the sea islands, overseeing labor on Port Royal Island later in the winter and frequently writing letters to Barton asking for more clothing or to thank her for her efforts.[61] While in a few cases white schoolteachers like Ellen Murray or wives of Red Cross volunteers like

Black women preparing potatoes for the Red Cross. While no photographs of the sewing circles were taken (or, perhaps, survived), this photograph of African American women preparing Irish potatoes suggests a similar dynamic: this was community work, done by Black women old and young, together. *Women Cutting Potatoes for Planting*, February 1894, *The Red Cross: A History of This Remarkable International Movement in the Interests of Humanity* (Washington, D.C.: American National Red Cross, 1898), 199.

Mrs. John McDonald and Admiral Beardslee's wife headed the sewing circles, the majority of women in charge of sewing circles were African American.

Nineteen sewing circles across the sea islands operated for months, mending thousands of garments. Dozens of women worked for each sewing circle. One hundred and one women bent over cloth and thread at Bennett's Point, twenty women on Sampson Island, sixty-seven at Seabrook, and many more elsewhere. On Hilton Head Island alone, the sewing circle repaired and gave away 3,400 articles of clothing, blankets, and mattresses to a population of 2,300.[62] The women agreed to sew without pay, though for the weeks that they worked in the sewing circles, they received double rations. This was the typical wage that the Red Cross paid to the Black sea islanders, both men and women, who volunteered labor for the organization. Indeed, the Red Cross deliberately paid the same to men and women. In an official report on the effort, one of Barton's associates wrote to defend the choice as self-evident: "Will some of my readers think that these women, some with large families to support, and all having some one depending upon them, should receive less than the men, because they were women?"[63]

Some of the sewing circles drew up constitutions pledging their labor to the Red Cross.[64] Though Barton had dispensed some basic instructions on the goal of the circles, each community sought to make the sewing circle a formalized organization. One such compact paraphrased the Gospel of Matthew from the Bible, in which its members agreed to cooperative work "in peace & harmony" while remembering "that He who numbers the hairs of the head sees all they do, and who cares for the sparrow cares for them and will provide for their needs & wants."[65] Beyond the much-needed extra rations, Black women may have found that the sewing circles provided a valuable outlet to cope with grief, to channel their sorrow into useful work, and to reconnect with other women in the community. They frequently sang call-and-response songs as they worked, giving voice to those sentiments.[66] Undergirding the songs' lush melodies and lyrical words lay the connection between labor and community that Black women on the sea islands had nurtured to overcome natural calamity and human atrocity alike over the generations.[67]

The Red Cross worked hard to make recovery a grassroots effort. However, its severe budget constraints and the expanding size of the field burdened the organization's ability to direct recovery. Furthermore, a second, more compact but powerful hurricane came ashore near Georgetown on October 13 and killed at least eighteen people, exacerbating the need for assistance farther north on the South Carolina coast.[68] While initially Gov. Tillman and Sen. Butler had only asked Barton and the Red Cross to oversee recovery on the sea islands, letters and petitions first trickled and then poured into the Beaufort headquarters from coastal and inland communities stretching from Georgia to North Carolina. They asked for food. They described their ruined crops. Rarely did farmers inland on the coastal plain face the same dire deprivations that sea islanders did. Few deaths occurred inland, and houses were more likely to have lost their roof rather than been swept away, but the hurricane had drowned crops and robbed poor farmers, both Black and white, of their livelihood that year. Many felt a growing panic with the onset of winter. One white farmer shook his head, "I never begged before, I wouldn't now only I can't see my children starve. . . . I've stood it as long as I can; if I can't get help nowhere, can't say what I'll do."[69] The schoolteacher Mather foreshadowed that, without sufficient assistance, "these desperate crackers and long suffering Negroes will combine," in an interracial movement for class solidarity that would "shake the proud old Palmetto State to her very foundation."

One letter written in mid-September by women who lived along the Broad River in South Carolina described "malarial fever and stomach troubles" among hungry people whose crops had been washed away by the hurricane.[70] The letter, penned by two white women, Anna Miller and Emma Mills, wrote of the "furiously destructive wind and terrible waves that swept over them . . . leaving helpless women and children in hopeless wretchedness" and claimed no distinction between poor African Americans and the few poor whites in the region, "whose condition in no way differs from their colored neighbors." "We women situated on the Broad River and its many tributaries," Mills and Miller continued, "in mass meetings on the battle field of Honey Hill do invite that angel of mercy and relief, Miss Clara Barton, to come." The white women living among the fields and forests of the coastal plain met on a Civil War battlefield to decide on their course of action—an evocative setting, given the likelihood that those women had husbands, fathers, or brothers who had fought for the Confederacy, to request aid from what other white South Carolinians considered a new invading force. They resolved to send a committee to meet Barton, as many others inland also did.

By November, Barton estimated that the original 30,000 sufferers had swelled to 75,000 as requests from mainlanders afflicted by the storm poured into Beaufort. She described them as "a mixed population of colored and white, equally destitute, equally storm sufferers, equally objects of charity and care."[71] Her characterization of mainland sufferers as "equally destitute" was not one she repeated often. Indeed, other Red Cross volunteers disputed the degree of mainlanders' destitution once white Democrats seized on the white farmers among those sufferers and began to accuse the Red Cross of neglecting this group because they were "negro lover[s]."[72] Those complaints would not reach their zenith until the late spring of 1894, but disgruntlement began to percolate among white South Carolinians once the Red Cross had conducted their work for a few weeks. These malcontents calculated that accusations of racial bias might prove a useful tool to dislodge the organization.

Politics aside, with no harvest and damaged homes, poor mainland farmers both Black and white did need material assistance. Yet such aid had not been within the purview of the Red Cross, and the state of South Carolina had not given the organization permission to tend to people elsewhere. The Red Cross was to be an invited guest only, so Barton was reluctant to impose the organization on a population without the blessing of state authorities. Indeed, the Red Cross was still a liminal organization: not integrated in state or federal governments but unable and unwilling to operate without their consent.

Barton did not ignore the appeals for aid, but the Red Cross also did not have the resources to tend to them along with the 30,000 wards already under their care. Furthermore, the Red Cross refused to see itself as a charity. Its purpose was not to collect donations and ask for money but to disperse appropriately the money and goods accumulated by community organizations and private citizens. Barton wrote that the Red Cross "cannot appeal to the governors of the various states for we never beg, we never even ask assistance. Ours is a benevolent organization, not a charitable one if you will recognize the difference."[73] And yet the Red Cross badly needed more money. Barton engineered a compromise in which she appealed for federal aid to support employment and infrastructure. "I thought to go to Washington alone," she wrote, "lay the matter before Congress, tell them the situation and ask of them the use of $50,000 to be employed in labor only to enable men to support their families through the winter."[74] On October 29, 1893, she traveled by train from Beaufort to the nation's capital and there lobbied for her cause.

When she founded the Red Cross, Barton intended for the organization to cut through government bureaucracy in times of crisis. The Red Cross, a "single independent, nationwide, permanent volunteer disaster-relief organization," would not be beholden to the gridlock and delays of the federal government.[75] More agile in its response to calamities ranging from war to hurricanes, the Red Cross could provide immediate relief while Congress bickered over the necessary appropriations. However, the Red Cross, as Barton began to realize, was alone on the field. During earlier efforts, the Red Cross had served in a supplementary role, alongside other charitable organizations as well as town, county, and state governments, and with the expectation of some federal relief. Not so in South Carolina.

Although Barton expressed reluctance about requesting money from the federal government, she nonetheless hoped that Congress might appropriate funds for the sea islands because of its increasing willingness to dispense aid to flood victims. In 1884 on the Ohio River, floods wreaked havoc in Cincinnati and other towns, and the rains that caused the river to swell also brought a tornado that ripped through even more villages. Congress appropriated $500,000 and sent the U.S. Army to assist in its distribution. Barton arrived in Ohio to find army boats there to rescue stranded victims and issue rations. After the Ohio River flood, Barton arrived at the scene, and her efforts catapulted private donations to $175,000.[76] But the economic depression of 1893 had closed wallets as northern businessmen and philanthropists alike dealt with their own financial struggles.

Congress, Barton hoped, would hear out her plea and pass a bill for the sea islands, no. 1149. A key ally, Senator George Hoar (Republican, Mass.), vociferously defended the bill after introducing it to the Senate floor. Hoar, an avid opponent of political corruption across party lines and a critic of American imperialism, argued that "this is a measure of very pressing necessity. Some 30,000 people are starving to death." He avoided racism or paternalism in his speech and instead proclaimed sea islanders both Black and white "our fellow citizens." Sen. Butler, crossing regional and political lines, supported Hoar and a handful of other senators. The senators appealed to the low cost, the frequency with which Congress divvied up money for disaster victims, and the humanitarian responsibility of the federal government to assist its citizens.[77] Hoar made an impassioned appeal. "I will do anything short of going down on my knees," Hoar pleaded, and read aloud a statement from Clara Barton describing the impact that the money would have on the working people of the sea islands. The "wet and sour" soil lay spoiled, with "drains and ditches choked up and useless"—unless Congress appropriated money to pay Black sea islanders to reclaim the sodden earth of the sea islands for productive farmland.

Ultimately, Hoar, Barton, and Butler's efforts were for naught. The bill failed. By November 12, 1893, Barton was back on a train to South Carolina. Discouraged though the supporters of the bill were, Barton had fixed her eye on the South Carolina General Assembly. She had convinced an unlikely, and likely reluctant, potential ally to hear her out: Governor Tillman. On her return journey, she stopped in Columbia to talk with Tillman about the possibility of state assistance for the relief effort, and Tillman agreed to read a "feasible plan" before the general assembly if Barton drew one up.[78]

Other Lowcountry allies had been pressuring Tillman to provide state assistance. Ellen Murray, in her position as a longtime schoolteacher at the Penn School on St. Helena Island, wrote to Tillman two months earlier that "the general Government, wealthy as it is, with its thousands to spare for the World's Fair, boasting of its prosperity and opulence, might and ought to assist."[79] After all, "these negroes have been law abiding . . . we have had two constables among six thousand people. No working class in the world have been more orderly, more self-restrained. Is it right, is it wise of the nation to force these people into crime by the pangs of starvation?" While white women like Murray and Barton tried to convince Tillman of the needs of Black sea islanders in part by construing them as harmless, Tillman did not see them the same way. And while the U.S. government may have had funds, South Carolina, a state with little tax revenue, did not.

In his yearly address before the South Carolina General Assembly on November 28, 1893, Tillman described the hurricane as "the most disastrous storm and tidal wave on the coast from Beaufort to Charleston of which our annals have any record."[80] But for private charity and the arrival of that "noble lady" Clara Barton and "her lieutenants" "laboring in the cause of humanity," thousands more may have died of starvation. Tillman acknowledged his uncertainty: that "the question of relief and how best to administer it is a difficult one; even the amount absolutely necessary to prevent starvation is unknown." However, "it is not the will of the people of South Carolina that any of her citizens, no matter how humble they may be—even the poorest negroes—should starve." True enough. The death of Black sea islanders would deprive white landowners and phosphate executives of a workforce. Tillman made a few recommendations. First, the general assembly should, along with the inspectors of the Phosphate Commission, send an emissary to collect information to "report fully also on the condition of those islands and as to the advisability of an appropriation to aid Miss Barton." Second, many coastal residents would not be able to pay their taxes, and the general assembly ought to make provision for them. Tillman asked that the general assembly grant the comptroller-general permission to "suspend the collection on all property within the devastated region in Beaufort, Colleton, Berkeley, and Georgetown, and to remit the taxes of all kinds" where appropriate.

That, however, was the extent of Tillman's request. He did not ask the general assembly for rations or money, even as payment for labor. Tillman, accepting that tax relief would be helpful, otherwise assigned busywork rather than doling out substantive measures of relief. Neither was his speech motivational. The general assembly, Barton wrote bitterly in her midfield report two months later, "at the holidays adjourned without having made the slightest provision for these sufferers."[81] Tillman's appeal, Barton had hoped, would "awaken the General Assembly to a clear sense of its duty," but the "guardians of the people's weal" only extended tax collection by a year, "thus making it possible for the State to collect from them next year what it could not possibly have done this." Perhaps it was no surprise that the South Carolina General Assembly refused to extend a helping hand to the sea islands. The majority Democrat body had likely sensed the growing disapproval of the Red Cross among its white voter base and had no real reason to humor Barton or to provide monetary assistance to Black sea islanders.

Barton, though discouraged and disappointed, had one final option to explore. She wrote a series of letters to the federal secretary of the treasury, John G. Carlisle, requesting revenue cutters to assist in the distribution of rations, clothing, and building materials around the islands. The shallow inlets and dangerous shoals that carved out the sea islands made distribution a difficult task, and the hurricane had swept away many of the nimble craft capable of navigating the treacherous intertidal waterways. The geography of the coast made delivering aid efficiently impossible, as Barton wrote: "These seventy islands are cut and crossed by rivers, sounds and creeks, often too narrow and too shallow to navigate, too wide and deep to ford, and again sweeping, swift and dangerous, like unto the open sea."[82] Not only that, but even if Black sea islanders still "had their little boats, in the long row of twenty to forty miles to come for their provisions, in the frost and cold, half clad and half fed, they would perish." Barton had already tried to develop a system of relief that accounted for this geography, through the creation of districts and the election of local committeemen. But for that decentralized system to function, the committeemen would still need a way to travel to Beaufort, and Red Cross district managers still had to travel between their districts and the Beaufort headquarters. Barton wrote urgently to Carlisle, "The nights are getting cold. We have no way of getting to the people the supplies they need. Unless a boat can be had soon, many deaths from exposure must result."[83]

A week later in early December, Carlisle had agreed to dispatch a few revenue cutters to the coast to assist. Yet even those were inadequate because none of the boats owned by the Treasury Department were of sufficiently "light draft" to navigate all of the waterways, even at high tide. Barton had no choice but to make do with the craft Carlisle sent. In her letter of thanks, she painted a picture of an entire coast on the brink of ruin and starvation, reporting three deaths that were the result of exposure: "Not one in fifty has a bed, blanket, or cover; not one in a hundred has food for two days save the remnant of the weekly issue of charitable provisions we can make, which is the pitiful amount."[84] The Red Cross had "less than 50 cents" per person "for the next six months."

Barton's letter, published widely in newspapers on the East Coast, indicated a level of despair she had not yet shown in public. She must have hoped that the letter would spur charitable donations or perhaps government action. "I bring these facts to you, Mr. Secretary, not to move you to greater pity, but to show you how needful a provision you have made," she wrote. She continued with a description of the desire of Black sea islanders for work. "Seven o'clock of every morning finds a gang of 150 to 200 men in front of the headquarters," she wrote, "waiting to learn if we have shovels or hoes to let them go to work, either ditching the land, building their houses, or preparing the ground for the next year's planting." Barton declared that "we could put 5,000 men at work in three days, under their own foremen, at 75 cents a day, payable in meal and meat for themselves" who would, by next summer, transform the sea islands into "the garden spot of the eastern coast. Hereafter let no one say that the sea island negro is not willing to work."

Even this plea was to no avail. The South Carolina General Assembly did nothing; the U.S. Congress did not reopen the question of federal appropriations. Just as the civil rights of Black southerners mattered less and less to white northerners in the late days of Reconstruction as political corruption, labor strife, and economic troubles distracted them, the sea island crisis paled in comparison to the Panic of 1893. Barton would have to formulate a new plan that did not rely on funding from outside sources, and Black sea islanders would have to find another way to survive until the summer.

Not every sea islander would make it that long. Jack Snipe, "a young man, almost a boy," worked as one of the foremen for a rebuilding project on Port Royal Island. From January 11, 1894, to March 2, Snipe led a crew of twelve men who dug 2,000 feet of drainage ditches, built five chimneys, installed 4,000 shingles on tattered roofs, and repaired eleven houses.[85] Dr. J. B. Hubbell inspected his work on July 27, 1894. Jack Snipe, he wrote, "was a hard-working, conscientious man, but not very strong physically." Despite the care of one of the Red Cross's nurses, Mrs. Barker, he succumbed to a fever in the late spring, soon after completing his work. Perhaps he died of malaria. Perhaps it was some other ailment exacerbated by the back-breaking work of building homes and digging ditches and the meager rations available to Snipe.

Though the Red Cross established a series of free clinics across the sea islands, each staffed by a volunteer physician who both paid home visits and ministered to the ill from an office, they could not save everyone. Dr. Egan, who provided medical care for sea islanders on Port Royal and Hilton Head from the Beaufort headquarters, recorded that two-thirds of his cases were "malarial fever."[86] He declared that these fevers were "undoubtedly aggravated by the condition of their wells" through saltwater intrusion (though doctors a decade later would know that while the brackish water could incur a variety of health problems, malaria was not one of them). But, six months later, Egan saw a marked improvement in the health of sea islanders. He credited this progress with the "cleaning of the wells, the digging of new ones, the clearing out of drains that had been choked up for years and the carrying off of stagnant water."[87] Those improvements could indeed have diminished habitat for mosquitoes carrying the malarial parasite. The work that Jack Snipe undertook may have saved the lives of others in his community, just not his own.

For the Red Cross, the improvement of the lives of Black sea islanders applied, too, to the improvement of the land. One Red Cross volunteer observed that, between the hurricane and the rain after, water "lies and becomes stagnant [on the sea islands], causing fever and rotting the crops. . . . [Draining] will improve the production of the land . . . in addition to making it more healthy."[88] The Red Cross began to consider how to engineer the Lowcountry environment to make it more healthful and productive, having seen how water-quality issues affected the lives of Black sea islanders. In January, the Red Cross determined that they had enough supplies to "pay" sea islanders double rations, the same that women earned if they volunteered for the sewing circles, on weeks that they worked digging drainage ditches, rebuilding homes, and constructing bridges. Barton had wanted to pay the sea islanders outright, but the lack of federal or state funding stymied this plan. Black sea islanders, though, were more than willing to engage in these building projects, which was perhaps no surprise given the enthusiasm among Black women for the sewing circles. The spring planting season, which comes early to the nearly tropical climate of the sea islands, was rapidly approaching. The time was right for the Red Cross to put into motion another part of its plan: to remake the sea islands into a healthful, fruitful agrarian landscape, and to deploy Black sea islanders as the agents of this dual process of improvement.

The amount of time and power required to carve out and maintain drainage ditches was often too much for independent Black farmers to complete on their own. It was difficult enough to plant their own crops each year without the additional burden of spending weeks clearing ditches. Because of the nature of the work and the constraints on Black landowners' time, drainage ditches on the sea islands were indeed choked with weeds and plants. At Gray's Hill on Port Royal Island, Dr. Hubbell reported that the boss, Sandy Haywood, told him that the ditching there had been neglected for over ten years.[89] The Red Cross, then, provided Black sea islanders with an opportunity. The Red Cross gave them rations, lessening the urgency of feeding their families while waiting for the crop to grow, and the tools with which to clear the ditches. Now Black sea islanders possessed an unusual advantage that freed them from their usual dilemma of having either to work on white-owned land or to starve.

As with the ration distribution and the sewing circles, the Red Cross not only facilitated the organization of labor at the community level, which in and of itself was not unusual, but they made it into a form of remunerative labor. African Americans elected their own foremen, or "bosses." The bosses led crews of twelve men on a weekly basis to refurbish homes, ditches, and wells. African Americans who had experience engineering irrigation ditches for rice fields were particularly in demand because they could build "flood gates, or 'trunks,' as they are called, and dams . . . to protect the opening of the ditches from the incoming tides."[90] They also coordinated their drainage ditches across property lines to ensure that a drainage system on one farm did not interfere with another.

The bosses traveled to the Beaufort headquarters to check out shovels, axes, hoes, and other tools, all marked with the organization's Greek cross, from the warehouse. (This procedure was already a contrast to how many African American tenant farmers had to buy tools on credit and become indebted as a result.[91]) They signed their names or made their mark and then brought the tools back once the work was completed.[92] Finally, they wrote reports documenting the labor completed. A few of the bosses, always African Americans, included the names of their crew. Rarely did Black workers receive recognition for their labor. But the lists of names bore witness and claimed a stake in the historical record for their labor and skill. It was a small but important act of pride in a political and economic system that told African Americans that only the products of their labor mattered, not their personhood or proficiency.

The two pecks of hominy and two pounds of bacon per week that the Red Cross paid in exchange for labor may not have been much, but the work was worth a great deal to Black sea islanders, who threw themselves into it despite sometimes being "without proper clothing, barefooted and hungry."[93] D. E. Washington of St. Helena, who worked as a boss leading a work crew, wrote in his report on the thirteen ditches at 34,331 feet that his crew dug had a "value to us" of "$2000."[94] Another boss agreed, scribbling a note on his report that "the people through whose land these ditches run, consider that they are of more value to them than if they had received $100 cash."[95] All told, Black sea islanders dug and cleared out between 250 and 300 miles of ditches crisscrossing Port Royal, Hilton Head, Lady's, Coosaw, Mussleboro, Sampson, Beards, Little Edisto, Edisto, Wadmalaw, and Johns Islands.[96] They rebuilt or repaired at least 200 houses and eleven bridges, the longest of which spanned ten feet by eighty feet.[97] The work crews of the sea islands were willing to work for double rations and because they recognized that the Red Cross's rations gave them an opportunity to refurbish the Lowcountry landscape, labor that they were often forced to neglect to ensure a survival crop.

However, they also viewed the Red Cross as a relatively sympathetic employer of sorts with whom they could negotiate. For all the Red Cross's noble rhetoric on personal and agricultural improvement, the fields of the Lowcountry were still worksites for Black sea islanders, as they had been for centuries. The Red Cross required African American labor to make their relief effort a success, and that was a position that Black sea islanders could leverage for more tools, more nails, different food, and extra clothing. Wesley L. Jackson, a carpenter and boss, reminded the Red Cross that they owed him twenty-four pounds of flour from the previous week and with that request included the names of the twelve men under his guidance in need of their rations. However, Jackson wrote, "I would be very glad if you would change the flour for me into something else."[98] He did not say what grain he preferred; perhaps his men preferred hominy or rice, two chief staples of the sea island diet, to wheat flour. Friday Smalls, a boss at Barnwell Place, wrote asking Barton for "a cuple lbs of nails" to help repair a bridge, following strenuous labor "in mud an water."[99] On March 2, 1894, a group of Black sea islanders living at Cedar Grove on St. Helena Island wrote to Barton and asked for tools "to dig out our ditch . . . for the plase is over flow with warter. The people here cannot git in there Corn land not until April. . . . The men here thought that they would come together and made out this list and send it in by these two men."[100] A list of twenty-six names followed the missive, though the author, William Dais, wrote his at the top in a clear, flowing cursive. This was collective action, no doubt, as Black sea islanders advocated for better goods in exchange for their valuable labor.

The Red Cross rations and tools did not just aid in survival. They gave Black sea islanders the time and freedom to conduct a large-scale project of infrastructure improvement that would make the soil more productive and their harvests richer. Throughout the Red Cross recovery, Black sea islanders tapped into traditions that they had developed during Reconstruction and that valued community-oriented cooperation. The Red Cross's resources prevented Black sea islanders from pursuing low-paying fieldwork with white landowners so that Black sea islanders and the Red Cross could apply themselves to the work of recovery. The Red Cross was attached to an ideal of yeomen independence emblemized by a small, family-run farm that rewarded hard work. Though Barton found the organization frequently frustrated by the halls of power and their refusal to mobilize money and aid for impoverished Black southerners, Black sea islanders transformed the Red Cross's hominy, pork, shovels, and axes into something much more powerful than Barton could have anticipated. No longer meager offerings, in the hands of Black sea islanders, these goods became tools of liberation.

Chapter 6

White Backlash

Not everyone was pleased with the relief efforts on the coast. A cadre of white South Carolinians launched a misinformation campaign, arguing that the recovery effort favored African Americans at the expense of suffering white coastal residents. One "anonymous eye witness" wrote to *The State*, demanding, "What about the white sufferers?"[1] He thought it a "mistake" to "expend . . . so much of the funds for provisions and over-stocking the negro population with eatables that in three months' time will be gone, leaving them no better off than they were before." Worse than that, he claimed that African Americans were behaving unlawfully: "It is a noteworthy fact that the very negroes whom I saw swarming the wrecks stealing and plundering everything they could lay their hands on . . . are now the loudest in their appeals for help." Using rhetoric that would be repeated for generations to come, he demeaned African Americans as looters, stealing firewood and food, and as beggars, thronging for help once those slim pickings ran out. He contrasted their behaviors with those of a young white head of the household who wanted not food but tools to take back up his trade—yet, like many other "white sufferers," he felt a "hesitancy" to claim aid. This supposed eyewitness treated his statement as logical fact rather than anecdotal evidence or outright fabrication, declaiming a simple interest in preventing "abuse of charity" among African Americans.

The anonymous eyewitness played into the white supremacist myth of African American men as brutish and, without the civilizing impact of subjugation through punishment, lawless. Many white South Carolinians wielded tales of white suffering to foment white solidarity.[2] These campaigns began soon after the hurricane and recurred well into 1894, always marshaling around the same refrain: that attention to Black suffering had not only superseded the plight of poor white South Carolinians but that the publicization of and assistance to Black sea islanders indicated a nefarious power grab on the part of Black South Carolinians and their white Republican allies. This rhetoric proved useful for the white elite, eager to consolidate white supremacy—or at the very least compel poor white men to permit wealthy white men to make decisions for them. These tales also served the utilitarian purposes of white landowners, who saw Red Cross aid as stealing Black farmworkers away from their fields.

Clara Barton and the volunteers for the American Red Cross believed so strongly in their vision for recovery that they underestimated how structural and political forces undercut their successes. The recovery became quickly enmeshed in South Carolina's fraught politics. Governor Tillman, who had invited the Red Cross to the state in the first place, soon turned the organization into a target for white Democrats' discontent. White farmers, politicians, and businessmen, when they saw how the Red Cross was working with Black sea islanders, repudiated the Red Cross as staffed by carpetbaggers intent on imposing a second Reconstruction. Wild accusations of corruption, fraud, and graft splashed across newspaper headlines and traveled far and wide through word of mouth.[3] White Democrats' hatred of the Red Cross was not dog-whistle racism but a blaringly loud clarion call to oust the organization and halt aid to Black sea islanders through evoking racist caricatures. White elites saw the devastation wrought by the hurricane as undermining a troublesome region of Black independence, and they maneuvered at every turn to accelerate the processes of destruction and subsequent dependence that the storm set in motion. They wanted to keep Black sea islanders in the condition in which African Americans found themselves after the hurricane: hungry, without shelter, and without a crop.

Most white South Carolinians did not want Black independence. Instead, they moved to consolidate white dominance, maneuvering to end Red Cross relief to Black sea islanders so that they could take advantage of the hurricane's destruction. They sought to narrow the divide between Black South Carolinians' losses as a result of the hurricane and Black South Carolinians' potential privations as an outcome of racial oppression.[4] They hoped to forge a Jim Crow environment in the crucible of the hurricane that would bring ruin down upon Black landowners and farmworkers. Within this environment, Black sea islanders would lose control of not only their harvest but their freedom rights too. White Democrats thus saw the Red Cross's presence and actions as an insidious threat to their nascent Jim Crow project to subjugate African Americans economically, politically, and socially.

Throughout the conflict, the Red Cross struggled to maintain a middle ground, even as whites worked to impress them with the impossibility of impartiality and as Black sea islanders tried to convince them of the immorality of neutrality. Many forces had pressed in on Black autonomy in the Lowcountry even before the hurricane.[5] The white elite hoped to use the hurricane to challenge Black autonomy in the Lowcountry. If starvation, disease, and exposure threatened African Americans' lives and livelihoods after the storm, so too did the rising tide of Jim Crow. Whether Clara Barton and the Red Cross would bear witness to and stand up against white supremacist challenges to the Red Cross's authority and Black autonomy was, for Black sea islanders, an open question that they hoped would be answered in their favor.

Stories of hurricane-induced suffering quickly became fodder for white supremacist propaganda. Many white South Carolinians readily parroted this rhetoric. On September 9, two weeks after the storm, *The State* published an appeal for "a good many white people on Wadmalaw Island" who were "on the verge of starvation." The remedy was swift. Joseph Barnwell of the Charleston committee immediately sent seventy-five dollars worth of provisions to those "white sufferers."[6] Two days later, a headline in *The State* cried out: "Distress on Jenkins' Island: Three White Families There in Destitute Condition."[7] A competing committee thus challenged the integrated Sea Island Relief Committee as these angry voices crescendoed. By September 9, a few white Beaufortonians had formed a Citizens' Relief Committee, which T. G. White chaired. There was no question as to the committee's purpose: it would ensure aid to white sufferers. *The State* introduced the committee as "embod[ying] the views of a number of our representative and prominent citizens" in the need for judicious hands to "dispose of any and all contributions . . . for the benefit and immediate relief of whites as well as blacks."[8] Thomas Talbrid, a probate judge; Dr. A. S. Gibbes, a physician from the politically active South Carolina Gibbeses; and Dr. A. P. Prioleau, a physician that the Sea Island Relief Committee refused to include, joined the colonel on the committee's board. Unsurprisingly, none of them were Black.

The Citizens' Relief Committee also had ready capital. The aptly named White had already accrued $975 and numerous clothing and supplies that the committee set about distributing. "This committee," *The State* emphasized, "is in no way antagonistic to the efforts" of the other relief committees— "except that it differs from it in its policy." That "policy," in which they rendered assistance to whites only, was intentionally divisive. Neither the Charleston nor the Beaufort committees had established any rule prohibiting aid to white South Carolinians who were hungry or homeless; nor were there any concrete allegations that they engaged in discriminatory distribution of aid. White's committee even seemed to have trouble finding enough white sufferers upon which to spend their donated largesse. A week after its establishment, White's committee announced that it was keeping money and clothing in reserve in case any white sufferers should need such supplies, despite their earlier claims that the plight of white sufferers was dire. The "white sufferers," who preoccupied the imagination of so many white South Carolinians, existed in greater numbers as rhetorical devices to advance a racist agenda against disaster relief than they did in reality.

The whiteness of these sufferers always mattered, a badge of their innocence and deserving, and the implications were always obvious to South Carolinians. The proportions of Black storm sufferers to white did not matter, given that those overwhelmingly killed, displaced, or ruined by the storm were African American. The plight of white storm sufferers rarely remained unheard for long. Many other white South Carolinians were all too eager to trumpet these causes in newspapers, in their rush to remonstrate any relief worker who forgot the supremacy of white suffering over Black. Alongside the narrative of destitute Black sea islanders in need of care percolated a more nefarious tale: that Black sea islanders were taking advantage of the hurricane's destruction to claim state aid and to usurp the rights of white South Carolinians in so doing.

In contrast to the stated fears of white South Carolinians, the Red Cross's mission was deliberately neutral and explicitly provided for dispensing relief after a "natural calamity." Indeed, the organization's purpose within the United States was to give succor to communities stricken by disasters of all kinds, and Clara Barton had deliberately crafted the Red Cross to respond to floods, earthquakes, epidemics, hurricanes, and so forth. After the Civil War, as Barton began to consider a national branch of the Red Cross, she wrote of calamities as a unifying force: "Our southern coasts are periodically visited by the scourge of yellow fever; the valleys of the Mississippi are subject to destructive inundations; the plains of the West are devastated by insects and drought, and our cities and country are swept by consuming fires. In all such cases, to gather and dispense the profuse liberality of our people, without waste of time or material, requires the wisdom that comes of experience and permanent organization."[9] These calamities were not particular to any one region of the United States and could therefore elicit sympathy from disparate communities across the nation. Barton appealed to a reconciliationist viewpoint, most of all in her language of "our southern coasts," which as one historian explained, "spoke to the emergent desire for healing between whites in the North and the South after the end of Reconstruction."[10] Her rhetoric of reunion would not overcome the rancor and racism of most white South Carolinians, who were stubbornly opposed to finding commonalities between themselves and anyone else. But Barton nonetheless conceived of the Red Cross as an organization that could serve diverse populations across the nation.

By early spring of 1894, as African Americans prepared the earth for planting using Red Cross supplies, white landowners had found an angle by which to undermine the Red Cross's relief effort and to reframe the plight of Black storm victims as a blatant power grab. They staged their attack against the Red Cross on two battlefields: Bluffton and Pocotaligo. Both were technically mainland communities barely apart from the sea islands. But in Pocotaligo, as was often the case on rice-growing lands farther inland, white landowners had a tighter control over Black laborers; and Bluffton remained a tiny town of a couple hundred dominated by a group of white elites. In South Carolina, it was astonishing what a difference being on the other side of an intertidal waterway could make. White landowners' choice of Bluffton and Pocotaligo was deliberate, calculated to maximize their success.

Bluffton, a small town situated on the May River, is closer to the ocean than Beaufort, but it is nonetheless on the mainland. Nestled in the two arms of the Harbor River's terminus on the coastal plain, Pocotaligo was once a Yamasee village and, by the 1890s, a rice community near the bustling Yemassee railroad junction. White South Carolinians' contention against the relief effort across the state followed a template: the Red Cross neglected white South Carolinians in favor of African Americans, and Black South Carolinians were undeserving and corrupt. That the Red Cross had requested money from private citizens, the federal government, and the South Carolina General Assembly to extend their field of recovery to include and tend to mainland farmers of both races meant little.[11] Indeed, the Democrat-dominated South Carolina General Assembly had refused to dispense state funds to the Red Cross in November 1893.[12] The white South Carolinians who led the charge against the Red Cross did not care to consider how they had already failed poor whites in the state. They were dedicated instead to a project of white supremacy. What the Red Cross saw as aid to encourage Black independence and self-sufficiency, white landowners saw as dangerous and misguided welfare that would steal African Americans away from their fields. Aid disrupted the property relation between white landowners and Black farmworkers, because white landowners continued to believe that they were entitled to the labor of African Americans, along with the land that they worked.

The first hints of conflict appear peppered in private letters and conversations. On November 7, 1893, Thomas Martin, a white landowner and Confederate veteran, wrote to the editor of the *News and Courier* J. C. Hemphill with a litany of accusations and slander against Barton. According to Martin, John McDonald, the Red Cross's representative to Bluffton, "did not call on a single white landowner or a representative white citizen."[13] Instead, McDonald "went around among the negroes." And not just any African Americans, Martin wrote, but Black Republicans, Union veterans, and crooks—arrogant malefactors who, Martin claimed, poisoned the minds of other African Americans. Martin alleged the rampant abuse of welfare perpetrated by Black South Carolinians and enabled by the Red Cross. He pointed to the demographics of Bluffton, arguing that "this peninsula has more white farmers and people than colored, and also more Democrats than Republicans, yet not the list of managers and the fact that no whites are helped, and it looks rather peculiar if not partisan." Martin implicitly, though not subtly, alluded to the Freedmen's Bureau and Reconstruction, claiming that "the negroes all say that Uncle Sam is going to feed them a year, and they will not wrok [*sic*] and are selling nearly all their corn." "The Red Cross," Martin declared, "is an inflammatory failure." Martin tethered recent memories and the meaning of Reconstruction to the Red Cross's recovery, eager to turn the politics of disaster recovery to the ends of white Democrats.

Feeling that such an account would "require a full and candid explanation," Hemphill sent this letter to John McDonald and asked him if he had a rebuttal for Martin's accusations. McDonald did indeed have a response and composed a powerful and rousing repudiation. He opened his letter with an anecdote about his arrival: "I saw three of the representative white citizens standing on the wharf,—one a landowner—and overheard the remark, as I was only fifteen feet away, 'That boat has brought the men belonging to that Cross concern which proposes to feed all these worthless n——.'"[14] McDonald built a case exposing the intersecting classism and racism of the white elite in Bluffton and refuted every charge laid at the Red Cross. He was disinterested in humoring the claims of well-to-do white men like Martin and expressed his abiding trust in the poor farmers, both Black and white, of Bluffton: "What information from alleged representative white citizens, who represent none but themselves! What could they have told me that would compare with what I saw? Would they have told me that they were threatening the people with ejection from their homes, if they did not get the rent? . . . I have in my book a list of such cases where hogs, horses, mules etc., have been taken away from these poor people by the white merchants, because the cyclone had rendered them helpless."

McDonald informed Hemphill that he was in Bluffton on the authority of two petitions sent by two different committees in Bluffton. Neither petition was signed by a "single landowner, merchant, or professional" white man, so McDonald did not feel bound to represent their interests. Instead, McDonald felt a responsibility to expose white Blufftonians' callous exploitation of their less fortunate neighbors. McDonald, a Scotsman, also refuted the charge of partisanship, saying that he had never cast a vote in the United States. Instead, he wrote, "Mr. Martin's spiteful cry of partisanship recoils upon himself, for he evidently wishes to make a political issue out of the trivial matter to measure out a peck of grits to starving people." Indeed, that was just what Martin wished. Martin and his friends had much to gain if their bluster frightened the Red Cross into retreat.

McDonald had seen enough to grasp Martin's goals. The gravest accusation that McDonald leveled at Martin was that white landowners, storekeepers, and merchants in Bluffton were colluding to stop aid distribution to "force these destitute people to purchase on credit or mortgage, from the stores." The economic stranglehold of the white elite on land, labor, and capital in the rural South was not unique to Bluffton. Martin and his associates were conspiring to replicate what they saw happening all over the South. The hurricane had forced poor farmers in Bluffton into a vulnerable position, and the kind of destitution that foments desperation now characterized their precarious economic condition. Martin and his men were waiting, ready to extend credit that would afford no real relief, only an endless cycle of debt and exploitation. The Red Cross was standing in their way and providing poor farmers with a lifeline that would foil elites' plans. The white petite bourgeoisie knew this. One "prominent merchant" complained, "This relief work is nonsense; the n—— don't want any more help, let them catch fish and crabs and help themselves."[15]

Ultimately, Hemphill did not print either letter in the *News and Courier*. But Barton knew of and took the accusations seriously. She recorded the trouble with extending the field of recovery to the mainland and pondered the Red Cross's exodus, originally scheduled for late March. Barton feared March was too early if she wanted to help both sea island and mainland sufferers through the growing season. In early December 1893, Barton, with the Sea Island Relief Committee, penned a letter to Tillman that was pregnant with despair: "This field, ever an exceedingly hard and perplexing one, has been made doubly difficult, owing to the great number of appeals from the mainland. Delegations, committees and single petitioneers from the mainland swarm around us in such vast numbers that, added to our island wards, well nigh dead-locks our relief effort. . . . If we are not speedily relieved, our supply will be entirely exhausted."[16] On January 9, 1894, she wrote: "We have great trouble with the main land people."[17] By mid-March, she had sent an emissary, Mr. Wisler, to investigate the needs of the "main land people," yet again. On March 27, she met with Wisler and commented, "This main land question is very hard." The next day, Barton had decided to extend the Red Cross's stay to address the concerns of the people on the mainland and to see the spring crops planted. Niels Christensen wrote to his wife, Abbie, in Massachusetts, expressing a mix of exasperation and gratitude for Barton's choice: "Miss Barton has awoke to the fact that there is still much to be done in the way of relief and has concluded to stay longer than she anticipated. It is luck that the spring started so early."[18]

Barton announced the postponement of the Red Cross's departure to midsummer 1894 in a circular distributed to local pastors to read to their congregations. The memo listed the Red Cross's efforts thus far: Barton counted out "350,000 feet of lumber, 500 kegs of nails, 600 hatchets, and as many saws besides," in addition to the distribution of clothing, garden seed, Irish potatoes, beans, and peas.[19] The Red Cross felt compelled to see the sea islanders through their harvest because, Barton explained, "you have come to be very near to our hearts, your welfare is our welfare, your interests our interests." The Red Cross had collected a store of White Flint corn so that a peck—enough to plant three acres—could be sent to every family. The circular contained reassurances as to the Red Cross's dedication to the field of recovery and as to their sensitivity to local conditions.

Rev. D. C. Washington, a Black minister, hosted an event at his church to announce the circular's message, attended by Red Cross volunteers.[20] He voiced thanks for the Red Cross, in such a way that hinted at what they could have suffered at the hands of both the storm and white South Carolinians: "If God Almighty did not send Miss Barton here, what would have become of us? . . . Your most noble works will be left for ever, in the best hearts that beat in the sunny South."[21] Pullman, a Red Cross volunteer who spoke as a guest, marred the proceedings with characteristic paternalism: "You are better, in that you are placed in a more self-reliant way; you will have learned a good and difficult lesson."[22] Another Red Cross volunteer, Mr. Tillinghast, continued in the same vein, warning Black sea islanders that "after awhile [Barton] will go away and you must then depend on yourselves; and are you able to do that? You must learn to work, to know the value of work. . . . You must learn that time has a value, a money value."[23] While Black sea islanders publicly expressed gratitude for the Red Cross and patiently worked with them because their supplies were invaluable, the unrelenting, condescending racism from Red Cross volunteers must have chafed, most of all when they lectured to Black sea islanders while guests in their holy spaces.

While Black sea islanders reconfigured their plans for the spring based on Barton's announcement, white South Carolinians redoubled their efforts to force the organization out, first in Pocotaligo in late March 1894 and then for a second time in Bluffton in May 1894. The Red Cross encouraged Black sea islanders to focus on their own crop and continued rations distribution to ensure that they would not borrow money during this lean time of the year. That course of action collided with the interests of white landowners, who relied on the hunger of Black farmworkers as their food stores ran low in the late spring so that they would sow, tend, and harvest crops on white-owned fields. White landowners in Pocotaligo schemed to deprive Black farmers of the rations that gave them this relative independence and to compel them back onto white property. In Pocotaligo, despite its mainland location, the Red Cross had been providing rations for months and had facilitated the organization of work crews, with William Grant serving as the community's committeeman.[24] The Black farmers of Pocotaligo were determined to plant and harvest their own crops, and the Red Cross rations had presented them with an occasion to claim greater self-sufficiency.

Red Cross assistance irritated white landowners in Pocotaligo who relied on Black labor to tend their fields, particularly Eugene Gregorie. Gregorie was a prominent local landowner, who, in 1882, was called forth as a witness on behalf of George D. Tillman, Ben Tillman's brother. Robert Smalls had sued George Tillman after the contested 1880 U.S. Senate election, which Tillman had stolen from Smalls through voter fraud and intimidation.[25] Members of the white supremacist paramilitary organization the Red Shirts, who backed Wade Hampton III in his seizure of the South Carolina governorship in 1876, had spent the election season of 1880 menacing and attacking Black voters, killing at least two, to prevent Smalls's election. Gregorie had been one of those men. Fourteen years and one destructive hurricane later, Gregorie resented the Red Cross's intervention because it emptied his fields and his wallet at the very moment that he had hoped to use the hurricane as a disciplinary tool to bend Black farmworkers to his will.

The hurricane created a new kind of opportunity for whites in South Carolina, one that distorted the line between the harm done by nature and the habituated disaster of racial oppression. Gregorie wrote to William Elliott, the powerful Beaufort politician who had facilitated requests for aid to the federal government a few months earlier, to convey a message to Barton about "the wrong she is doing to the planters by her indiscriminate issue of rations to the lazy negroes of this section."[26] Gregorie claimed, "My foreman tells me that the negroes about Pocotaligo boast that they are not going to work as long as the RC issues rations and clothing. . . . Unless it is all stopped before May we had just as well not plant." White landowners across the South tried to keep Black farmworkers in a state of perpetual need, with the hopes of retaining a large and cheap labor pool. Gregorie clearly recognized that the Red Cross's rations threatened white control over Black labor and, in response, fomented the origins of a powerful racist argument against welfare provisioning to protect his own economic well-being and to prevent Black independence.

Somehow, the Black farmers of Pocotaligo learned of Gregorie's accusations. Soon after Gregorie's letter arrived on Barton's desk, Black farmers in Pocotaligo organized a letter-writing campaign to appeal to the Red Cross as sympathetic mediators. They wrote seven petitions protesting Gregorie's stance, clarifying the deal that white landowners had offered to them, and explaining the beneficial effects of Red Cross rations.[27] Although these Black farmers were not members of the Colored Farmers' Alliance, which had grown in previous years to become a substantial organization representing the interests of Black farmers in the South, the language of their petitions nonetheless reflected one of the central tenets of that organization. They stated that collective action "is the only thing that will give us protection for our labor and crops."[28] Just like members of the alliance, the Black farmers of Pocotaligo believed they had to rely on themselves to protect their property and their interests.

But these particular rice workers had another point of inspiration: many may have been veterans of the 1876 Combahee River strikes in May and August to September, during which Black rice workers in the region petitioned for higher wages paid in cash, walked off the fields, and rallied together to vie for their labor rights amidst a pivotal election year.[29] The resonance of the strikes can be felt in the Pocotaligo petitions. The seven petitions from a small farming community on South Carolina's coastal plain were an exercise in popular democracy at a time when such protests were increasingly dangerous for African Americans generally, and particularly for those in the rural South.[30] Legacies of labor resistance were neither dead nor forgotten in the Lowcountry.

The petitions exposed the insidious, entwined goals of labor control and white supremacy and the gendered nature of both. Their authors were unequivocal and unflinching in their condemnation of white landowners. They also employed forms of legalistic language, just as the sewing circles had used in their constitutions. The first petition opened with the phrase "This is to certified [*sic*]," and continued, "that we was informed that the contrey farmers had Reported to you saying that we wont not work for them even when they offered Wages that we are to lazy to work for our living we will denie it it is not the truth. . . . Any Bodey whosoever say that the People of Pocotaligo is lazy and wont work for themselves is telling untrue it is a fals Report and I can Priase it by the vegitable that is growing in the garden and by the cultivation of soil in the Field."[31] The proof of their hard work was in the earth, planted for themselves, material evidence to refute the white landowners' claims. Other petitions elaborated on other points of exploitation and abuse. A second opened in similar fashion: "We lernt that Parties have Sent in Repor that we will not work for No Price on account of being seplied by the RC. That Statement . . . we denie."[32] However, this petition, signed by nine men, added more information to counteract Gregorie's letter: "We also lernt that the farmers said that they offer the labourers from 75 cents to one dolla Per day. That statement we also denie." Instead, white farmers were giving around forty cents for a day's work, and not in cash but "white papers" that could be exchanged for goods at a single store in Pocotaligo whose prices were double or triple their market value. Four other petitions confirmed this figure.

One of the petitions, with eleven signees, pointed out a gender discrepancy: "The [white farmers] have willfully exaggerated the wages paid to women are 30 cents and to men from 40 to 50 cents per day and only give us 4 cents per task for ditching and who can live at such wages as that?"[33] For the same

On The Combahee.

On The Ashapoo.

Distribution at Chehaw.

On the Combahee, *On the Ashepoo*, and *Distribution at Chehaw*. These three photographs, from Rachel Mather's slim book about the effects of the hurricane, are of African American communities along three Lowcountry rivers, the Combahee, the Ashepoo, and the Chehaw (a branch of the Combahee), that were centers of rice production. Absent photographs of the Black farmers from Pocotaligo, these are a useful representation of similar communities in rice-growing areas after the hurricane. The men and women pictured were almost certainly rice workers; many may have been on the two days' system or confined to some other form of tenancy. The storm surge from the hurricane, as discussed in Chapter 2, rushed down Lowcountry rivers and was devastating for Black rice workers, who were deliberately placed by white landowners in low-lying areas vulnerable to flooding. Any supplies that the Red Cross or other local organization provided would have been much needed. *On the Combahee* (opposite p. 68), *On the Ashapoo* (opposite p. 100), and *Distribution at Chehaw* (unnumbered first page) in Mather, *Storm-Swept Coast*, courtesy of Beaufort County Library, South Carolina.

labor, Black women were paid less than Black men—not that either wage was sufficient. Another letter agreed that the wage was untenable: "Dear sir who can live on forty cents per day with a large family?"[34] In a system where men and women worked the fields, that difference, between forty and seventy-five cents, mattered. The white landowners of Pocotaligo had attempted to turn the community into a kind of agricultural company town, depriving Black workers of other opportunities for employment or recourse.

Some petitions boldly named their oppressors. The act of putting names to deeds was an act of deliberate and daring protest. One sworn and notarized testimony contained a statement that the author had asked Gregorie for work, and Gregorie had turned him away because he claimed to have enough hands.[35] Another author agreed that Gregorie was rejecting laborers: "I will name one man that went to him for work for it and he refuse to give him any thing."[36] A few singled out Gregorie as a perpetrator in this scheme, and others named John Frampton and a Mr. Schnider. They even indicted the behavior of John J. Dale, the Beaufort businessman whom Tillman attempted unsuccessfully to force onto the Sea Island Relief Committee but who later consulted frequently with Barton and the Red Cross. While Dale "claim that he is helping the People," "when one old Collerd Person came to him," he only handed over "4 quarts of grits." But when "a white Person came to him he gave to one man 10 lbs of Bacon one Bushel of Peas One Bushel of Corn and a Plow." He also exploited their need for cotton seed, the primary cash crop available to them. While Dale would sell cotton seed, "you have to contract to gave him in the Fall 12 lbs of seed cotton to evry Bushel of cotton seed witch he gave you." With cotton seed weighing thirty-two pounds per bushel, and ten pounds of cotton seed planting an acre of cotton, that was a heavy cost to Black farmers in the Lowcountry, whose plots of land were rarely more than five to ten acres.

The final petition resulted from widespread community organizing. Signed by sixty-five people after "a mass meeting," it included numbered grievances, resolutions to demonstrate their gratitude to the Red Cross, and sections to "condemn . . . the statement made by the planters."[37] The five-part petition broke down their case: first, the landowners only offered "employment to not more than 12 person in one week," with pay not in cash but in white papers at the nearby store: "Whether we want the goods or not we are compell to take it." Second, they called out the exorbitant prices at the store: "Take the white paper [of] $4.00 to the farmers stor we are paid in goods from the same which will amount to about $1.50 for the week work." Third, they pointed to the job scarcity: "There is nothing for the mass of the poor people to do. And these Farmers know it. What little work there is not one third of the people git any thinge to do."[38] Fourth, the petitioners explicitly described the system of racism that entangled and constrained their lives and labor: "The White Farmers is pregeded agence the poor colored people. . . . The know if we can git a little Helpe that we can stay on our own place and plant our crops and work there. The depend on us to make there crops and as long as there can make falce statement to the World agence the Black Man to git you to withdraw from us the Beter."[39] This remarkable petition, the result of community organizing, declaimed and named the white supremacist project of the Pocotaligo landowners as they tried to harness the hurricane's destruction to Jim Crow. The means of exploitation were manifold: pay too low to support a family, too few jobs, the country version of a company store. The Black farm laborers of Pocotaligo understood all too keenly the stakes of hurricane recovery and the politics of disaster. The Red Cross presented them with an opportunity that white landowners were desperate to deny them.

Finally, the authors of that petition thanked the Red Cross for their past assistance, assuring them that "no matter what is said By the Southern white People we Belive that god sent you here."[40] They also promised to use the rations wisely and played to the Red Cross's concerns over their diligence: "We are eager to plant and do all we can for our self." Then followed the names of twenty-three Black farmers, seven of them female heads of house, each including how many members their household contained, for a total of sixty-five. The other petitioners ended with similar sentiments requesting that the Red Cross continue rations distribution so that it would be possible to plant their own crop so that they would not rely on white landowners for a paltry wage. "We Beg that you will contrive to help us a little longer so that we may Be able to make our crop as we are making every Effort to do So," one group wrote; another, a short statement signed by eight-two Black laborers, assured the Red Cross that all they wanted was to "make our crops and secure provisions for the Future."[41]

The Black farmers of Pocotaligo succeeded. Barton continued sending rations to Pocotaligo. William Grant, the community's committeeman, sent a letter of thanks to Barton for a shipment of four wagons containing seventeen sacks of provisions, including eight of corn, nine of grits, and eight of seeds, divided among 509 people.[42] These farmers had won a victory, uncommon for the time and place. Their community efforts also belied any myths of passive Black compliance with early Jim Crow and demonstrated that African Americans in the Lowcountry understood the power of and were ready to engage in labor organizing at the first pragmatic opportunity for success. This flurry of organizing and protest had triumphed for two reasons: the vigor of the effort on the part of Black farmers and the Red Cross's increasing realization that white South Carolinians intended to make the organization's job harder through deception and exploitation. The attempts of white South Carolinians to stonewall recovery from the hurricane and Black independence would not, after all, end with their loss in Pocotaligo.

Less than two months later, in May 1894, a second controversy flared in Bluffton, South Carolina. There, white landowners focused not on spreading racist stereotypes of African American workers but instead on the Red Cross's supposed neglect of poor whites on the mainland whose livelihoods had been hurt by the hurricane. Months earlier, Barton estimated that the original 30,000 sufferers had swelled to over 70,000 as mainland sufferers flocked to Beaufort.[43] Poor white mainland farmers were indeed hurting—but, as before, Barton was reluctant to move her operation beyond the sea islands because Tillman, leery of any perceived attempt to encroach on his own power, had limited the Red Cross's effort to the islands. Nonetheless, Barton did distribute many rations to the mainland sufferers, but the ostensible line drawn around her field of care allowed white Democrats to attack the Red Cross's perceived lack of care for white mainlanders.

The trouble began after the *New York Daily News* made the sensational announcement that 600 white residents from the Bluffton area were destitute and on the brink of starvation. The *Columbia Register* in South Carolina immediately dispatched a reporter, W. W. Price, to Bluffton to investigate the claims. Price penned a hit piece entitled "CURSING THE RED CROSS!"[44] The exercise in inflammatory slander emphasized the differences between the Beaufort enclave and the rest of the state and implicated the Red Cross in a discriminatory scheme: "Not a dozen white men in Beaufort County, unless it be in the towns of Beaufort and Port Royal, have a good word for the society. It is roundly cursed and abused and the charge is openly made here that the suffering and destitution of the white people in this township are due to the discrimination of the RC or some of its managers." He charged that Black sea islanders and white Republicans in Beaufort assisted the Red Cross only to advance their own political agenda.

Furthermore, Price alleged, the Red Cross had only lent a hand to "the worthless class of negroes, who never had anything in their lives, who had nothing to lose by the storm, and who were as well of[f] after the storm as before" while ignoring "industrious negroes." In the minds of men like Price, material wealth was equivalent to respectability and industriousness. Absent those markers, a Black sea islander was surely of poor character. At the end of his diatribe, Price appealed to pride in a mythical heritage and history, prized rhetoric among white southerners eager to claim a clear and exclusionary lineage: members of the "Anglo Saxon race were on the verge of starvation."

Gov. Tillman, smelling blood in the water, charged forward. He dispatched an appeal "on behalf of the white residents of Bluffton township," and newspapers across the South picked up the story.[45] Tillman leveraged the incident as a moment to encourage white racial solidarity, describing that he had only recently learned that "great destitution existed among the people of our own color" at Bluffton. These poor whites had lost their crop and had subsequently "exhausted all means of credit in an effort to support themselves and to plant anew." *The State* reiterated Tillman's call for charity for white sufferers and made explicit the underlying charge: "these white people have been strangely neglected by the Red Cross, while the negroes on the coast have been well provided for," proving "the discrimination against white sufferers by the Red Cross."[46] Monetary contributions poured into Tillman's office from private citizens, white churches, and a few Farmers' Alliance organizations from South Carolina. Many of them echoed the same sentiment, sometimes emphasized with a thick line of ink: "please distribute among the White storm sufferers"; "for the benefit of the destitute white citizens of Bluffton"; "for the storm stricken white sufferers of Bluffton"; "for the Bluffton white sufferers."[47] Tillman donated $125 from his own coffers to the white sufferers of Bluffton.[48]

The members of the Sea Island Relief Committee, who had administrated relief in the area before the arrival of the Red Cross and had worked closely with Barton since, shot back a response in early June 1894. They sent an outraged letter to Tillman, which they also had published in newspapers, that did not deny that white farmers at Bluffton were in trouble but did remind Tillman that the Red Cross had only been asked to administer to the sea islands.[49] They felt "impelled by a sense of duty, and in simple justice to an organization which came to our relief at a time when their advent was regarded as a most fortunate event, to say, that we have been deeply impressed with the integrity, impartiality, economy and unswerving devotion to duty" of the Red Cross. The "unfounded and unprovoked," "numerous, ill-considered criticisms in the public press, some of which have verged upon the borders of libel," could not stand unchallenged. Barton had gone out of her way to assist tens of thousands of people, "white and colored," on the mainland. The relief committee members, including Robert Smalls, typed their names under a curt *Signed.* Barton followed up with a letter to the head of the General Relief Committee, Joseph W. Barnwell, who had slighted the Red Cross's recovery effort. The *News and Courier* reprinted her address:

> You will recall . . . that on the 30th of March last the Red Cross, through me, presented to your honorable committee the reports then beginning to circulate relative to alleged suffering and destitution on the mainland of the state adjacent to the sea islands. It was pointed out that an appeal

to the country at large would be distasteful to the citizens and an injury to the fair name of South Carolina, that with very slight pecuniary or material assistance all danger would be averted and the situation relieved. No action was taken. You will recall that no criticism has been made by us on that non-action.[50]

Barton, who rarely showed anger in public, here betrayed some of her frustration. She refused to let the Red Cross take the blame for negligence in a state where no authority had stepped forward to provide aid. She especially could not let such accusations pass when they came from politicians who had refused to advocate for their own people, against her recommendations.

That 600 poor white farmers were destitute in Bluffton was not a point of dispute, especially after June 16, 1894, when a letter of thanks signed by dozens of whites from Bluffton appeared in *The State*.[51] The proportionality of their suffering when compared to the other 30,000 on the sea islands, however, could have been contested. That the Red Cross had not practiced racial discrimination did not matter. It did not matter that the Red Cross distributed 20,000 garments, rations for 41,509 people, and double labor rations for 6,500 to the mainland alone.[52] What mattered was that white South Carolinians believed that white northerners were meddlers with an insidious agenda. They were right to a degree, if by "insidious" they meant that the Red Cross was willing to bolster African American landed independence.

By late May, as the Bluffton controversy swelled, Barton did not regret her decision to stay until summer, but the onslaught of slander made the responsibilities of managing the recovery weigh more heavily upon her and her organization. The Red Cross wound down its relief efforts in preparation for their departure in June 1894. Dr. Egan, with Black Beaufortonians, cleared out the Beaufort warehouse. Volunteer doctors completed a final round of home visits and began to close the makeshift clinics that had once seen forty to fifty patients per day.[53] Red Cross representatives scoured the countryside, speaking to Black sea islanders, taking extensive notes, and writing reports on the improvement in their houses, fields, and lives. McDonald wrote of thriving crops, thanks to the hundreds of miles of ditches, which "reclaimed and rendered tillable . . . a large area of otherwise waste land."[54] H. L. Bailey described Black sea islanders as more "prosperous and happy than ever in their new rebuilt homes, with their newly fenced gardens, wells cleaned or new ones dug, three hundred miles of newly made ditches that reclaimed the best lands they had."[55] "It gives me pleasure," he wrote to Barton after surveying Edisto, Wadmalaw, Kiawah, and Johns Islands, "to tell you that these people are grateful for what has been doing for them and are to be commended for their excellent behavior and patient attention. . . . I have every reason to think that they can now take care of themselves."

As a final initiative to reinforce the Red Cross's recovery effort, Barton launched a lecture series on "lessons in practical economy" for Black sea islanders.[56] The talks were well attended, with Barton writing that between fifty and 400 would crowd in to hear the lessons imparted. The topics included "Owe no man anything. How to keep out of debt. Don't sell cotton before it is picked. Plant more vegetables, and why. Divide cottages into rooms. Don't mortgage."[57] Dr. Hubbell threw himself into the lectures with vigor and inflected them with practical instruction. He lectured on "seeds, crops, gardens, planting, cultivation, fruit trees, and keeping of wells and ditches clean, giving practical and actual examples of making tin graters from a tin pan or a tomato can, and grating the new corn into meal while it is too soft to grind."[58] Dr. Hubbell hoped Black sea islanders would use these talks "to cultivate thrift, ingenuity, enterprise, develop prosperity, with the view of breaking up the 'credit system' of business among them and making them independent, progressive and prosperous." The Red Cross, again and again, misunderstood what constrained Black sea islanders: not their individual character but a creeping system that conjoined economic, environmental, and political oppression.

To the last, Red Cross volunteers were not veiled in their rhetoric. They adopted racist language and used slurs that echoed tropes about the happy slave or those seen in minstrel shows. One Red Cross volunteer commented on the "happy, careless, cheery manner" of the "industrious, grateful class" of Black sea islanders to whom they tended.[59] The Red Cross's general field secretary, George Pullman, as he reflected on the effort, demeaned Black sea islanders for not planting Irish potatoes before. To a *New York Times* reporter, he recounted, "We were told that you couldn't get a n—— to have anything to do with an Irish potato," but once distributed by the Red Cross, "they were like children with a new toy," noting "each inch of growth with shiny eyes and grinning faces."[60] Some volunteers suggested that the deadly hurricane, which killed thousands of Black sea islanders, had been a strange kind of blessing to them: "They did not know how to live," one said, but "the Red Cross lifted those people and placed them years ahead of where they would have been had it not been for this cyclone."[61]

From the perspective of Barton and the Red Cross, their recovery efforts had been a success because they had directed African Americans to renew the agricultural landscape of the sea islands and had conveyed lessons in frugality and self-sufficiency. Indeed, Dr. Hubbell compared the improvement in Black sea islanders' character to the growth of their crops, suggesting how connected the two missions were for the Red Cross: "There has been a great deal of mental seed planted and germinated among these people and it needs a little more attention and cultivation before it is 'laid by' to harden and mature. We want a good harvest and an abundant one."[62] Black sea islanders tolerated the Red Cross and attended the lecture not because they were ignorant of how to plant gardens or grow a healthy crop but because they were eager to stave off the debt that so many white landowners were forcing onto them. These were lessons that Black sea islanders were compelled to take, whether they wanted them or not.

Despite the constant stream of vitriol from certain quarters, many Lowcountry residents on both sides of the color line agreed that the recovery was a triumph. Even a significant cadre of wealthy white Charlestonians lined up to congratulate Barton. Charleston's Vanderbilt Benevolent Association, of which both Wade Hampton III and Sen. Butler were members, awarded her with an "exquisite gold badge" in the shape of a star.[63] The association's president, the prominent Charleston businessman A. C. Kaufman, presented it to Barton on July 9 and gave a speech extolling her and the Red Cross: "This heroic band of workers in the cause of the 'needy and him who hath no helper' illustrated their profession by their practice, for little less than one year, in caring of the thousands on the storm-swept coast of our State." "It was," Kaufman announced, "a herculean undertaking from which even these stout hearts were inclined at first to shrink, but duty beckoned them onward, and to that call there could be no refusal. Humanity the world over owes them a debt of gratitude for their magnificent management, and pre-eminently so the people of South Carolina."[64] Kaufman concluded by stating that the governor and general assembly ought to issue a resolution of thanks (no such resolution was ever forthcoming, a reminder that most of the white elite felt differently). Mayor George Holmes and Commodore L. A. Beardslee of the Sea Island Relief Committee issued a statement, published in the *New York Times*, which extended "our most heartfelt and sincere thanks and high appreciation of the gratifying results of their labor. . . . We deem it a pleasure to be enabled to place upon record our confidence in the Red Cross."[65] Even Niels Christensen, despite his earlier criticism of Barton as too rigid, wrote to his wife, Abbie, that "Miss Barton sent for me the other day and we had an affectionate good bye with each other."[66]

Black sea islanders penned numerous letters to Barton thanking her and the Red Cross. Several missives used the language "We the People" or similar wording to show their collective appreciation.[67] Others wrote of the Red Cross's fair treatment. S. J. Pinckney, apologizing needlessly for his spelling and grammar, wrote, "I feel like I could work with you all my lifetime for the way you all have treated me."[68] M. C. Black wrote, "I could not have been treated any better. It will be a lasting regret to part with you."[69] The Women's Aid Association of Beaufort, of which Annie Smalls was president, conveyed their "heartfelt thanks and inexpressible appreciation for your noble and unselfish work."[70] Eighty-three African Americans living on Bonny Hall, the very rice farm where the August 1876 Combahee River strike began, thanked Barton and told her, "We are satisfid with the Committee finding that we will have justis of what you deliver in too his Hand."[71] A group of Black Beaufort lawyers, professionals, and pastors also drew up a statement of thanks.[72] The letter described Barton as a "faithful, inestimable, and Christian lady" who was "wise and impartial dealing with all the sufferers alike."

Black sea islanders had pushed the Red Cross to recognize their worth and equality through their labor, their collective action, their dedication to their communities, and the trust that they placed in the organization. Barton had once said that the purpose of the Red Cross's mission was "to fit [Black sea islanders] for the citizenship which, wisely or unwisely, we had endowed them with."[73] The irony, of course, is that the Red Cross was "fitting them for citizenship" just as Jim Crow was poised to take it away. The letters were perhaps a final attempt to convince Barton and the Red Cross of something that was, to them, self-evident: that Black sea islanders, too, were indeed part of "We the people."

The anger that Barton faced from white landowners had jarred her sensibilities, no matter how gratifying Lowcountry residents' general response to the effort. The Red Cross was supposed to be an organization that worked for the good of the nation, maintaining a moderation and neutrality that would allow them to work effectively no matter the political context. The hurricane relief effort in South Carolina had challenged that framework. White resentment of the Lowcountry's traditions of Black autonomy and landownership was amplified by the Red Cross's presence and policies. The ground that white Democrats had gained over the past ten years, between the disenfranchisement of African Americans across the state and the rise of Tillman, had thus far stopped at the Lowcountry. The Red Cross–guided recovery effort was further bulwark against the incursion of Jim Crow white supremacy, and African Americans deftly allied with the Red Cross to strengthen their defenses. Within that context, the attacks that whites launched against the Red Cross and African Americans were shot through with vindictiveness.

The effort took another kind of toll on Barton. She traveled to Charleston from Beaufort in July 1894 and, once there, collapsed of exhaustion "resulting from overwork."[74] She had clocked fourteen-hour days almost consecutively for ten months, which would be a strenuous undertaking for a person much younger than Barton's seventy-two years. Barton rested for two weeks. At the end of her convalescence, she met with the *News and Courier* to conduct an exit interview at a warehouse on East Bay Street in Charleston that a businessman had lent to the Red Cross for collecting goods. The newspaper, one of the local few kindly disposed toward Barton, listed the number of houses rebuilt (5,000 of 6,000 destroyed), the 245 miles of ditches dug, the five tons of garden seeds, 1,800 bushels of seed corn, and 1,000 bushels of Irish potato seeds given out, and the rations distributed on the islands (30,037) and to the mainland (41,000).[75]

The journalist also described Barton in effusive terms that emphasized her feminine grace, which suggested that some white South Carolinians saw Barton as a convenient white savior figure to whom they could give credit for the hurricane recovery. Clad in an elegant black skirt of velvet and silk and a black blouse held closed by an amethyst and pearl brooch, Barton was notable too for her "exquisitely modulated voice" and how "to every subject . . . she lent that deep sympathy and warm-hearted impulsiveness which have been such factors in shaping her own career."[76]

Barton was sharper than the paper's depiction of her might suggest. She did not accede to failure on any point, despite the smarting memory of Pocotaligo and Bluffton. "The people of the sea islands with whom we have dealt," Barton declared, "are to-day better clothed, better fed and in a better condition than they have ever been before. They have more vegetables than they can eat or sell." With the assistance of the Red Cross, Black sea islanders were now a "prosperous and self-helping people." The Red Cross accomplished this with no thanks to the South Carolina General Assembly: Barton commented tartly that "no response was given" to the Red Cross's pleas. Barton defended her organization and extolled the victories of the recovery. She also acknowledged the divisions she had witnessed and become embroiled in. Barton may not have known, upon her arrival on the sea islands ten months earlier, how trying the recovery effort would be because of the "internal dissension" among South Carolinians.[77] It might have given her even greater pause, uncertain as she had been whether the Red Cross could successfully manage such a large field of relief.

As Barton left South Carolina to return to her home in Washington, D.C., a few whites aimed parting salvos at her. A group of the agitated white landowners from Bluffton, infuriated by the *News and Courier* interview, had the paper's editor publish a rebuttal (though the editor gave it the unflattering headline, "Berating Miss Barton: A Bitter Letter from Bluffton's Relief Committee").[78] "This woman," they fumed, "venomously slanders our people, as well as ourselves, and make them appear by outrageous, untruthful, studied and purposely planned statements as either liars or rogues, and simply to carry out her unworthy, selfish ends." "Who is she that drives the streets of Beaufort with Bob Smalls, and sends out agents that sleep, eat and drink with negroes to succor the wants and cares for our poor and proud white people?" they asked, insinuating an improper association between Barton, a white woman, and Smalls, a Black man, that could have dangerous consequences for Smalls especially. They demanded of Barton, "Who is this woman, anyway, that throws reflection on our people, and dares to insinuate that they would, without the direst necessity, ask for help?" White landowners would not soon relinquish their wrath. Anything Barton said in defense of her work was a grave insult to white landowners and only proved her and her representatives' unseemly connections with Black sea islanders. White anger increased because of Barton and the Red Cross's outsider status. Just as white Democrats abhorred the legacies of Reconstruction and treated the events as an open wound, so they found the Red Cross culpable in deepening the cut.

In the months and years after she left South Carolina, Barton did not forget about the people of the sea islands. In late February 1895, Barton wrote to Black sea islanders that she hoped that they "have been doing the preparatory work of ditching for the raising of good crops."[79] "Get the neighbors to join together," Barton suggested, "and clean out the old ditches, make all the new main ditches and canals that they can, and then make the smaller ones to connect with them; this will help to give them better health, less fever, larger crops and better ones." They should "strive to raise the best of everything," "keep out of debt," and persevere in "the general continuance of this work of improvement"—as though it were that simple. Those tasks would prove increasingly difficult, as the political climate of South Carolina turned to the new season of Jim Crow. During the rest of 1894 and the spring of 1895, Black sea islanders watched their crops mature, tended their gardens, and held their families close after the catastrophic, tragic losses of the hurricane. One year after the storm, it appeared to be a fruitful year already. By August 1894, farmers were growing the season's second crop of White Flint corn, a type that the Red Cross introduced, a first for the sea islands since their previous varietal took the entire planting season to mature.[80] Others reported that their crops were producing twice as much per acre as usual.[81]

The Red Cross–administered recovery effort was an unusual event in Lowcountry history and an unusual case in the history of disaster relief. This is not intended as a naïvely rosy parable about communities coming together after a tragedy or a glibly triumphal tale about the victory of the recovery effort, because either would be inaccurate. Instead, this is a historical narrative about hard-won possibility. Black sea islanders knew that they were hemmed in by hostile whites waiting to strip them of their franchise, their economic independence, and their land, to storm the enclave of Black political power whose walls the hurricane eroded.

Black sea islanders had long seen their traditions of autonomy as a buffer against white supremacy and applied the same tactics to hurricane recovery, with the Red Cross as a critical mediator. They understood that they could ally with the Red Cross to support their struggle against South Carolina's white elite, for the organization gave them time and resources to regroup and to rebuild as best they could under the circumstances. Through their alliance with the Red Cross, they achieved meaningful, though ultimately limited, successes for their communities that promoted their independence from both white landowners and an oppressive political system that was steadily gaining ground. The recovery that Black South Carolinians sought would remedy not just the grievous wounds of the hurricane but the gross injustices and privations of an ascendant white supremacist regime.[82] Landownership and political self-determination should not be particularly radical goals. But in South Carolina in 1894, to advocate for those goals was to challenge white supremacy itself.

White landowners had instantly recognized the danger of the alliance between the Red Cross and Black sea islanders, of the autonomy that Black sea islanders could sustain if they were not entirely cut off from sympathetic allies. They had seen it during Reconstruction, during a fusion ticket of Black Republicans and moderate Democrats in the 1880s, and, most recently, during the recovery effort. White landowners thus fought the Red Cross whenever possible. For white Democrats, the Red Cross effort seemed less about restoration to prehurricane conditions than a dangerous experiment in social and racial engineering that would forestall Jim Crow. The Red Cross believed that hard work could translate into economic independence, and with that logic, the Red Cross justified the extension of aid over many months to avoid Black sea islanders' long-term dependence on welfare.

But long-term dependence was precisely the goal that white southerners had for Black southerners, and the only hard work that white southerners recognized was the hard work that African Americans did for whites. Anything else held no value. In an increasingly frantic series of attacks that depicted white South Carolinians as the true victims of both the hurricane and outside intervention, white Democrats repudiated the need for aid at all, accused the Red Cross of racial favoritism, and plotted to dislodge the Red Cross from the Lowcountry.

The conditions of Black sea islanders after the hurricane reminded white Democrats in the state of the conditions of poverty and dependence that they endeavored to inflict on every Black South Carolinian. The hurricane had interrupted this white supremacist project in unexpected ways, allowing three different constituencies—Black sea islanders, white landowners, and the Red Cross—to grapple with the hurricane's destruction and to implement their own visions for Black dependence or independence. And yet the recovery effort carved out a meaningful interlude during which African Americans, using the assistance of the Red Cross, worked to strengthen their communities. This, then, was a rare southern story about the failures, however temporary, of white backlash against carefully preserved Black autonomy. African Americans' recovery was bittersweet, entwined with lingering grief. But for a short time, a sense of tenuous achievement blossomed delicately across the sea islands.

PART III

Cascade

Chapter 7

Draining the Black Majority

Each time hope sprouted, soon after came the scythe. Clara Barton, in her final address to South Carolinians upon her departure from the Lowcountry, pointed to the roiling politics of the state that had so inflected the recovery effort:

> I realize the throes of internal dissension which are rocking you from centre to circumference. . . . I see how impossible it will seem to you to take note of any lesser matter till this seething bubbling pot of political discord is still; and yet, brothers and sisters, it will be for your credit that you turn a single glance at this little strip of misery, degradation and want hedging you like a sea wall, and take some concerted, and elevated step to wipe the spot out. Don't let politics touch it, they have had enough to do with it already, but give the reins of thought and action for one day to a little humanitarian effort.[1]

Barton recognized that white South Carolinians were poised to institute a political system that would have a detrimental impact on the rights and livelihoods of Black sea islanders. She appealed to white South Carolinians to stay their hand rather than dealing a death blow. In drawing connections between the hurricane's lingering effects and the institution of Jim Crow laws, Barton suggested that the timing and the voracity of white South Carolinians' efforts to diminish the political power of Black coastal residents may not have been incidental. At the very least, Barton made it clear that she found the push to consolidate white supremacist power over African Americans still struggling from the hurricane callous, even vindictive.

The story of Jim Crow's rise in the South Carolina Lowcountry is necessarily a complex one because of the unique autonomy and demographic dominance of the region's African American majority—a majority so overwhelming and an autonomy so embedded that it defied white southerners' normative vision for Jim Crow. What could Jim Crow segregation look like in counties like Beaufort and on rural islands such as St. Helena, where over 90 percent of the population was African American?[2] The coastal fringe of the state presented a challenge to white South Carolinians' conceptualization of a post-Reconstruction regime as a pervasive, one-party system of violence, disenfranchisement, economic subjugation, and segregation operating through state support. The price to remedy the disaster of slavery was Reconstruction fulfilled—a price that the white South refused to pay and the white North lost the will to enforce.[3] Another disaster, which rose from the ashes of Reconstruction and gained strength from the hurricane, loomed: Jim Crow.

White Democrats had caught a glimpse of what a Jim Crow could look like on the coast after the hurricane, in which formerly independent African Americans were forced into conditions of deprivation and made reliant upon white largesse. While white South Carolinians did not publicly express their desire for a mass death event like the hurricane, the Quaker minister Joseph Elkinton voiced his suspicions after his January 1894 tour of the coast, saying that "Southern Democrats thought, as did their representatives in Congress, that there were too many 'n——s' and it would be well to let them die."[4] When a few thousand died in the deluge in August 1893, there were fewer Black voters. If tens of thousands more survived and recovered nonetheless, then white supremacists would have to find other means of diminishing their presence. The stalwart persistence of the region's Black voters and politicians, despite the bloody state coup of 1876 and a flurry of voter suppression acts in the 1880s that ravaged the Black vote elsewhere in the state, was a perpetual source of frustration among white Democrats. While the Red Cross had dimmed their vision for Jim Crow, white South Carolinians were emboldened by Ben Tillman's racist exhortations to prevent "negro domination" and could not relinquish the idea of a subdued Lowcountry.[5]

The hurricane had started what white Democrats considered a necessary process of reducing the coast's Black autonomy and population. They would do all they could to finish it in the following years. A main point of white Democrats' obsession was thus demographics. To institute Jim Crow in South Carolina, the Black majority could not stand. The Lowcountry, as the stronghold of the Black majority, had to be crushed not only through traditional political means, like a new constitution to replace the one from 1868, but a wholesale attrition of the Black population. Tillman (who moved from his position as governor into a U.S. Senate seat in 1895) and South Carolina Democrats fixated on population statistics. Whether at the ballot box or in the fields of the Lowcountry, African Americans outnumbered whites.

Jim Crow in the Lowcountry thus required a built environment to support and supplement state structures of dispossession and disenfranchisement.[6] While white Democrats poured effort into crafting a new political regime for the state, they also looked to projects of environmental engineering and racial control to make the Lowcountry landscape whiter. Even as Black politicians grounded their contentions for African American citizenship in the soil of the Lowcountry, white landowners and politicians sought to develop an architecture of white supremacy fitted for the region. The range of strategies that white supremacists employed to force Jim Crow upon the Lowcountry demonstrates that Jim Crow was a contested, varied, often highly localized project with environmental as well as traditional political and economic dimensions. The ultimate effect was a Jim Crow regime tailored to the Lowcountry, built in the wake of the hurricane.

In many parts of the Jim Crow South, one historian observed, "interracial proximity meant that social relations had to be negotiated and renegotiated each time a person walked down the street."[7] But in the Lowcountry, the substantive Black majority altered the dynamics of Jim Crow. The uneasy closeness of white and Black southerners living in large numbers within the same community, the rule for large parts of the South, did not characterize the entire region. Instead, white politicians found the Lowcountry difficult to manage because so few whites lived there at all. The Black majority that persisted not only in the Lowcountry but in the entire state until 1930 fostered a unique character of long-lived Black autonomy. White supremacists seized upon the Black majority to position themselves as the victims of an imagined "dark wave of corruption, misgovernment and well-nigh anarchy" that Black and white Republicans supposedly foisted on the state during Reconstruction and to whip up paranoia about the potential for "negro supremacy" in the future.[8]

The Lowcountry, where Black South Carolinians successfully defended their autonomy through the storms of politics and nature for decades, was at the heart of these fears. To succeed in imposing Jim Crow on the Lowcountry, white Democrats had to confront carefully cultivated Black resilience on political, economic, and environmental fronts.[9] One highly valued target was South Carolina's 1868 Reconstruction constitution. A convention of seventy-six African American delegates and forty-eight whites wrote the constitution, which provided free public schools for all races; gave married women full property rights; allowed the electorate to choose the state's presidential electors; established a poll tax dedicated to primary education; laid the basis for a normal school, an agricultural school, and an institute for deaf and blind students; and required maintenance of the state university.[10] Most objectionable to white supremacists in the state, it ensured Black voting rights and had no provision against interracial marriages.

Even after the end of state-government-led Reconstruction in South Carolina in 1876 and after the passage of a series of laws in the 1880s that reduced the Black vote elsewhere in the state, African Americans along the coast maintained their political power due to rights that the constitution empowered.[11] They deployed this power through fusionist tickets, numerical power at the ballot box, and the savvy politicking and tenacity of Black politicians like Robert Smalls, Thomas E. Miller, W. J. Whipper, James Wigg, and George

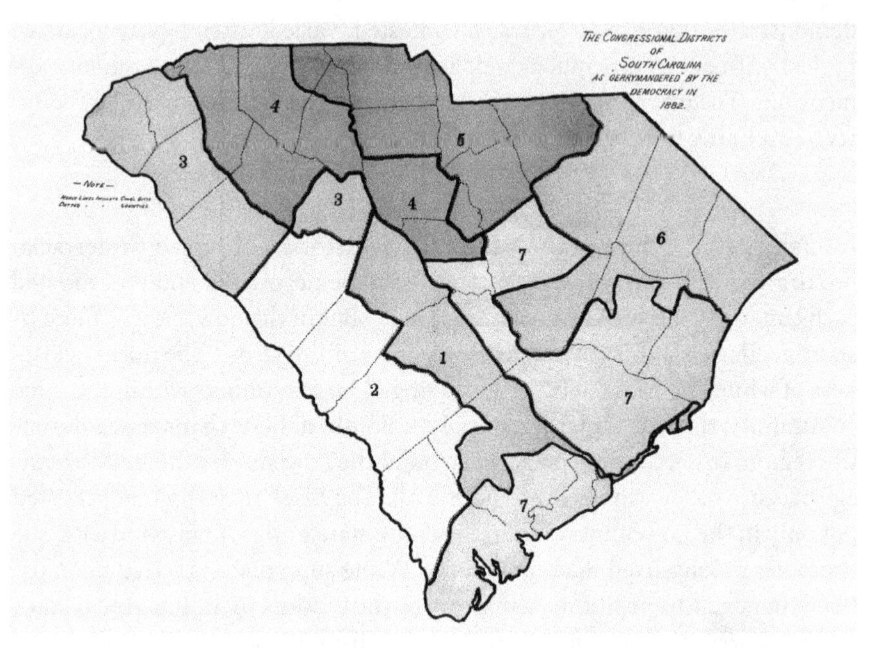

South Carolina congressional districts, 1882. This map shows the consequences of redistricting by white Democrats in South Carolina in the 1880s. South Carolina legislators drew the lines of the new Seventh District specifically to crowd as many African American voters as possible into a single district and to thus dilute their influence in U.S. House of Representative elections. *The Congressional Districts of South Carolina as "Gerrymandered" by the Democracy in 1882*, Library of Congress Geography and Map Division (LCCN 2015588077).

Washington Murray.[12] The Lowcountry's Black voters stuck in the craw of Democrats. Their independence—whether in the context of their landed autonomy, their modest recovery from the hurricane, or their political mobilization—presented a barrier to white supremacy.[13] The proponents of a new constitutional convention could not imagine Black political participation, no matter how limited, as anything but "negro supremacy."[14] The wispy specter of "negro supremacy" justified the blunt, hard-nosed program of white supremacy.[15]

This perspective said more about how dimly white Democrats regarded the practice of democracy than anything else.[16] White politicians saw the project of disenfranchisement as a numbers game, first to unite white Democrats and then to devise constitutional provisions that would defeat the state's Black majority. To "prevent a bitter factional fight" among Democrats at the convention, leading conservative and Reformer Democrats signed a pledge in February 1895 that vowed to support "such qualifications [in the constitution] of the suffrage as will guarantee white supremacy."[17] Conservative and Reformer

Democrats, while divided on economic questions and polarized by Tillman's abrasive governance, could always find common ground if they turned to their racism. This fusion, as white politicians saw it, was necessary within a state where 100,000 white men of voting age faced 140,000 Black men of voting age.[18] Tillman had no doubt what the Black majority meant for the state: "Can the 100,000 white men in SC energize and marshal the 140,000 negro voters in this State and lead them to the polls, contending for mastery, without destruction to all of our business interests and a paralysis of every industry? Can this vast horde of ignorant and debased voters participate in government without corrupting and debauching the public service? . . . If we had a white majority in the State, this burning question of negro supremacy or quasi supremacy need not cause so much alarm."[19] Indeed, "negro domination" because of the state's Black majority was a favored rhetorical bogeyman of Tillman's. He knew that white supremacy in the state ran upon the carefully tuned mechanisms of law and practice and that fear-based, racist politics had a long tradition of success among the white South Carolinian electorate.

Black South Carolinians faced the convention as yet another battle to be fought rather than a predestined victory for their opponents.[20] In the months before the election for the convention's delegates, the Republican Party registered 10,000 new voters in the state.[21] They managed to sweep all five seats for the constitutional convention delegation from Beaufort and one of the three from Georgetown, the latter on a fusionist ticket. All six of the Republican delegates—Robert Smalls, James Wigg, Thomas E. Miller, W. J. Whipper, Isaiah Reed, of Beaufort County, and R. B. Anderson, of the Georgetown delegation—were Black, though they had clashed with each other over their many years in politics.[22] Those six Black delegates took on the responsibility of struggling against odds that they knew were, by design, against them.[23] It is doubtful that they believed they could alter the course of the convention. The six men keenly knew the odds against them.

The convention began on September 10, 1895, in the capital of Columbia, just two years after the hurricane. There, white delegates worked to dismantle the legitimacy of the 1868 constitution and to block Black delegates from meaningful participation. The Black delegates nonetheless launched a passionate defense for the African American vote at the convention. In their speeches, the delegates rallied around arguments for African American citizenship grounded in a vision of American history that centered Black contributions. African Americans had been brought to the shores of South Carolina unwillingly but had nonetheless been vital to building the country as it was. Indeed, appealing to history and constructing a narrative that countered the one that white supremacists had crafted was central to the Black delegates' strategy. White politicians wielded their memories of slavery, the Civil War,

Robert Smalls, 1895. August Kohn took the portraits of every delegate to the Constitutional Convention, including, of course, the eminent Gullah statesman Robert Smalls (b. 1839, d. 1915). *Smalls, Robert, Delegate from Beaufort*, from the August Kohn photograph collection South Caroliniana Library, University of South Carolina, Columbia, South Carolina.

and Reconstruction as a weapon against Black South Carolinians and their rights, just as they had in the wake of the storm.[24] In their speeches, the Black delegates provided a corrective to white supremacist mythologies and traced their vital role in American and Lowcountry history.[25]

Significantly, much as proponents for relief for Black sea islanders had done after the hurricane, the delegates constructed a defense of African American voting rights on the basis of a Black, working-class agrarian identity. Smalls called Black rural workers "the bone and sinew" of South Carolina.[26] Wigg demanded of the white delegates, "Tell me whose labors bring this promise of contentment and joy to the homes of South Carolina?"[27] Miller spoke of interracial class solidarity, arguing that the convention was "for the disenfranchisement of that class of people, whose chief lot has been to toil, toil, toil," "to disenfranchise the negro in the rice fields and his poor, uneducated white brother," transforming them into "nonentities in the government."[28] The six delegates, aware that the convention might strip them of their rights, looked to Black labor and its transformative power over the southern landscape, and the Lowcountry in particular. White delegates might refute the intelligence of Black South Carolinians or their ability to govern wisely and use that repudiation to strip them of voting rights, but they could not deny that they had, for better or for worse, made Black labor the bedrock of their economy.

Despite these efforts, the white delegates prevailed. They approved property and literacy qualifications, which required either that a potential voter

Benjamin Ryan Tillman, 1895. Kohn's portrait of Benjamin Tillman (b. 1847, d. 1918) captures a stern expression. As was his preference, Tillman's missing eye is in shadow. He lost it to a cranial tumor as a teenager, which prevented his plan to serve in a coastal artillery unit for the Confederacy during the Civil War. *Tillman, Benjamin, Delegate from Edgefield*, from the August Kohn photograph collection South Caroliniana Library, University of South Carolina, Columbia, South Carolina.

must have paid taxes on and demonstrate that they had property in the state "assessed at $300 or more" or could "both read and write any Section of this Constitution submitted to him by the registration officer."[29] Tillman admitted the danger of the clause: "The property qualification straight would disenfranchise thirty thousand people, white men, who own nothing in their own names. . . . I said last night that the chalice was poisoned."[30] However, Tillman averred, "Some poisons in small doses are very salutary and valuable medicines." The poisoned chalice would be a dangerous prospect to accept. But Tillman drank deep and blamed the brew on the state's African American population, saying that the property and education qualifications were still a compromise, given the federal government's constitutional protections of African American civil rights. Tillman revealed the authoritarian nature of white supremacist rule: white supremacy mattered more than voting rights, even for whites.

The convention lasted for another month after the passage of the suffrage clause. On the evening of December 4, 1895, the final day of the convention, Evans called for a vote on the final ratification of the constitution with 116 yeas and 7 nays.[31] The Black delegation voted against it. When the Beaufort County delegates were called on to sign the constitution, Smalls asked, on behalf of his comrades, to be excused.[32] They could not put their names on a document abhorrent to their politics, their histories, and their people and so solemnly filed out of the hall. With the Black delegates gone, Governor Evans, the president of the convention, spoke to the white delegates in solidarity: "Fond memories of this Convention will ever arise, while watching you make a law upon which the hopes of our people for supremacy and prosperity are based. . . . Let me wish you a merry Christmas and a happy New Year."[33]

It was just as well that the African American delegates were not present to hear Evans's beaming approbation of the convention, a mockery of what they had lost and sacrificed. It is unlikely that the Black delegation looked forward to the cheer of the holidays, no less because Smalls was grieving for reasons beyond the political losses of the convention. His brilliant, civic-minded wife, Annie, who had been such an active volunteer during the Red Cross recovery effort and a leader among the women of Beaufort, fell seriously and unexpectedly ill two days after Tillman launched a lengthy attack upon her husband on the convention floor. Smalls rushed from the convention to her deathbed in Beaufort.[34] She died three days later, on November 5.[35] Smalls returned to the convention nine days after to watch the constitution take its final, terrible shape, with his fellow delegates by his side.

The constitution signaled a shift in the Lowcountry, consolidating the white population around a Jim Crow agenda of Democratic dominance of state politics. Whites living in the region were undoubtedly emboldened by the ascendance of the white supremacist constitution, having gained energy in mounting their resistance to the Red Cross recovery effort. In 1896, the Democratic Party held its first "camp meeting" in Beaufort County since the Civil War, where white Democrats rallied around a platform of an all-white county government that propelled the party to victory.[36] The few whites who did live in Beaufort County—while split for decades between northerners, usually Republican, who moved to the county in the years after the Civil War, and the southerners who had held onto inherited property—began to unite under a common ideology. Even so, in Beaufort County and on Black-majority sea islands, African Americans continued to participate in municipal elections for nearly twenty years after the constitution passed, despite the fact that Democrats controlled the registration apparatus in the state and thus clamped down on their participation in all other elections.

Black South Carolinians smarted under the losses to their civil rights and from the symbolic significance of the constitution, even those whom state laws and other forms of voter repression had functionally disenfranchised over ten years earlier. White South Carolinians frustrated by the successful defense that African Americans had mounted for decades even in the wake of the most destructive hurricane in the state's history had devised a way to bring down that stalwart wall: to shift politicking from the local level to the state level, where they could use their power to weaken the Lowcountry's Black bastion. The Red Cross recovery effort had highlighted the extent to which Black political and demographic power in the Lowcountry derived strength from local politics and interracial alliances. In response, white Democrats literally moved the region's most powerful Black politicians away from the Lowcountry to the capital of Columbia to begin the process of destroying Black political power along the coast. In so doing, white supremacists had hit upon another strategy that they would deploy in the coming years: draining the Lowcountry of its Black majority.

The backlash to the recovery effort and the new constitution were only the beginning of white South Carolinians' efforts to reclaim the Lowcountry. Indeed, white South Carolinians' plans took on a surprising dimension: some began to advocate for ridding the Lowcountry of African Americans entirely. Not all agreed, of course. Inland counties lobbied successfully for state laws prosecuting emigrant agents who lured Black workers to employment out of state, whether in New York as domestic laborers or to Florida turpentine camps.[37] Some descendants of enslavers who operated coastal labor camps hoped to retain a Black labor force on their dying farms. But some of them, like Harry Hammond, the Charleston judge Theodore Jervey, and the editors and founders of *The State* Ambrose and Narciso Gonzales, came to concur with politicians like Tillman. The African American population along the coast, with their engrained traditions of landed yeomanry, were of little use to whites, scarce in the rural Lowcountry anyway.

Jim Crow in the region, therefore, would hinge on the reduction of the region's Black majority as much as the shackling of African Americans to exploitative economic systems like sharecropping that dominated elsewhere in the South. After all, as Hammond had argued just two weeks after the hurricane struck, the ruination of Black communities in the Lowcountry was a sign that it was time for the population to leave.[38] White South Carolinians had recently equated Black emancipation with white oppression; as George Tillman, Ben's older brother, had argued during the constitutional convention, "If we were free, we would have negro slavery."[39] Unable to reinstitute slavery, white South Carolinians devised new forms of bound labor and sought to "solve" the so-called "negro problem" with a wide variety of policy proposals and practices.

White politicians regarded the Lowcountry with continued suspicion and feared even a disenfranchised Black majority, and they wound the constitution into a larger set of goals with the purpose of "redeeming" the state from its African American population.[40] The constitutional convention, so close on the heels of the hurricane, amplified their fears that the region's Black majority was an ongoing threat to white rule. The paranoia that African Americans in the Lowcountry could mount a political rebellion against the Jim Crow constitution drove their plans, since there they had the advantage of sheer numbers. White South Carolinians also sought Black dispossession in the Lowcountry because there African Americans owned land and formed a threatening yeoman class that had proved resilient against many of the tactics that white supremacists had deployed elsewhere. The South Carolina legislature thus attacked Black rural laborers with a lien law in 1897 that fined or imprisoned people who received advances in money or supplies and then did not perform "the reasonable service required of him by the terms of the said contract."[41] White landowners and lenders used the law to target African Americans. Though the state Supreme Court found the law unconstitutional in 1907 because "it required involuntary servitude" in violation of the "13th and 14th amendments," the law demonstrated the determination of white South Carolinians to break the will of Black landowners, tenant farmers, and sharecroppers and to compel them into new forms of servitude to white bosses.[42]

To complement the political tactics of oppressive state laws, many white South Carolinians challenged the existence of the landed Black majority in the Lowcountry. They advocated for white immigration into the Lowcountry, for Black emigration from the region, and for methods of environmental management through drainage that would lure white farmers to the Lowcountry. Given the recent, Red Cross–supported drainage campaign, whites' interest in drainage seemed to be a distorted echo of those community-oriented efforts: whereas Black sea islanders worked together to revitalize their fields and their livelihoods after the hurricane through digging and refreshing drainage ditches, white powerbrokers seized upon drainage and Black convict labor as ways to suffocate Black autonomy in the region. Control of the Lowcountry environment meant control of the region's African American population.

The desire to rid the Lowcountry of its Black majority took slow root in the weeks after the hurricane in the furious clashes over Black sea islanders' fitness for relief. Arguments for or against often hinged on what interlocutors saw as Black sea islanders' suitability, or lack thereof, for citizenship. Senator Hoar of Massachusetts opened his appeal on the floor of the U.S. Congress by emphasizing Black citizenry: "The people who are suffering are our fellow citizens."[43] In contrast, whites who opposed the Red Cross's relief effort cloaked *themselves* in the language of citizenship, using the term differently to bolster their accusations against Barton and the organization: one headline read, "Strong Statement about the Red Cross Woman. A Responsible Citizen Makes Charges which Show Her Up Badly."[44] Harry Hammond also disputed Black sea islanders' rights to land and liberty in his call for African

Americans—who had in his view wasted the rich Lowcountry land and, because of the high rate of deaths during the hurricane, proven themselves unable to survive in the climate—to depart the region.

Thomas R. Heyward, a white landowner in Bluffton who had angrily protested the Red Cross's intervention, echoed Hammond's sentiment, suggesting that the Red Cross ought instead to force Black sea islanders into economic subservience. "It might be well for the State Immigration Society," Heyward argued, "to get the Red Cross to fill our lands with tenants."[45] Heyward trumpeted a common attitude among white southerners: that African Americans in the agrarian South were useful as sharecroppers or wage laborers for white profit and less so as freeholders. Throughout the Red Cross's recovery effort, white landowners challenged the organization's presence and their apparent goal to fit Black sea islanders for independence. But as it became clear that the Red Cross's recovery effort could not be stopped by the slings and arrows of angry white landowners and that the Red Cross's assistance had only bolstered Black sea islanders' autonomy, many whites decided that the central flaw with the Lowcountry was that too many African Americans lived there.

This rhetoric, as it developed between the 1890s and early 1900s, was an extraordinary reversal for white South Carolinians. Black labor built the Lowcountry. In the vast rice fields of the region during the eighteenth century, wealthy whites profited from enslaved labor to build enormous fortunes that made Charleston the richest colonial city in North America for whites. Charleston was the beating heart of the Confederacy, a hallowed ground for the South's white aristocracy, who chose to tear their country apart rather than relinquish their death grip on enslaved labor. There, Confederates crafted a nation that enshrined slavery as a natural right and fired on an undersupplied federal fort in Charleston harbor in the first shots of the Civil War. Without forced Black labor, none of this history would have been the same.

The white South Carolinians who glorified the gentility of the antebellum South and sighed wistfully over the grace and beauty of Lowcountry labor camps now advocated for something shocking. The Lowcountry should be drained of its Black population. The white elite would rather pursue plans of environmental management and dispossession and thereby risk finishing off the region's struggling, hurricane-damaged economy than allow the region's Black majority to persist there as they had for centuries. Since the hurricane had not immediately destroyed Black autonomy, and if the stringencies of Jim Crow could not sufficiently strangle the Lowcountry's Black population, then it would be better for them to leave entirely. This plan even united whites once divided over the future of the Democratic Party in the state, as evidenced by Tillman and the Gonzales brothers' rare agreement on the subject, ordinarily bitter enemies.

Harry Hammond was soon not alone in his vision for emigration of Black sea islanders in the wake of the hurricane. The state's key problem was easily identifiable, as the Gonzales brothers, who were themselves of the white Lowcountry gentry, saw it: "There are too many negroes in South Carolina."[46] The brothers, two of the strongest proponents of Black emigration from the Lowcountry, wrote in their newspaper *The State*, "We would be greatly gratified to see a general emigration from South Carolina. . . . Through the white man the negro is oppressed, and through the negro the white man is made to suffer even more grievously."[47] The Gonzales brothers contended that the presence of African Americans was "an obstacle" to white South Carolina's economic, political, and moral "progress."[48] "The presence of the negroes," they wrote, "enables the meanest qualities of the meanest white men to be developed and to triumph over the better thought of the race."[49] This argument mimicked antebellum rhetoric that slavery was more harmful to enslavers' morality than it was to enslaved Africans.

The path to the African American exodus from the South, the Gonzales brothers admitted, was not without impediments: "The southern farmer wants negroes enough on his farm to do some of the plowing, hoeing, fodder-pulling and cotton-picking. He is willing, doubtless, to have the negro majority reduced, but he wants his neighbors' hands to go—not his own." They pointed to the hypocrisy of the "avowed advocate of deportation" who nonetheless "appears to except from those whom he would send out of the country enough darkeys to do little things for him."[50] The logistics of emigration posed yet more problems, for because of this stubborn attachment to Black subservience at the expense of white improvement, "wholescale deportation is out of the question. . . . The next best thing would be the better distribution of the negro population over the rest of the state. In South Carolina, for instance, negro-ridden counties like Beaufort, Berkeley, Georgetown and Fairfield might be relieved of a large portion of their black people, with great benefit to all."[51]

The Gonzales brothers found the African American population of those counties particularly objectionable because of the predominance of "idleness . . . in the negro-ridden sections"—language that called to mind the objections of white supremacists who claimed that Red Cross aid had encouraged indolence among Black sea islanders. In contrast, the Gonzales brothers argued, counties where the proportion of African Americans to whites was more balanced and where they were "a laborer or the tenant of the white man" cultivated a diligence in African Americans. Within the mind of white supremacists, labor for white wealth, not labor for Black autonomy, was the only real measure of hard work. If African Americans could not be deported entirely, then the Gonzales brothers felt that transforming the Lowcountry's autonomous African American population into laborers for whites in the upstate was an acceptable compromise.

As the Gonzales brothers saw it, the Lowcountry was wasted on its African American residents because they no longer produced wealth for white South Carolinians at the same level as they had under slavery. "We believe," the editors elaborated a couple years later in 1898, "that one of the greatest blessings South Carolina could attain would be the transfer of these rich farming belts [in the Lowcountry] from the occupancy of negroes to the occupancy of whites. . . . The salvation of the low country must come through the overflow of the white population from the Piedmont region to the rich lands south of them, to be bought at low prices."[52] The Lowcountry, *The State* argued, must be cleared of its African American yeoman farmers to allow the immigration of white farmers to the region and the flowering of progressive modes of agriculture that would weaken the state's all-consuming reliance upon cotton.[53]

The emptying of the state's Black population would not only provide a solution to the "race problem," but it would also pave the way for modernization of the state's agricultural practices. No longer would monoculture deplete soils and render farmers vulnerable to the vagaries of the market: instead, white farmers could move toward diversified agriculture and food production. In the Lowcountry since the Civil War, "big plantations have been rented to negroes, with the result that white labor has been kept out and a slovenly all-cotton policy of farming has been continued."[54] The departure of African Americans would not ruin the state's economy but would in fact improve it, for "every negro who goes lifts the burden of a bale or more of unprofitable cotton from our agriculture and makes us turn the more to crops that a white man can make."[55] Given South Carolina's history of reliance on Black labor for profit, the Gonzales brothers' boosterism seemed wildly optimistic, a profound misunderstanding of what motivated white landowners in the state: inertia rather than innovation was the dominant force.

The brothers' reductionist perspective pinned the blame on African Americans for the Lowcountry's economic troubles rather than considering the matrix of factors that had contributed to the region's decline over the past few years. The Gonzales brothers willfully mischaracterized African Americans as the key to the Lowcountry's economic downturn, brushing away the harms of the hurricane, external competition, and political oppression. It was easier to point the finger at African Americans than it was to admit that a white supremacist system that fundamentally devalued Black freeholders might not be an adequate economic model to bring prosperity to the region—especially since valuing free Black labor might mean ceding profits to African Americans rather than consolidating them for whites. And, finally, it was more convenient to ignore the long-term impact of the hurricane on the Lowcountry economy than it was to place responsibility for the region's economic diminishment on the shoulders of the very population whom the hurricane had hurt the worst.

Tillman was also an avid proponent of white immigration to the Lowcountry, and he seized on this tactic with pugnacity.[56] If the constitution was sufficient to guarantee white supremacist rule in South Carolina, then he would have precious little left to do in his political career. Thus, the Black majority in the Lowcountry became a new target for his ire in the years after the constitutional convention. He monitored population growth like a hawk as he fomented strategies to transform South Carolina into not only a Jim Crow state but also a *white* state.[57] Tillman was not a segregationist, though that was the state of affairs for which he could grudgingly settle. No, he yearned for not only all-white public spaces but an all-white South Carolina. He grumbled that African Americans "covet segregation and the condition of independence from the white man" and that segregation made "the negro labor in this State . . . more and more worthless, more and more uncontrollable, more and more resentful of white supervision."[58]

Tillman did not think that segregation inculcated obedience or even guaranteed white supremacy: the mere existence of African Americans alongside whites prompted the death of the white race. For years, he tracked the demography of the state obsessively to determine the exact numerical disadvantage of whites to African Americans in the state. Throughout the 1890s, he carried with him small leather notebooks, covers worn flexible and thin with use, filled with newspaper clippings of census reports on the Black and white populations of each county in the state—penciling check marks next to each county that had a white majority.[59] He must have glared with his habitual choler at the Beaufort County entries in his notebook, which, in 1890, listed just 2,563 whites to 31,553 African Americans. Given his frequent tabulation of population figures and the racism that was as vital to him as breathing, it would not be surprising if he privately welcomed the hurricane, which caused roughly the same number of deaths of African Americans as there were whites in all of Beaufort County.[60]

Paranoia that African Americans would overcome the restrictions of the constitution drove Tillman's action in the years after the constitutional convention. The victories of the constitution, to Tillman's mind, were fleeting.[61] "The relief," Tillman warned, "is only temporary." To the South Carolinians who said "that there is no race problem; that we have solved that by the Constitutional Convention of 1895," Tillman had a harsh rebuke: by the early 1900s, more African American children were enrolled in school than white, so "are we so blind as not to see . . . that the number who can become voters by complying with the registration laws is increasing day by day?"[62]

Tillman saw a sharp, irreconcilable divide between the races, with the "Caucasian race" as "the superior race on the globe; the flower of humanity; the race responsible for the history of the world. . . . There can be no doubt about the attitude of every South Carolinian on that point. I mean every South Carolinian worthy to be called a South Carolinian."[63] With the gains that Black South Carolinians made daily in population and education, Tillman asked, "Is it not apparent to any thoughtful man that the first thing we ought to set about to accomplish is to get rid of the negro majority by reinforcing the white race?"[64] "The struggle for mastery as between a majority of negroes and a minority of whites," Tillman warned, "is bound to come."[65]

As a result, Tillman opined that immigration from western and northern Europe ought to supplement the state's white minority, and the white "people of South Carolina should use every legitimate and proper means in their power to encourage white immigration—to obtain men of our own race here."[66] White immigration of Europeans from England, Ireland, Scotland, and Germany would be the solution to what Tillman thought of as the Lowcountry's insidious Black majority.[67] Europeans from the continent's west and north, Tillman asserted, were preferable because they had much in common with the whites who had colonized South Carolina in the eighteenth century: "The foundation stock of South Carolina were first the English on the coast, then the Huguenots, later the Scotch-Irish . . . later a large influx of pure-blooded Scotch; pure-blooded Irish and Germans here and there; pure-blooded Dutch."[68] A new infusion of this "white blood" into the state would hold strong white supremacy.[69]

In contrast to the supposed purity of these potential immigrants, Tillman pathologized the coast's Black majority as a "cancer" that threatened the longevity of South Carolina's "Anglo-Saxon" race.[70] Blood-letting might be a cure that the state's whites ought not falter from—"we will make this State red before it ever becomes black," Tillman swore in 1908—but until then, he sought other cures.[71] White immigration was one dose. Environmental management of the Lowcountry was another, and Tillman paired the two. In the years after the hurricane, white South Carolinians other than Tillman had grown interested in the kinds of large-scale drainage projects that the federal government undertook in the American Midwest to dry out the land for the commercial cultivation of wheat and corn.[72]

A major contributing factor to this interest was because of the news from scientists and physicians that a parasite *Plasmodium* that thrived in female *Anopheles* mosquitos caused malaria, which had long plagued the region and which had wracked thousands of Black sea islanders after the hurricane. Once experts had confirmed mosquitos as the agents of the disease, white South Carolinians heralded drainage, which would empty the Lowcountry of mosquito habitat, as a preventative. James Cosgrove, a white Charlestonian and an advocate for drainage in the Lowcountry, wrote in 1907 that, "aside from the reclaiming of the land," drainage "would rid that section of the dreaded mosquito and the bite that carries with it the dread malaria."[73] But white South Carolinians, including Tillman, imbued this project with racist rhetoric that linked the threat of malaria with the specter of Black control of the Lowcountry. Ridding the Lowcountry of the mosquito would also dispossess Black residents of the coast—both desirable goals for white supremacists.

Wealthy white southerners had long feared and abhorred the region's swamplands.[74] The marshy land north of Charleston had been a barrier to agriculture, and for generations white Charlestonians associated the tangle of wetlands with a dangerous stronghold that harbored Black fugitives. Whites characterized wetlands across the South in the same way. Not only did white southerners consider swamps cesspools of ill health because of the "miasma" that they assumed caused malaria, but swamps were also impenetrable tangles of untamed nature, the antithesis of the gentled, beautiful, and organized plantation that white southerners imagined defined their society. Enslaved African Americans often made swamps their refuge, forming maroon communities across the South, particularly in the vast wilds of wetlands like the Great Dismal Swamp in eastern Virginia and North Carolina.[75]

Even in the post–Civil War era, when quaint folktales of the South grew in popularity and softened the swamp's nightmarish reputation into a setting for picaresque adventures, the swamps across the South remained best avoided by respectable whites.[76] In the postwar Lowcountry, whites continued to closely associate swamps with rice, whose fields were submerged half the summer in fetid water in which mosquitos thrived. Malaria had kept whites away from Lowcountry labor camps much of the year, which kept the white population in the region low and provided a degree of insularity for enslaved workers to nurture a unique set of cultural and social traditions.[77] But with the postwar decline of rice and rise of white supremacist regimes, and with the discovery that mosquitoes bore malaria, new possibilities opened to white supremacists who resented the region's Black majority and, due to the changing economy, no longer had as much to gain by keeping so many African Americans there.

Plans for drainage to accomplish this end usually overshadowed the outcome.[78] Many counties around the state worked with the federal and state governments to form drainage districts, with results that were often limited. The Sanitary and Drainage Commission of Charleston County was the only body to implement a large-scale, government-driven drainage project. This may have been in part because Charleston officials were old hand at drainage:

it was necessary to prevent tidal overflow into streets, and for centuries city planners had directed laborers, often enslaved, to infill marshy areas and creeks and to improve the sewage and drainage systems within the city grid.[79] White planners in Charleston County did point to citizens who were delinquent in their taxes as a way to unroot Black landowners, arguing that "idle territory" forfeited to the state could become "prosperous regions" tended by "industrious and excellent young white settlers."[80] Tax schemes could supplement drainage projects to dislodge Black landowners.

Perhaps unsurprisingly, while white South Carolinians boosted drainage as a key step to creating a white-dominated agrarian landscape, they used Black convict labor to experiment with and carry out these drainage projects.[81] Though the exploitation of Black labor for white wealth was nothing new, the desire to deprive the Lowcountry of its Black population was a change, and its boosters hoped that drainage schemes would supplement the gains of the 1895 constitution to establish Jim Crow in the region. Draining the land and promoting white emigration to the region, which would eventually squeeze out Black sea islanders or outweigh their population gains, might ensure a white supremacist regime for the long term.

The project of draining the Lowcountry emerged first in Charleston. In an initiative spearheaded by James Cosgrove, a member of the South Carolina General Assembly from Charleston, the Sanitary and Drainage Commission (SDC) of Charleston County was established in 1899.[82] Cosgrove promoted drainage because "the agricultural resources of Charleston County have not been developed by the white man" due to "the fear of malaria," which, "owing to a lack of any systematic drainage," chased them off the land.[83] Cosgrove opined that immigration was "a subject closely allied with drainage, for our immigration movement can not assume the proportions it should until our waste lands are in a condition to receive immigrants in large numbers."[84] *The State* agreed that with the careful application of drainage, "disastrous overflows in times of freshet" might diminish in their damage to coastal crops and would "be the means of reclaiming valuable lands in the State."[85] Through a proper system of drainage, the "hinterland" around Charleston, SDC officials claimed, could "support a population of 100,000 frugal, industrious white farmers."[86]

Unspoken in this drainage boosterism was that African Americans had long lived on and owned Lowcountry land that, as Black sea islanders during the Red Cross recovery had shown, also benefitted from drainage, even as the constraints of the economy prevented African Americans from conducting regular maintenance of drainage ditches. But in the minds of the drainage boosters, African American landownership was a waste of the region's perceived natural fecundity and accessible location, which had "lands as fertile as the Delta of the Nile, with a climate unequalled in salubrity, with the largest markets of the country only a few hours distant."[87] In a distant echo of settler rhetoric that derided Indigenous land use practices, whites in the Lowcountry demeaned Black yeoman farmers as undeserving of their property because Black landowners did not pad whites' wallets as much as sharecroppers or wage laborers. Hammond himself had explicitly made this argument in the month after the hurricane, asserting that Black sea islanders had abused the abundant gifts of the Lowcountry's "genial climate" and "fertile soil" and were thus no longer fit inhabitants of the region.[88] Once again, white supremacists chose not to acknowledge the array of factors diminishing the livelihoods of Black sea islanders and seized on a more facile explanation that found inherent fault in Black sea islanders themselves.

Tillman pounced on Cosgrove's desire to attract white settlers to this swampy region that eluded white control and remained the bulwark of Black landownership in the state.[89] He declared that politicians should "start in every way practicable about getting more white men and white women to come to South Carolina."[90] Tillman looked at Cosgrove's drainage project in the Lowcountry and saw an opportunity for white supremacy. Tillman agreed that within the Lowcountry lay "the richest in its possibilities of all the land we have, but even the frogs have chills and fever down there and malaria is rampant."[91] He hubristically asserted that "all it needs is a little canaling" to tame the land—a statement of such ignorance that it would surely have made the ghosts of the first enslaved Africans who cleared Lowcountry cypress swamps for rice fields laugh bitterly. According to Tillman, the Black majority posed the larger problem, for there "the sons of Africa" lived "in swarms," pests like the mosquitoes themselves, "ready to compete and act as a deterrent" to white immigrants.[92]

To achieve true demographic change, then, "the negroes could be rooted out" to transform the environment from a "wilderness of bog and swamp into a garden."[93] That African Americans were the barriers to the Lowcountry's productivity, despite the not-so-distant history in which Black labor had built the fantastic colonial and antebellum wealth of Lowcountry whites, was incredibly perverse logic. But within the mindset of white supremacists who were determined to redeem the Lowcountry from Black landowners, the drainage of Lowcountry lands paired harmoniously with the dispossession of African Americans and the "immigration movement" to improve the region's economic future for whites.[94]

Before this white utopia could come to fruition, though, the ditches had to be dug, and drainage boosters looked to the state penitentiary and county jails. The idea of extracting forced labor from people who were imprisoned also had traction among a network of professionals, politicians, and engineers. The

Yale professor S. E. Baldwin, originally from South Carolina, presented a paper at the International Prison Conference in Budapest, Hungary, "in which he advocated the employment of convicts on great works of public improvement, such as irrigation and drainage and the reclamation of waste lands." *The State* gave thorough coverage to his speech, in which he urged policy makers to remove forced convict laborers from road projects where "the convict is constantly in the public eye and feels himself degraded and scorned" and to "pioneer work" like drainage, which "would be more uplifting."[95] Theodore Jervey, a Charleston judge responsible for locking up many thousands of African Americans in the 1890s, argued that this forced labor was a solution to the "Race problem": "As a portion of the race is lifting itself, however, another portion is sinking, and for this portion here, where they are in such numbers, compulsory work is the only salvation, and of its reformatory effect I can personally testify."[96] Slavery, in its essence, was the solution.

Cosgrove thus found his labor force in the Charleston County Jail. The Court of General Sessions for Charleston County in Charleston delegated Black men who had been convicted of misdemeanors to Cosgrove's Sanitary and Drainage Commission to serve as forced labor.[97] While the records for the Charleston County jail are spotty and incomplete, extant records provide some insight into the sort of crimes for which people living in the area were arrested. In the late nineteenth century, Charleston's police department regularly recorded over 3,000 arrests each year, and in the year that Cosgrove requested forced labor for his project, the police chief W. A. Boyle recorded 3,455 arrests.[98] The vast majority of people convicted were found guilty of misdemeanors related to regulating behavior, often with an eye to policing African Americans in particular: as a case in point, 2,079 Black men were arrested in 1900 out of 3,455 total arrests in Charleston.[99]

The Charleston police arrested African American men for these misdemeanors at wildly disproportionate rates. For example, of 574 people arrested for "disorderly conduct," 312 were Black men and 156 were Black women, compared to twenty-two white women and eighty-four white men.[100] The court sentenced 943 people total to the city chain gang: white men made up 137 of that number, and Black men the other 806.[101] Within the Charleston chain gang system, Black men bore the brunt of forced labor at highly disproportionate rates to white men. This was intentional on the part of the city's leadership. Police Chief Boyle mentioned, for example, the success in implementing a recent vagrancy law, another tactic used against African Americans as a form of labor discipline to push them into unfavorable labor agreements with white bosses. White Charlestonians rarely hesitated to admit it: in 1898, *The State* commented that in Charleston, "the negroes show signs of chafing under the enforcement of the [vagrancy] law."[102]

While this data is from the city of Charleston rather than the county, it nonetheless shows how the use of forced Black laborers derived directly from Jim Crow–era policing practices that criminalized African Americans as a means of social control, in full force in Charleston by 1900.

Cosgrove's first project was a 15,000-acre plot of land north of the city of Charleston, and his hope was that the land could "be thrown open to occupancy by white men."[103] According to Cosgrove, the land suffered from a dual infection. He wrote that the land "has been called 'Hell's Half Acre,'" due not only to malaria but also because "the section has been inhabited almost altogether by negroes." Historically, Cosgrove argued, malaria had profoundly damaged the future of the white race in the Lowcountry. "For centuries it has been believed that no white man could live in the summer months in the swamp lands of Carolina, without contracting malaria," he said in a speech at the 1908 National Drainage Congress.[104] Those white families who attempted the stay "became invalids for a great part of the year, and their offspring grew to manhood and womanhood handicapped with disease that unfitted them to become industrious and useful citizens."[105] In a display of disregard for Black landowners, he hoped that "when we have drained these lands, this fear [of malaria] may be banished" so that "white settlers may come into the territory with the assurance that health may be had there, and the lawless element will be either driven out or made to pay its quota in support of the general good."[106]

The work began on April 1, 1902, on a 3,000-acre section, with eighteen Black forced laborers digging canals and ditches delineated by an extensive topographical survey.[107] Over the course of nine months, fifty prisoners, all African American, rotated onto the SDC's chain gang. Wading through "waist deep" water, with spades, shovels, and wheelbarrows as their tools, the prisoners did the dirty, difficult work of ditching.[108] The civil engineer in charge of the project complained that "some time was required for proper organization" of the laborers, before the supervisor and guards were able to "get . . . the men properly broken in."[109] The forced laborers, with one superintendent and three guards during the day and one at night, were "quartered in the stockade at the old Edisto Phosphate Mill," a site that may have been truly unhealthy given the toxic legacies of phosphate manufacturing, which contaminated the soil with arsenic and lead for generations.[110]

Within the mill, the forced laborers may have been kept in the iron-slatted, wheeled cages that typically housed chain gangs in South Carolina in the early twentieth century and that the SDC used throughout the 1910s.[111] Whether the forced laborers were restricted to cages at night, the supply list that detailed every item that the SDC purchased and used during the project did reveal that these men must have been confined physically: six padlocks, two fifty-foot chains, two ball and chains, and thirty-four pairs of shackles and chains, as well as five guns, one rifle, and one pistol, ensured the captivity of the laborers. Despite these shackles, over the duration of the SDC's first experiment in drainage, one man escaped. Another was murdered by guards for an unstated reason.[112]

When the forced laborers were done nine months later, they had dug a matrix of 30,000 feet of open ditches and canals and installed underground terracotta tile drains, 8,000 feet of which were furnished by local landowners.[113] They also "cleaned out, reshaped, and releveled 20,000 feet of County ditches," improving drainage outside of the 3,000-acre section.[114] Cosgrove and city officials declared the project a great success. They reported no malaria among the prisoners or white guards. And once the A. C. Truxbury Lumber Company moved onto a lot on the site soon after the drainage was completed, the SDC publicized that the company likewise had no incidents of malaria among its employees.[115] The land, the "breeding place for thousands of mosquitoes of the malarial type, and where but a short time since naught was heard but the hoot of the owl and the buzzing of the mosquito," Cosgrove crowed, was now filled with "the busy hum of machinery."[116]

The use of a large piece of the land by a corporation suggests that the white South Carolinians who promoted drainage as a route to landownership for working-class whites was perhaps an empty sentiment, or an implicit admission that attracting white farmers to the Lowcountry was not as easy as they had hoped. Grateful white residents who did live nearby wrote to Cosgrove that the work of the forced laborers "is truly making this country what it has never been before, a fit country for a white man to live in."[117] The SDC subsequently purchased at $1.50 an acre some 75,000 acres of wetlands from the county and estimated the increase in value per acre from two dollars to, in some cases, $250.[118] The county was so impressed with the commission, and the use to which it put Black convict laborers, that they granted the SDC permanent oversight of the county chain gang and of the construction of roads and bridges in Charleston County.[119] The SDC continued to operate the county chain gang and oversee the construction of roads and bridges in the county until at least 1927. In that year, imprisoned men, thirty-five per day on average, constructed the River Road on Johns Island, topped a road from Adams Run to Dawho with gravel, and widened and raised the Dawho Causeway, which ran through a broad swath of salt marsh.[120]

The zeal for drainage spread beyond the swamps north of Charleston, with city and state governments taking on drainage projects for the sake of the public good and the improvement of private property—as long as they benefitted whites. The city of Charleston, which had managed a chain gang since 1892 for city maintenance, put convict laborers to work on drainage projects at the same time as Cosgrove's experiment.[121] Forced laborers from the city jail, serving sentences between five and thirty days, dug drainage ditches along roads and through public spaces at least until 1922.[122] Two of the most notable projects completed with the use of convict laborers included Chicora Park and the grounds of the South Carolina Interstate and West Indian Exposition, which ran from December 1901 to May 1902 and was intended to showcase South Carolina's resources and progress.[123] White leisure and progress ran on forced labor extracted from African Americans.

The state government took note of Cosgrove's drainage project and his exploitation of Black convict laborers. *The State* newspaper commented approvingly that "experiments with drainage have been made in Charleston county, convict labor being used, and the results have been so gratifying that the matter is worthy of the consideration of the entire State."[124] In 1901, the South Carolina General Assembly passed an amendment to the state's constitution to permit state-run drainage in the state and passed a bill that required that the state hire convict laborers from the state penitentiary to counties for road work and drainage.[125] The amendment allowed for "the condemnation . . . of all lands necessary for the proper drainage of the swamp and low lands of the State."[126] Furthermore, the general assembly ruled, "all able bodied male convicts to hard labor . . . may be required by the counties to work on the public highways or the sanitary drainage in said counties."[127] But the new amendment did not, contrary to politicians' expectations, spur a flurry of drainage projects in the state. Furthermore, some whites found drainage laws controversial and disapproved of their reach into the uses of private property. One man wrote to *The State* in 1908 declaring it "unconstitutional" and complaining that it "smack[ed] of federalism" because if the government "has the power to drain the low lands in the low part of the State, it has the same power to level hilly lands in the upper part of the State, or to do any other paternalistic thing it sees fit to do."[128]

Nonetheless, the general assembly enacted the Drainage Act in 1911, and the law set more drainage projects in motion across the state.[129] The state government then allowed the U.S. Office of Farm Management under A. G. Smith, the "father of drainage," to establish a "Coast Station," unimaginatively dubbed "Drainland," on 300 acres near Summerville donated by the Southern Railroad.[130] Convict laborers drained the acreage and produced "as fine crops as were to be found in South Carolina" on lands "previously considered unfit for agricultural purposes."[131] State officials embraced those results as proof that "drainage and proper fertilization will result in splendid crops."[132] The experiment at Drainland had demonstrated that drainage "will eventually add millions to the wealth of South Carolina by increasing immensely its productive area."[133] A second experimental farm, in Marion County on land belonging to Judge C. A. Woods, used "a gang of fifteen negroes" to lay tile drainage, in which clay tiles buried in the soil would run along a slight downward angle to an open drainage ditch.[134] There, too, the drainage trial was successful. The continued promotion of drainage projects as a benefit to whites when carried out by forced, unpaid Black laborers revealed the desire of white South Carolinians to reconcile progressive farming practices with a new version of enslaved labor.

The Drainage Act reached beyond the scope of the Drainland and Marion County farms, though. The general assembly instituted a Board of Drainage Commissions, which oversaw local drainage districts "to promote the public health and welfare by ditching and draining" the two million acres of wetlands in South Carolina.[135] With the reclamation of this land, lawmakers wrote, massive acreage could be "brought into agriculture, rendered healthy, and made to add to the annual wealth-producing agencies of the State," and potentially could contribute "several million dollars" to the taxable values of South Carolina.[136] To promote the formation of drainage districts, the U.S. Office of Farm Management organized informational meetings for white farmers throughout the state and reported that 5,000 attended in total.[137] The benefits of drainage should be confined, state official suggested, to white landowners; Black landowners in the Lowcountry were not deserving of this access; nor did white politicians wish to give them any opportunity to improve their economic prospects.

Tillman, Cosgrove, and the federal and state governments envisioned a state drained of its wetlands through a complex of ditches that would have rivaled even the antebellum system of rice cultivation, the newly fruitful farmland attracting hundreds of thousands of white settlers. Realities, however, fell short of expectations. Ten years later, less than 1 percent of South Carolina's land area, or 154,687 acres, had been organized into drainage districts, at a cost of $618,083.[138] Despite the boosterism of the state government, the state's "impoverished" finances and the convoluted procedures stymied drainage from flourishing.[139] The five steps necessary to simply establish a drainage district—not even to implement the drainage itself—were involved, and it is easy to imagine that a community would have difficulty coming to a consensus.[140] Even the most ardent white supremacist might find such an initiative trying, especially since the state provided little assistance in the organization and funding of the drainage district. The element of the state's erstwhile drainage plan that did thrive was the state's continued use of chain gangs to build roads and dig drainage ditches.[141]

White supremacists intended that the drainage of the Lowcountry would rid the Lowcountry of excess water and an excess population of Black coastal residents. Having witnessed the cultivated resiliency of African American landownership and political operation on the Lowcountry even after the ravages of the hurricane and the 1895 constitution, white South Carolinians pivoted in their tactics. Rather than maintaining the antebellum landscape that extracted profit from a Black majority for white enslavers, whites maneuvered to forge a new environment that would supplant the Black majority to benefit white health and wealth. In a cruel twist, they levied convict labor toward this project of elimination. Its success was limited, for widespread drainage never came to fruition. And while African Americans did migrate in increasing numbers away from the Lowcountry, the immediate cause was obviously not because of drainage projects. However, it contributed to the array of political, social, economic, and environmental factors that made life for African Americans in the Lowcountry more difficult.

The Lowcountry drainage projects were never extraordinarily successful in the sense that they neither drained the entirety of the Lowcountry nor immediately chased away African American landowners. Regardless the measure of success, though, the obsession with demographics and drainage revealed the virulence of white supremacist attitudes toward African American residents of the Lowcountry, the need for white supremacy to constantly evolve and modernize even as the regional economy stagnated, and the environmental valences of white supremacy under Jim Crow.[142] And through the SDC, the Charleston City government, and county governments, chain gangs around the Lowcountry continued the hard labor of digging and draining in the context of road construction and smaller drainage projects for generations. That system *was* long-lived and crushing for thousands of African Americans in the Lowcountry.

White supremacists saw the hurricane as lending fuel to their Jim Crow project, beginning a process of dispossession and demographic diminishment that they wanted to further. Jim Crow in the Lowcountry possessed traditional political, economic, and social dimensions—but whites also leveraged the region's unique demography and geography in forming their grievances and goals. They conceived a new scheme that underlaid not only the new constitution but the drainage and emigration initiatives: that African Americans in the Lowcountry were of less use to them, politically or economically, and that the region would be better off without the populous Black labor base that had driven the economy since the late seventeenth century. So fixated on establishing white supremacy that they were willing to rid the state of what Robert Smalls described on the convention floor as "the bone and sinew of your country," they might indeed, as Smalls then prophesied, wake up one day to find the African Americans of the Lowcountry gone.

Chapter 8

Unmooring the Regional Economy

On Sunday, August 27, a hurricane that had ravaged ships in the Atlantic and avoided detection in the Caribbean struck Tybee Island with little warning.[1] The storm brought high winds and a surging ocean. From Savannah to Georgetown, it devastated Black homesteads, the dwindling rice and sea island cotton crops, and commercial waterfronts.[2]

The year was 1911, not 1893. But the powerful 1911 hurricane, falling on the same day of the year and week as the Great Sea Island Storm of 1893, invited comparison. Residents of the Lowcountry had developed a particular rhetoric of response to hurricanes, and they cycled through them in a predictable cadence. "The terrible storm," Principal Rossa Cooley of the Penn School wrote, "swept over the Sea islands" and "resulted in great suffering," which required communal cooperation to feed and clothe Black sea islanders.[3] The *News and Courier* trumpeted of the resilience of Charleston's commerce, headlining "Hum of Business Goes Merrily On—Charleston Suffers No Loss from Storm—Optimism Keynote of Comment."[4] And white landowners beat their favorite drum: those near Charleston asserted, perhaps too firmly, that "there is absolutely no need for relief measures of any kind."[5]

But Black sea islanders remembered past hurricanes too. While white landowners, politicians, and businessmen hurried to leverage the hurricane to downplay Black suffering and to promote their economic interests, Black sea islanders understood that hurricanes could present them with an opportunity to advocate for themselves. On Wadmalaw Island, bound by creeks and rivers, hidden from the sea by Johns Island and Kiawah Island, Black farmworkers took a stand. The hurricane had not destroyed the cotton crop there, and the cotton bolls erupted the week after. White landowners offered Black farmworkers the usual sixty cents for every hundred pounds harvested. Instead, to white landowners' frustration, "no hands are coming forward. They are striking for a dollar the hundred."[6] As white landowners reported, "they are taking this attitude largely because they have an idea that the Red Cross or some other organization intends to send them supplies and provisions, wherefore they can afford to be lazy." Black sea islanders asserted their value as workers and insisted on fair treatment for hard work immediately after a powerful hurricane. White landowners twisted this labor action and worker solidarity to justify their own victim complex, complaining that "the people who have suffered most from the storm are the white people, not the negroes." The *News and Courier* did not follow up. Whether the Black farmworkers or the white landowners prevailed is unknown.

From the vantage of 1911, what had changed since 1893? Looking out at flooded fields, damaged cottages, and wrecked docks, taking their same stances, residents of the Lowcountry may have wondered what had improved and what had remained relentlessly the same. From the late 1890s to the 1910s, the Lowcountry was a region with an unclear future. The Lowcountry was increasingly the insignificant southeastern fringe of a nation that had rapidly expanded its capitalist and imperialist ambitions over the last few decades. The system of racial apartheid that white supremacists had gradually extended across the region was their primary accomplishment, but it did little to enrich them as much as the exploitation of wage laborers in northern factories had Gilded Age industrialists. Through the 1890s, Charleston and the region were struck by the double punch of the 1893 hurricane and a national depression, which "marked the lowest point in Charleston's postwar doldrums since Reconstruction."[7] Though Jim Crow had reinscribed white supremacy in the state government and cast its shadow on land management and dispossession schemes in the Lowcountry, it had done nothing to revive rice, sea island cotton, or phosphate. Many white South Carolinians resented the economic growth of much of the rest of the country. To compete, innovation-minded white landowners and politicians formulated new solutions of drainage and immigration to the old problem of the Black majority, which had been of their ancestors' making.

Yet while many of the white elite were attracted to those solutions, they also clung to the tatters of family property, feared large-scale economic changes for their potential to destabilize the racial hierarchy, and sunk into a white supremacist nostalgia for the lives their grandparents had had. This was as true in the hinterlands as in Charleston, where "chronic economic stagnation" greeted the few "failed efforts" to pull the region out of its decline.[8] Rice and cotton—or the regime that they represented—still called the names of the white gentry. Not the resistance of Black farmworkers nor the roar of the ten hurricanes that struck the Lowcountry from 1893 to 1911 could quite drown out the allure of what white landowners already knew. Stagnation held racial hierarchies and old economic practices suspended in place—or so whites hoped.

Stagnation in a region of flowing creeks, ocean tides, and shifting shorelines required targeted intervention. White landowners in the Lowcountry had to work to forestall or ignore, as it benefitted them, the environmental forces of the region. They also saw as their enemy the ongoing, laborious efforts of Black sea islanders to maintain what communities they could against their white bosses, a series of devastating hurricanes, and a grim economic landscape. It was 1911, not 1893. White South Carolina had passed their Jim Crow constitution and invested in the wages of whiteness.[9] White politicians and landowners looked for new battles to win against the Lowcountry's Black residents. An unprecedented wave of hurricanes had battered the coast for eighteen years. Rice, sea island cotton, and phosphate were slipping away for whites grasping at power and for Black workers searching for a wage. Black sea islanders were forced to develop new strategies to either adapt or migrate. And yet so much remained the same. The recurrent hurricanes in the first two decades of Jim Crow mirrored the cyclical, spiraling nature of change in the Lowcountry.

Though none approached the Great Sea Island Storm of 1893 either in ferocity or death count, nine more hurricanes between 1893 and 1911 served as potent reminders that hurricanes would inevitably come to the coast. The Great Sea Island Storm's younger sibling in October 1893 damaged the Georgetown waterfront and the shore north. A September 1896 storm hit northern Georgia and the lower South Carolina coast, killing at least thirty-six and ruining the year's crops. In late August 1898, a hurricane made landfall on Cumberland Island in southern Georgia and trailed destruction up the coast, flooding the sea islands and leaving Black sea islanders on the lower South Carolina coast hungry.[10]

Another hurricane a month later with a substantial storm surge that *The State* reported as comparable to that of the "great tidal storm of 1893" ruined the rice still grown on the Combahee, Pon Pon, and Edisto Rivers. The storm killed at least 139, despite initial fears that there would be a "heavy loss of life" much greater than that—perhaps because the hurricane's strike occurred during the day rather than during the middle of the night, and revenue steamers patrolled the sea islands, rescuing dozens if not hundreds of stranded Black sea islanders.[11] A hurricane in 1904, two in 1906, and one in 1910 damaged harvests but left no dead behind, at least that were widely broadcast. And in late August 1911, the storm described above slammed the coast near Savannah, Georgia, with winds just over 100 miles per hour, killing 17 and doing serious damage to the crops.[12] After this hurricane, an onlooker noted, a "migrant" from the Lowcountry "stated that four left from her home and that hers was not an exceptional case."[13]

Nearly two decades of regular and severe storms—an unusually intense cluster on the South Carolina coast, where two decades can pass with barely a brush from a hurricane—wore out Black sea islanders, especially since the cycle began with the horrific devastation of the 1893 hurricane. Not only did hurricanes increase the labor of the agricultural season, with fields requiring new drainage and clearing, but they also stripped Black sea islanders of their subsistence and saleable crops and emptied the wallets of land-rich whites who otherwise would have hired Black laborers at crucial times in the planting season.

The Great Sea Island Storm of 1893 helped set into motion an era of decline that the subsequent hurricanes compounded. Phosphate had been the great hope for white South Carolinians itching for economic modernization: it combined Black labor and the Lowcountry's natural resources with industrial production and capital investment, and it symbolized to boosters what the Lowcountry could become.[14] The hurricane wrecked phosphate infrastructure, and political maneuverings ensured the industry was not revived. The storm surge tore apart drying sheds, warehouses, and processing plants up and down the coast.[15] Ten out of eleven dredges had sunk beneath turbid waters, the eleventh, the *Columbia*, only weathering the storm because it was being repaired in a shipyard.[16] Coosaw lost nearly all the 30,000 tons of rock workers were processing when the storm hit. The storm surge and winds had scattered its five dredges, sixty-eight lighters, and forty floats along the creeks and rivers around Beaufort.[17] Phosphate workers who lived near the riverside factories were especially vulnerable to the storm surge, with the settlements near Charleston wiped out and a high death count at phosphate works near Beaufort.[18] River-mining companies' loss in property hit dismal lows. Coosaw and Carolina lost $150,000 each; Beaufort, $25,000; and Farmers, $20,000.[19] Only Carolina had any insurance, and that was on the dredge *John Kennedy* for $100,000. These numbers were significant to the region's 5,000 phosphate workers in that it portended either a great deal of repair work or the failure of some phosphate companies to recover to their former levels of production.

Devastating as the toll from the hurricane was, it took human actions to prevent phosphate from rising from the wreckage. Then-governor Tillman and the South Carolina state government's Phosphate Commission refused to spare the phosphate companies. Tillman resented the phosphate industry because it represented a resurgence of Lowcountry economic dominion and it held so much sway in the region, a corporate defiance of state (and thus Tillman's) control.[20] The phosphate companies banded together to request a reduction of 50 percent of the royalties that they owed to the state for the next year. Tillman and the commission, almost certainly with the awareness that the phosphate companies had provided 25 percent of the state's tax revenue the prior fiscal year, instead ordered the companies to fulfill a bare minimum of $75,000 in royalties for 1894, after which point the companies could mine rock royalty-free for the rest of the year.[21] After 1894, the royalty would be reduced to fifty cents per ton rather than the usual dollar until the companies recovered to their prehurricane production levels.[22] These terms would be distributed equitably among the mining companies, requiring each company under bond to contribute, whether they resumed mining in 1894.

The companies found the state's offer stingy, knowing that they would be both hard-pressed and unwilling to match the royalty while workers rebuilt the phosphate industry's infrastructure. Jacob Paulsen, Moses Lopez, Paul Felder, and F. Brotherhood, heads of the Beaufort Phosphate Company, Coosaw, the Farmers Mining Company, and the Carolina Mining Company respectively, wrote a letter to Tillman in October 1893, urging him to show greater leniency. The companies had 31,164 tons ready to ship when the storm hit, with no expectation of mining further rock in 1893, which would only make a small dent in the required $75,000 royalty. Carolina and Coosaw, the presidents wrote, would not be able to resume operations until June 1894 at the earliest. They pointed to the other dire pressures on South Carolina phosphate and noted that Florida's fifty-cent royalty on their burgeoning phosphate industry already put them at a disadvantage. They protested the mere existence of the royalty at all, informing Tillman that "had this cyclone not occurred it was only a question of time when the Companies would have had to stop, unless the State so reduced the royalty as to enable them to keep on."[23] The companies, they assured Tillman, wanted to cooperate with the state regulatory board but found the terms unacceptable. The phosphate executives were troubled by the hurricane and Tillman but still assured of their own influence and in the wealth they brought to the relatively impoverished state coffers. They thought they could negotiate with Tillman.

Tillman, however, saw the chance to crush an enemy. He shot back a response four days later, arguing, "The river mining companies . . . seem disposed to drive a hard bargain with the State, and demand concessions which we cannot give." With poor foresight, Tillman dismissed Florida as a future competitor, saying, "The royalty in Florida has nothing to do with the royalty here, and the river rock there cuts a very small figure in the market. . . . We had just as well leave the rock in the river as to give it away."[24] Tillman also miscalculated by bringing the matter before the general assembly with the assumption that they would force the phosphate companies into compliance. Instead, the general assembly seized control of the matter, some legislators even vying for the state to drive away the phosphate companies and manage the mining themselves, employing convicts as the industry had until the high death rates compelled the state to shut the leasing program down. They settled on a more temperate approach, reducing the royalty to fifty cents per ton for five years.[25]

The next year, the river-mining companies set their workers to rebuilding company infrastructure. By late May 1894, Farmers had "recovered its entire

Phosphate laborers at the Pacific Works, 1889. This photograph of the Pacific Works, a phosphate mine in a Beaufort County salt marsh, gives some sense of the exposure and difficulty of labor in Lowcountry phosphate. Dozens of African American workers and their family members perished in the storm surge here, just four years after Conrad Donner visited the mine to photograph it. Conrad Munro Donner (1844–1916), *Pacific, Excavator Trench. Includes Workers and the Donner Brothers*, Lowcountry Digital Library, 1889–95, courtesy of Beaufort County Library, South Carolina.

plant and is making extensive improvements at the works. . . . A large force of hand pickers has been put to work by the company."[26] Coosaw had purchased a new deep-water property for ships that could draw twenty-two feet of water as a distribution point. Carolina was still rehabilitating the *John Kennedy* and instead had put 100 men onto the river in flats to dive for rock. In 1894, they managed to ship 114,281 tons of phosphate rock for a royalty of $57,140—a quarter of what the state had received the previous year.[27] Though weakened, companies had not given up after the hurricane, deploying Black workers toward a recovery effort of their own.

But it would not last. Headlines throughout 1894 excitedly declared the discovery of rich phosphate deposits in Tennessee and Algiers as well as rising production at the Florida beds. In Bradley County, Tennessee, a particularly rich vein reportedly stretched sixteen feet wide and nine miles long.[28] Algiers was even more threatening, with a layer of phosphate between twenty inches to ten feet thick and 60–70 percent in concentration rather than in a thin, diffuse ribbon of phosphate like in South Carolina.[29] In December, *The State* ran a headline "POOR SOUTH CAROLINA," as Florida had claimed phosphate as the centerpiece of their display for the 1895 Cotton States and International Exposition in Atlanta. There, state officials showcased an enormous phosphate pyramid 100 feet square and 50 feet high that the exposition's 800,000 visitors could admire.[30] Florida's declaration of its ascendance in the industry and its rapidly growing export of phosphate rock eclipsed South Carolina's production and captured the domestic market for phosphate for the next century.[31] Though land rock mining continued apace, its production fell into a slow decline. River-mining production tumbled quickly too. By the time a United States Geological Survey geologist surveyed the state's phosphate in 1914, river mining was a fossil: "In 1893 a disastrous cyclone destroyed the plants of practically all the river companies and paralyzed the industry from some time. The richer deposits had by this time been exhausted, and the competition of the Florida, Tennessee, and Algiers phosphate had become keen; the cyclone therefore came at a particularly critical time and injured the river mining beyond hope of recovery. The industry rapidly declined during the next ten years."[32] Coosaw, the biggest river-mining company, had shuttered its operations in 1904, with smaller companies littered in its wake.[33]

The hurricane, then, worked as an erosive force that state governance, new sites of production, and storm after storm assisted. George Brown Tindall, an anthropologist and historian, noted in the 1920s that South Carolinians talked about the phosphate industry as "paralyzed by the blow, and never recovered in that area."[34] Another commenter wrote that "phosphate mining in South Carolina was becoming a sick industry before the advent of the storm, but the hurricane hastened its decline" and "injured the river mining beyond hope of recovery."[35] Capital dried up once it became clear the Lowcountry was a risky investment, lashed with hurricanes and guarded by resentful politicians. Thousands of Black workers found themselves out of work. The state government became poorer than ever. Miners opened richer veins of phosphate further south and west. The production of phosphate fertilizer using imported rock continued on the Charleston Neck, but phosphate mining in the Lowcountry was done.

White South Carolinians described the impact of the Great Sea Island Storm and its younger siblings on rice in similarly catastrophic terms. In the days after the storm, reports rolled in of rice fields swollen with saltwater; dikes, ditches, and dams broken; and drainage systems rendered useless. Thousands of acres of rice remained underwater for days, rotting the year's crop. Because of the storm's timing, Black rice workers had already harvested the earliest planting of rice from mid-March, one-eighth of the crop;

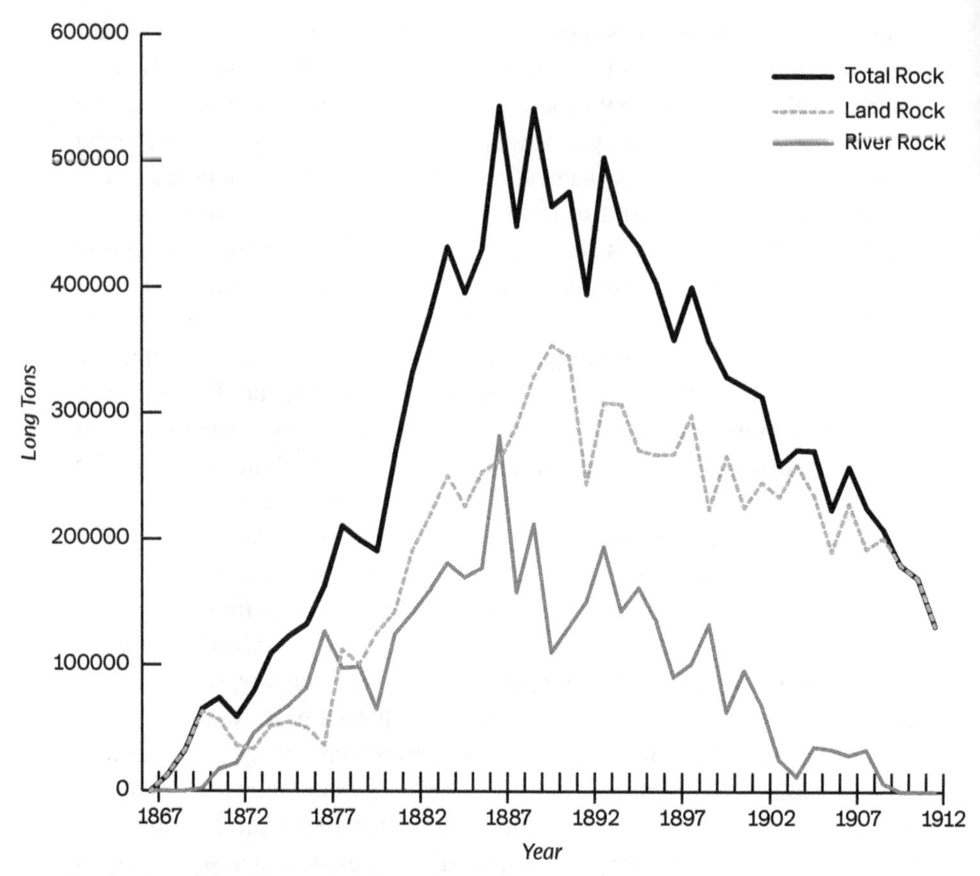

Production of phosphate in South Carolina in long tons, 1867–1912. Data from Rogers, "Phosphate Deposits."

three-quarters of the crop, however, was ripe in the fields, and the other one-eighth, the June rice planted later in the season, had not yet matured.[36] On the Combahee and Ashepoo Rivers near Beaufort, 9,500 acres lay submerged for five days, as were the homes of Black sea islanders who worked in rice, inundated by the storm surge. On the Edisto and Santee Rivers south of Charleston, some 6,000 acres lay in the same condition. Rice grains blackened in their hulls, rotting on the stalk under the saltwater and the hot sun.[37] The saltwater had intruded twenty miles inland and would not soon flush from the rivers or the adjacent fields.[38]

White landowners panicked, describing their tribulations with little mention of the Black rice workers who perished on their properties. Langdon Cheves, the grandson of the ardent secessionist and Speaker of the U.S. House Langdon Cheves, wrote to his wife, Sophia, on September 13, 1893: "The account from Combahee and Pon Pon gets worse everyday. . . . Heyward has lost nearly all his rice & has had breaks everywhere. The other Heyward places . . . lost almost everything. Also, I fear, Willie Haskell & Frederick Blake & Middleton. Pon Pon is just as bad or worse. The Allstons' place was under water at every tide up to 2 or 3 days ago. Young Willie Haskell's much the same or more."[39] Later in the month, he expressed final defeat, lamenting the constant rain in the month since the hurricane and admitting that most white landowners would be lucky if they could save any rice at all. Cheves's foreman wrote to inform him that the rain had kept rice hands from harvesting the remnants and that the only salvageable rice might be a portion of 180 acres of rice that was ripe at the time of the storm.[40] Berkeley Grimball directed Black workers to harvest whatever rice they could three weeks after the hurricane, salvaging rice whose waterlogged stalks had collapsed and shook loose just half of the grains in each sheaf.[41] Poor harvests were once troubling for white landowners' coffers in previous decades. With rice already on the decline, a bad harvest was disastrous.

Saving some of the rice harvest, however, was only part of the labor that awaited Black rice workers. Many white landowners found that the storm had wrecked the rice mills, both those who had one on their property and those in Charleston.[42] The warehouses and wharves on Charleston's waterfront, filled with recently harvested (now mostly waterlogged) rice, fared no better.[43] Harvested rice was useless without a mill to process it for market. Furthermore, of course, the most momentous damage was not to the crop but to the rice fields themselves. R. P. Sanders, Berkeley's foreman, wrote, "Very few of the trunk doors are left; they are gone. They were whipped off by the waves."[44] Breaks in the dikes that separated fields from each other and from the river made the irrigation systems useless. The rice fields were also clogged with detritus from floodwaters, and the soil was spiked with salt from the storm surge. Black rice workers were not numerous enough to complete all the repairs that white landowners wanted, and the white landowners had neither sufficient disposable income to put toward the repairs nor the willingness to pay Black workers fairly. John Screven, a white landowner of a Savannah River rice farm, summed up his situation as "woe to the vanquished," declaring "Vae Victis!" a month and a half after the hurricane.[45] Screven and other white landowners doubted their ability to muster the capital to pay for the materials for the extensive repairs needed to restore the fields to production.[46] They were not wrong to do so. No help was coming.

The commercial production of Lowcountry rice was at another turning point in its fraught history. James Henry Rice, thirty years after the storm, memorialized the industry and recalled the hurricane as the point after which it was clear that production of rice as a means of enriching white South

African Americans hoeing rice. This oft-reproduced stereograph of Black rice workers at the twilight of commercial-level Lowcountry rice production is a rich yet slippery depiction of agricultural labor. The composition suggests an underlying order that would have flattered white landowners, Black men and women maybe even placed deliberately by the photographer to create such a striking image. Several workers wield long-handled hoes, a characteristic tool adapted by enslaved West Africans from traditional implements for the new Lowcountry context; rice workers would use the hoes when water was drawn off the rice fields to dig up weeds and aerate the soil. Keystone View Company, *Hoeing Rice, South Carolina, U.S.A.* (Meadville, Pa.: Keystone View, ca. 1904), Library of Congress Prints and Photographs Division (LCCN 00652861).

Carolinians was doomed. He conceded that "rice planting slowly recovered as the years went on," after the Civil War and as Black farmworkers returned to the fields under labor contracts, "but on August 27–28, 1893, there was a hurricane that destroyed crops, backed up salt water and created havoc everywhere. . . . After a few hectic efforts rice planting perished."[47] The hurricane, which struck as the mechanization of rice culture in Texas and Louisiana began to outpace the production of South Carolina rice, resulted in "the slow crushing of the rice planter": "The recent storm, coming at a time when rice was low," Rice wrote, "finished the business."[48]

After 1893, though many landowners held on to their remaining arable acreage, "the industry has been speculative," Rice commented, while it ascended in Louisiana and Texas. There, Rice mourned, "rice planting became reduced to a science. . . . Machinery was employed. Water companies with abundant capital attended to irrigation," and South Carolina was forsaken.[49] Its marshy soil could not support mechanization, whereas "the buoyant prairie

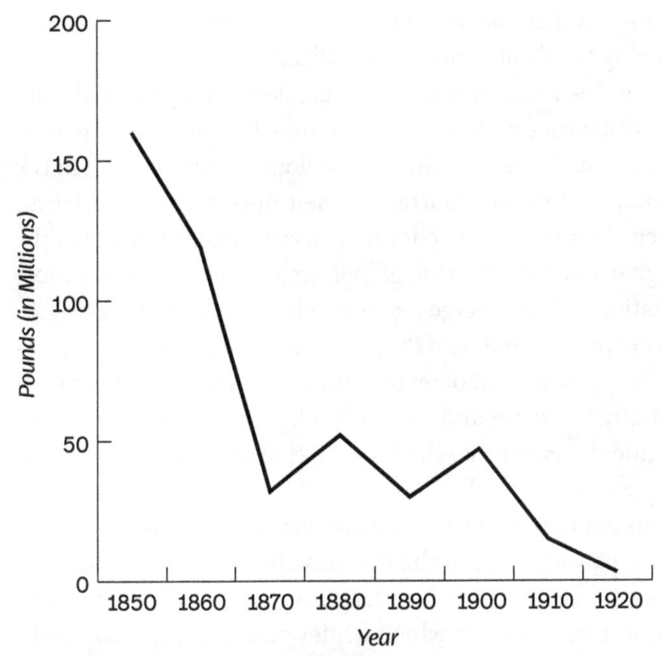

Rice harvest by pounds in South Carolina, 1850–1920. Rice data gathered from the 1850–1920 USDA censuses of South Carolina. These censuses can be found online at the U.S. Department of Agriculture, Census of Agriculture Historical Archive, courtesy the Albert Mann Library, Cornell University, accessed February 14, 2022, http://agcensus.mannlib.cornell.edu/AgCensus/homepage.do;jsessionid=3F2582F065D5E5034D6A5903D07791E4.

soil" of Louisiana and Texas "makes the use of machinery possible."[50] Black farmworkers were scarce and turned away from the "hand labor" that Rice derided as "the poorest and most unreliable labor in the world"—failing to realize that that racist attitude toward Black rice workers and their expertise might be precisely why maintaining a labor force was increasingly impossible for whites.[51] As Louisiana and Texas drove prices of rice down and attracted investment, South Carolina's rice production sank further. The two decades of hurricanes only worsened matters. Elizabeth Allston Pringle reflected on the legacy of not only the 1893 hurricane but also those that followed. "The rice-planting," she wrote, "which for years gave me the exhilaration of making a good income myself, is a thing of the past now—the banks and trunks have been washed away, and there is no money to replace them."[52]

The white elite of the Lowcountry shook their heads at the exodus of Black laborers from the ruined, eroded rice fields because it meant the decline in their own fortunes and in a social position—the "planter," though they planted nothing themselves—that they had jealously guarded for centuries. The white coastal elite, mourning the diminishment of a Black workforce and their illusions of control over the land, connected the erosion of the rice fields and the subsequent reforestation on the coast to the slow fading of the environmental remnants of slavery. The rewilding of millions of acres of land in the Lowcountry, as white landowners contracted their operations and African Americans withdrew their labor from commercial rice fields, symbolized the emptying of the region and the withering of their imagined idyll of the peaceful, gracious plantation.[53] The emergence of truck farming and logging in the early twentieth century did not hold the same cache for the wealthy whites of the Lowcountry.[54] The busy hum of textile factories in the upstate, fed with short staple cotton grown inland and mostly the labor of poor whites, provided few opportunities for either the white coastal elite or impoverished Black workers.[55]

The hurricane was not the only factor that had shaped, diminished, or otherwise affected phosphate and rice in the decades after emancipation, but it was a powerful moment that reinforced the fragility of the Lowcountry economy, especially the sectors of it from which whites profited. This was an elision of other causes of decline that coalesced with hurricane damage, which included white exploitation of Black labor, the increasing rigidity of Jim Crow, and the inability to adapt to changing economic circumstances or attract outside investment. But white South Carolinians found a force of nature easier to blame for hard times than either structural issues or personal flaws.

After "this tremendous storm" in 1893, "There was a loss of the cotton industry and the phosphate industry. So my father conceived of the idea of buying land," Herbert Keyserling recalled of his father, William.[56] In the late nineteenth century, William Keyserling worked on St. Helena Island, first as a cotton gin machinist and then as a manager for one of the grocery stores owned by the MacDonald-Wilkins Company. Before the hurricane, Keyserling operated the Corner Store, the commercial center of Frogmore, a pocket-sized but consequential town on St. Helena near the Penn School.[57] At the Corner Store, Black sea islanders purchased provisions and negotiated for cotton seed and other necessaries for planting their small crops of sea island cotton. After the hurricane "salted the cotton fields," the company lost money and searched for new ways to increase profits. Keyserling took initiative and purchased a property on roughly "fifteen, eighteen miles north of Beaufort." There, on the mainland, he "transported a lot of black people to that area—built shelter for them and so forth—and started large-scale farming." He brought business partners on board and named the operation the Dale Farms

Corporation after J. J. Dale, the cotton factor of dubious reputation among African Americans but of solid standing among whites.

The farm became profitable for its white founders and employed a large number of African Americans, where they "grew potatoes and cabbage and, I guess, some corn and so forth" as fodder for the farm's forty mules. The farm thrived, and the white-owned business expanded to cover 3,200 acres.[58] Keyserling viewed the operation as a pragmatic business decision: "If people weren't earning money, they couldn't survive either," his son explained. "So he put people to work to support his business."[59]

But Keyserling was not acting through generosity, and his business model revealed the through line of Lowcountry history: there was no white profit without Black labor. African Americans may have sent select family members to work on the mainland farm to earn extra money for the rest of the family, as they had since the Civil War. Perhaps families who had lost too much in the hurricane to continue farming their own land found employment at the farm. It also could have been an employment opportunity for young people desperate to assist their families during the difficult years after the hurricane. The farm was nonetheless a site of labor exploitation and of white bosses and Black workers. Three elements served as reminders of what the farm was: its name, after a cotton factor whom Black sea islanders mistrusted; its modus operandi, moving Black sea islanders from their homes onto a company farm; and its origin, in a company that controlled not only access to food staples but also to the annual cotton crop, a necessary supplement to Black farm incomes.

Indeed, the owners of the MacDonald-Wilkins Company had deliberately made their capital central to Black life on St. Helena. Control over resources like food and cotton came to include control over land: property begat more property. In an annual cycle repeated across the South, the company would "finance for seed and fertilizer; make advances, keep it on the books. When cotton was sold, [farmers would] settle up. They'd pay their bill and get what was left."[60] But unlike the tenant farmers and sharecroppers who became beholden to these systems of credit, these Black farmers mostly owned their own land. The company found a way to leverage that property. "Sometimes," Keyserling recalled, "people put up land as security. I don't think everybody was required." Instead, the company constructed a hierarchy of those who were deserving and those who were "of questionable character," whom they compelled to put their land up as "a security for the advances they got." The company demanded and received control over the property titles of Black families whom they deemed untrustworthy, a loaded, racialized descriptor in the Jim Crow South. While Keyserling was sure that the company always returned the deed, which may or may not have been true since the memories came from his childhood, that system nonetheless placed the company in a position of power over Black families and reminded them of how white-owned capital could seize control of their livelihoods and their future. For Black farmworkers on the mainland Dale Farm Corporation, the reminder of white control over their labor must have been a palpable presence in their daily lives.

Black farmers in the turn-of-the-century Lowcountry had to contend with the barrage of hurricanes, the political attrition of Jim Crow, the pressure of white politicians, landowners, and businessmen jostling for their labor and their subservience, and the whittling away of the variety of economic activities that they had used to dodge white domination. Some African Americans looked to community connections and the federal government to bolster their livelihoods in the wake of these accreting disasters. On the sea islands, the U.S. Department of Agriculture's (USDA) project in "Cooperative Extension Work in Agriculture and Home Economics," which began its work in the Lowcountry by 1909, became a valuable program for Black farmers in the besieged region. While the USDA's relationship with Black southerners soured over the decades as the agency increasingly adopted discriminatory practices, Black residents of Beaufort County deployed their skill in adapting progressive, white-led aid programs for their own needs.[61] The program, which lasted through the Great Depression, was especially significant for Black farmers in the Lowcountry in the 1910s and into the early 1920s as they dealt with the environmental catastrophes of hurricane and boll weevil, the political and economic constriction of Jim Crow, and the demographic changes diminishing their local communities.

The USDA's overarching goal was to introduce and disseminate improved agricultural practices among small yeoman farmers, most of all through diversifying crops and introducing new crops to replace soil-exhausting monoculture. In South Carolina, under the USDA's direction, the state agent for the Farmers' Co-Operative Demonstration Work, Professor Ira Williams, organized a network of segregated county extension agents, both white and Black.[62] Williams, one of many in a line of benevolent-minded whites eager to instruct Black sea islanders in proper methods, expressed a vested interest in engaging Black residents of the Lowcountry in the extension service. But since Black sea islanders already grew subsistence crops alongside cash crops, the extension agent's goals in Beaufort County, at least, came to focus on transitioning their cash crops away from cotton.

From the beginning, USDA officials recognized how racism in the South might hamper their programming and adjusted their agenda to avoid angering southern whites. Williams advocated that the USDA hire African American county extension agents, in part to segregate the USDA's workforce to ease the department's relations with white powerbrokers in the state and because he felt that Black farmers might be more responsive to and trusting of Black agents. Williams was also fearful of backlash from white farmers should any Black farmers appear to receive preferential treatment. In his initial report in 1909 and in subsequent years, he frequently cited the statistic that over 50 percent of the state's farms were operated by African Americans, to impress upon the USDA the importance of including Black farmers in South Carolina in extension work.[63] Black farmers clamored for assistance from the extension agents in ever-increasing numbers; Williams worried that the "agents are asked to take so many demonstrators that it is getting impossible for the white men to take the negroes, because they could not well refuse white people," which would "make our work unpopular."[64] As a result, Williams recommended that the USDA hire Black agents.

Williams also acknowledged the unique difficulties of arranging any form of education or instruction in South Carolina for African Americans. While agricultural demonstration could be beneficial to white and Black farmers alike who were "anxious to receive" assistance and instruction, Williams wrote that whites in the state "are prejudiced against negro book education and any man who attempts to help them would be put out of a job."[65] In his 1911 report, Williams implicitly suggested that the USDA's extension work could ameliorate the conditions of discrimination under Jim Crow. He referred to the reports of the six "negro agents" currently under the USDA's employ that show "how badly the work is needed" and quoted one agent who wrote that "negroes have suffered so much at the hands of organizations and individuals which sought to filch from them their hard-earned wealth."[66] Williams's report revealed how fraught establishing a federal program of agricultural assistance in Jim Crow South Carolina could be. Adjusting agricultural practices and policy was never a neutral proposition.

The USDA and the Black farmers who remained in the Lowcountry had not relinquished the possibility that the sea islands could become a thriving agricultural zone once more. In his first report to Dr. S. A. Knapp, the special agent in charge of the USDA program, in 1911, Williams explained how the Penn Normal, Agricultural, and Industrial School on St. Helena, founded as a school for African Americans during the Port Royal Experiment, had already "placed its farm under our direction," with one hundred of its acres at the disposal of the USDA.[67] Indeed, students at the school, working with a county extension agent, had planned and planted fields of corn, cowpeas, cotton, and garden vegetables that season. Rossa B. Cooley, the principal of the Penn School, wrote for the school's annual report that "the farmers who followed the [USDA's] instructions more than doubled their crops."[68] Williams also hoped that the agents, working with African Americans, could make use of the "thousands of acres of land on the coast that have heretofore been devoted to the profitable growing of rice" that "have been in recent years abandoned."[69] Like many before him, he pointed to drainage as a potential solution, wondering if "these fertile lands that are fertile as the valley of the Nile can not be made profitable again to their owners."

In this case, however, Williams did not envision a Lowcountry drained of its African American population but a Lowcountry newly fruitful and profitable both for African Americans and through their labor. Williams "could not but think of Holland . . . where artificial drainage has to be utilized, that they have made those lands profitable with livestock." An agent had planted an experimental plot of ten acres in pasture grasses, as "a number of gentlemen . . . are willing to begin the development of the livestock industry on these lands."[70] Williams's vision of a pastoral idyll of cattle grazing the Lowcountry was an interesting reversion to white settlers' use of the land in the early days of English colonization before the introduction of large-scale rice cultivation and the importation of an enormous enslaved population—though Williams's concept would not come to fruition. As Williams considered the Lowcountry, he took a view that incorporated and respected Black sea islanders' own agricultural preferences and traditions rather than looking at the region as the vehicle for white wealth that it had been for so long.

In Beaufort County, the USDA extension agents also operated in conjunction with local farmers' societies, especially the St. Helena Cooperative Society on St. Helena. The cooperation between the USDA and local organizations in the late 1910s and early 1920s proved especially significant in supporting hurting farmers.[71] The African American agents over the years frequently mentioned interactions with the St. Helena Cooperative Society, founded after and in response to the 1911 hurricane.[72] Black farmers on St. Helena Island obtained a loan of $2,000 from the Penn School as seed money for the organization and worked with J. E. Blanton, the superintendent of the Penn School and the African American USDA agent for Beaufort County, to establish a Committee of Management and the overall structure of the nascent organization.[73]

The society attracted a substantial following instantly. Cooley wrote in 1913 that the society had loaned out $780 in the previous year, with an average loan of fifteen dollars and an annual membership fee of one dollar, and that the Committee of Management met each month and held an open meeting with its members only once or twice a year.[74] This was "a very important point" for the farmers, Cooley noted, so that "it does not take them off their farms nor away from their work."[75] The society provided a measure of protection for Black farmers already coping with economic hardship through collective buying and selling arrangements. Through hurricanes, the boll weevil, and Jim Crow, Black farmers continued to experiment with how to make a living when they could find enough resources to escape the urgent press to grow only crops that could garner a cash payout. Though they had lost numbers since the 1893 hurricane, those who stayed continued to turn to the same kinds of community-oriented, and sometimes federally supported, palliatives that they had for decades, adapted to cope with the specific problems of Lowcountry life during Jim Crow.[76]

Together, the society and the USDA extension agency helped carry Black farmers in the Lowcountry through yet another crisis. Unfortunately, a new enemy began creeping their way in the 1910s: the boll weevil. The boll weevil first arrived in South Carolina in 1917 at the state's far southeastern tip, after beginning its ravenous travels east from Texas in the 1890s.[77] In that year, farmers in Beaufort County still produced 1,093 bales of cotton, and 1,688 the next; in 1919, farmers reported on 167 bales, and none in 1920.[78] It is unclear whether farmers stopped growing cotton because they feared that the boll weevil would ruin the crop or the weevil ate up the crop that they attempted to grow.

South Carolinians knew that the boll weevil was coming: it had been making its way east, field by field, for decades. In theory, there had been time to plan for this slow-motion disaster. And some preparations had been made. The South Carolina government established the South Carolina Boll Weevil Commission in 1915 and released pamphlets in that year and in 1916 filled with information about the weevil.[79] The pamphlets provided instructions on poisoning the creature—calcium arsenate in the form of a dry powder, applied to plants at night, moistened with dew, in three applications every four days. They also contained widely accepted science on the "climatic factors" that allowed the weevil to flourish: "minimum winter temperature and maximum summer rainfall."[80] They also made dire statements about the future of South Carolina's booming cotton economy, predicting that "the coming of the boll weevil will put an end to the present advance system of the cotton crop now so generally practiced in South Carolina."[81]

At times, the pamphlet's primary author, Walter Riggs, president of Clemson University, connected the pressures of a white supremacist system of tenant farming with the boll weevil and the Great Migration:

> In Louisiana and Mississippi a large number of the young and able-bodied Negroes left the State to seek employment elsewhere. This emigration was chiefly due to inability to get credit, and therefore inability to make a living under the tenant system of farming. There seemed to be a general lack of appreciation on the part of the white people of the importance of retaining their Negro labor. With no capital and no credit on which to make another crop nothing remained for many Negro farmers but to move away and seek employment in other agricultural sections or in other lines of business.[82]

Riggs stopped short of suggesting policies that ensured assistance for cotton farmers, regardless of race, or protected Black farmworkers from exploitative creditors. But, fearful of losing the state's labor pool, he did urge whites in South Carolina to "by helpfulness and consideration seek to retain" Black farmworkers. Without them, and with the boll weevil, he worried that "the values of farm lands" would become "greatly depressed." Some of Riggs's prescriptions followed progressive farming tactics, solid advice for southern soils exhausted from monoculture regardless of the boll weevil's presence. He recommended, much as the Red Cross had in 1894, that South Carolina farmers "raise all provisions, keep out of debt, and cultivate cotton by improved methods on a restricted area of the best lands," as well as use "intelligent diversification and proper rotation of crops."[83] Riggs's suggestions would be easier said than done for cash-strapped Black farmers, especially those who did not own their own land. The South Carolina Boll Weevil Commission's publications demonstrated the deeply felt anxiety over the future of the agricultural South, especially among white policy makers and farmers—an existential crisis that did not begin or end with the boll weevil.[84]

In the Lowcountry, the boll weevil was yet one more of a cyclone of amalgamated factors weighing on the lives of African Americans, those affected most by predatory lending practices and environmental disasters. Help in any meaningful sense was not coming. Only in Beaufort County, where the USDA cooperated with existing infrastructure like the Penn School and Black community networks, did Black farmers find aid. As the needs of Black farmers grew, the St. Helena Cooperative Society met those challenges by giving them access to larger loans than before the boll weevil. In 1920, a high-water mark of the crisis, the average loan that farmers requested from the society grew to forty-three dollars, well up from the loans of fifteen and twenty dollars that they had applied for in earlier years.[85] The extension agency encouraged farmers to shift away from cotton and toward sweet potatoes and peanuts.[86] Beaufort County's history of interracial alliance, however uneasy, had found ways to endure.

By 1920, Black residents of the Lowcountry had experienced decades of devastating hurricanes, the arrival of the boll weevil, the ravages of white supremacy's new guise, and the decline of the economic autonomy that they yet fought to maintain. Black sea islanders, motivated to maintain independence and sustain their families, had constructed an economy around the sustenance of themselves and their loved ones rather than the enrichment of the white southerners who had so frequently profited at their expense. However, the resilience that African Americans had woven into this economy, while designed to absorb a hit here or there, began to lose its elasticity after 1893. The hurricane-wrought damage, only worsened by an onslaught of calamities, set into motion a rearrangement of the lives of African Americans living on the coast. Uneven and intermittent support from either northern or federal interlocuters did not ultimately provide long-term solutions. The flexibility of engaging in a variety of economic activities could only succeed as a strategy to bolster autonomy if there remained a diversified economy and familiar environmental conditions to which they could adapt. The political pressures that white South Carolinians applied to coastal African Americans fused with the economic hardships and environmental changes of the turn of the century, catching the Lowcountry's Black population in a vise grip that they would have difficulty escaping in the coming decades.

White South Carolinians had trapped themselves between their race hatred for the region's Black majority and their need for Black labor. Even as their own wealth declined, and as they began to lose their labor pool, they calculated the value of white supremacy as a currency of its own. But it was increasingly clear that they had not sufficiently anticipated the costs of environmental instability. The economic success of whites in the region depended on three factors: a tractable nature, a labor pool that they could brutalize, and the maintenance of a familiar economic regime. They had rarely had a firm grasp on all three at once, and by 1920, they were unmoored. The nature of the Lowcountry threatened to pull loose the racial hierarchy they had tried to secure through both legislative actions and economic control over Black South Carolinians.

Chapter 9

Jim Crow Lowcountry

On Ossabaw Island, a barrier island south of Savannah, a community of African Americans had grown in the decades since emancipation. Some had been relegated to sharecropping after broken federal promises following the retraction of Gen. William T. Sherman's Special Field Order No. 15, and some had been able to secure land titles in the two years immediately after the Civil War.[1] There, they grew sea island cotton and garden vegetables, fished, hunted, and harvested oysters. They founded Hinder Me Not Baptist Church in 1874.[2] But then, the Great Sea Island Storm swept the island on its route north. Another hurricane in September 1896 caused yet more damage on Ossabaw. Black sharecroppers, already beginning to search for an alternative to their lopsided labor agreements, were "forced . . . onto the mainland of Georgia" because of this "series of Atlantic hurricanes."[3] In 1896, unable to stand the storms and the sharecropping any longer, a few of these families purchased cheap land along Shipyard Creek from Judge Henry McAlpin, on the marshy fringes of a subdivision of summer retreats for wealthy whites from Savannah.[4] As Hinder Me Not's congregation plummeted, its members made a new home at Sweetfield of Eden Baptist Church. The Pin Point community, as it is known, thrived for decades, its residents fishing for shrimp, crabs, and oysters that they sold at the Savannah Market and later working for the A. S. Varn & Sons Cannery.[5]

The Lowcountry has long defied easy narratives of either progress or declension. There, both can happen at once. The blossoming of Pin Point, amid the turmoil of the 1890s, represents the adaptability that Black sea islanders were compelled to develop to survive. It also illustrates that environmental and political forces could not be disentangled in the region: Black sea islanders' experiences of work linked the two. The economic deprivations that drove African Americans out of the Lowcountry cannot be separated from the political conditions of the era, because white control over the Lowcountry economy and the Black labor that fueled it was intrinsically tied to white supremacy. The history of Pin Point, though a markedly brighter story than that of other Black communities in the era, only reinforces those connections.

As many African Americans characterized it, the Great Sea Island Storm brought the beginning of the end of their time in the Lowcountry. Over the decades, little in the economy or government improved in the lives of many

Lowcountry oyster cannery workers. While flexible wage labor remained scarce in the Lowcountry, oyster canneries were one of the most common and viable options for Black sea islanders. While these cannery workers lived on St. Helena Island, a comparable scene would have been familiar to the African Americans living at Pin Point: from the giant pile of discarded oyster shells, the dimly lit clapboard cannery, to the jumble of boxes for unloading unshucked oysters, along with the briny scent from the oysters and the marsh alike. Marion Post Wolcott, *Cannery Workers Going Home, St. Helena Island, South Carolina*, ca. June 1939, Farm Security Administration—Office of War Information Photograph Collection, Library of Congress Prints and Photographs Division (LCCN 2017754138).

African Americans. The storm became an erosive force for most African Americans' livelihoods, a "slow violence of delayed destruction."[6] Its harm to the region's economy lasted for decades and deepened the wounds that Jim Crow incurred. For African Americans, the process of realizing how their world had altered in the years after the hurricane was painful and protracted. Historical change does not always occur in a sudden shock, and when jarred and traumatized by the direct hit of a deadly hurricane, people do not respond with lightning-fast reflexes. Abiding ties to home, to land, and to a community are not easily severed. Coastal African Americans had to bring themselves to accept the changes in pieces and over time.

The hurricane, though the most devastating in the lifetimes of Black inhabitants of the Lowcountry, could be explained away for a time. After all, as the sociologist Clyde Vernon Kiser wrote, referring to the reaction of Black sea islanders from St. Helena Island to the hurricane of 1893, "catastrophes come without warning and affect many individuals who have given little thought to migration."[7] African Americans forged communities on the sea islands from the wreckage of stripped, exhausted Lowcountry lands, buying plots already infused with their blood and sweat because they could not find willing sellers elsewhere and because the Lowcountry was the place in which their identities and families were embedded. African Americans did not want to relinquish land to which they had a legal claim and in which they had knotted roots centuries deep.

Hurricanes, a meteorological phenomenon with which they were intimately acquainted, were not in ordinary times an impetus to move—partly because there was nowhere else that African Americans could move to and still retain their landed independence. But the hurricane had dislodged and rotted some of their roots there. Black sea islanders found their livelihoods hollowed out as the kinds of labor to which many were habituated in rice, phosphate, and subsistence farming no longer provided a tenable living. Many African Americans, first young people and later families, migrated to cities—some as near as Savannah, others as distant as New York City.[8] Others worked to ameliorate the losses to their livelihoods through turning to community aid and their own traditions of solidarity. In the years after the hurricane, they had to make decisions about how to cope with changing circumstances and how to survive within a political, economic, and environmental climate that had turned on them. African Americans struggled with their increasing economic vulnerability at a time when they had much to lose and little to spare.

Examined over decades, the hurricane had protracted, unpredictable effects. The whites who resented the region's Black majority began to see their wish of a white-majority state realized as African Americans began to move away from the region. Disenfranchisement, the removal of political power to the state level, and a pervasive rhetoric of dispossession combined with the hurricane's lingering impacts to severely restrict the futures of African Americans. The attrition of the Lowcountry's population had a deleterious impact on the region's economy. Whites had toyed with the trade-off of a weakened economy for a white majority state for years and would witness that hypothetical come to life in the Lowcountry with a vengeance throughout the first decades of the twentieth century: the sea islands underwent a half century of economic depression, in part because of the long-term impacts of the hurricane; and at some point in the 1920s, South Carolina went from a Black majority to a white majority state even in coastal areas like Charleston County. White elites in the Lowcountry, simultaneously relieved at the attrition of the region's Black population and resentful of the diminishment of their wealth, thus began to construct a Lost Cause mythology of the Lowcountry's history that erased Black contributions and exalted the labor camp and the white enslaver. The demographic, economic, and political emptying of the sea islands fulfilled some of the dreams of white supremacists, and those cruel victories came at a high price to African Americans.

Cracks that began after the hurricane in the 1890s had widened by the 1920s. Frederick Christensen, the son of the Beaufort businessman Niels Christensen, commented on some of these fissures as they opened on his return to Beaufort after completing his education in Massachusetts: "What a fine county this is where eleven-twelfths of the people are regarded by the influential twelfth as no better than apes (I've heard this several times since I've been here) and where every other man carries a pistol and all celebrate Christ's birthday by getting drunk. Mr. Johnson promises to show me an unarguable article to prove that N—— have no soul."[9] And while lynchings were not common in Beaufort County, the one recorded lynching in the county's history and another attempt occurred both in 1901. In that year, a gang of masked white men dragged a Black man arrested for trespass out of the Port Royal jail and beat him severely. They claimed that they then ran him out of town, and the injured man made his way to Savannah, but Frederick Christensen confided to his journal that he "and a good many others thought he had been lynched."[10] Soon after, whites in the town of Port Royal accused William Cornish, a Black sailor, of attacking several white women and shot him; African Americans in Beaufort County conducted an exhaustive search for his body but never found his remains.[11]

White supremacist organizing only increased with the nationwide rise of the Second Klan in the 1910s and 1920s. As one Beaufortonian summed it up, "There was a Ku Klux Klan element in the area. There always has been in the South."[12] Reverend Marshall of Beaufort's Episcopal Church, as Herbert Keyserling recalled, "was an active member of the Klan."[13] With the Klan increasingly integrated into the town's political structure, its members felt empowered to hold marches in the downtown in the early 1920s. Keyserling, then a young Jewish boy, recalled, "The Klan used to parade in Beaufort. We used to stand on the street as they came by and identify all of them, either by a limp or how fat they were."[14] In such a small town, the Klan's membership was no secret. When the Klan *did* attack Black residents in Beaufort County, they usually did so on the mainland. Albert Ullman recalled one such case from his youth in the 1920s, in which "a very nice black man in Bluffton" "spoke to a white woman" in the town post office.[15] On the basis of that innocuous exchange, "the Klan came in from Savannah [at night], pulled him out of his house. . . . They tarred and feathered him."

Though physical violence of that level was less common in Beaufort than elsewhere in the South, the Klan's strategic attacks, their parades, and the avid participation of white community luminaries made a strong statement about the new regime in Beaufort County. The shift in the region was also signaled with the deaths of Clara Barton in 1912 and Robert Smalls even more so in 1915 as a "cantankerous old man, quick to anger," perhaps embittered and saddened by the progress lost in the twilight of his life.[16] Unfortunately, even as Smalls's death symbolized the passing of an era, Benjamin Tillman's death in 1918 did nothing to dampen the harm that he had fomented.

Many African Americans found ways of holding on to their land and their autonomy for years afterward. Those who had survived the hurricane gritted their teeth and tried to make a living through the avenues that they had developed over the past three decades, but they found those routes increasingly trying. Laura Towne and Ellen Murray of the Penn School wrote in 1898 that the economic losses of the storm hit Black farmers hard for years, with some unable to pay their taxes and others entering into crop lien agreements for the first time with white cotton merchants.[17] They saw the hurricane as an undeniable force in changing the fortunes of Black sea islanders forever. "The self-supporting, self-respecting farmers were no disgrace to any race or nation," they wrote, though they spoke "in the past tense for in 1893 a terrible calamity swept over the island and ended its prosperity. One awful night made a ruin of the thirty year [*sic*] of work."[18]

The hurricane exhausted many African Americans, and there was no cure under Jim Crow. Towne and Murray wrote that though Black sea islanders "have worked hard," some farmers even exhausting themselves "till they had to sit down dizzy, and faint in their furrows," with "several d[ying] in their fields," they "ha[d] not yet regained their former prosperity."[19] The hurricane had worn away rice and phosphate, depriving African Americans of two pillars of their livelihoods. Those losses contributed to their entrapment in the kinds of debt that commonly plagued Black farmers in other parts of the South, a significant change in a place where African Americans had leaned on their landownership and demographic majority to shield them from such practices. They would eventually find the changes too great to remedy, no matter the ministrative recovery effort of the Red Cross or the past flexibility of their economic practices. Some African Americans found it too difficult to even envision a brighter future there. One young man of twenty, the oldest of five and son of a Black woman recently widowed, told her, "'I must find work or I shall die, I can't see you all suffering this way,' and so left home weeks before and had not been heard from since."[20]

The hurricane was a transformative event that had imprinted itself deeply on African Americans' collective memory, and many attributed the storm as the gateway to change in the Lowcountry. Black sea islanders who migrated to Harlem recalled three decades later that whereas "in normal times the proceeds from cotton, the 'work-out' money, and the home-grown food supplies enabled the Islanders to get along," the hurricane "reduced many families to actual need."[21] In the 1920s, the sociologist Clyde Vernon Kiser interviewed 170 St. Helena Islanders who moved to Harlem starting in the 1890s; he found that Black sea islanders were reluctant to abandon their homes on the island. No large out-migration, Kiser's interviews showed, had occurred from St. Helena until the decade beginning in 1900. Remnants of old economic practices and continued high rates of landownership around the islands of Beaufort County helped Black sea islanders hold on to their livelihoods and delayed their departure for years. Furthermore, the hard-won fight for emancipation and landownership was a clear memory for Black sea islanders in the 1890s, and they likely did not wish to concede the ground to white South Carolinians nor flee from their homes too quickly.

But even just a few years after the turn of the century, white South Carolinians noticed the out-migration. The 1907 South Carolina *Handbook*, which cribbed much of its material on history from Harry Hammond's 1883 edition, published numbers on the "drifting of the negro problem to the Eastern, Northern and Middle Western States."[22] The 1907 handbook, like Hammond's tome, adhered to a white supremacist perspective, viewing the departure of African Americans not as a judgment on white treatment of Black sea islanders but instead as an opening of "opportunities for white settlers and laborers."[23] Its author, the Charlestonian Robert Goodwyn Rhett, pointed to the slowing growth of the state's Black population: a huge jump in the decade from 1870 to 1880 of 45 percent, then to 14 percent the next, then 13 percent in the 1890s, and finally from 1900 to 1905 with growth of just below 6 percent.[24]

Beaufort County also showed only creeping increase. In 1890, its population stood at 34,119, and by 1900, despite high birth rates in the rest of the state, the population had added just 1,300 people.[25] Charleston County over the same period, for example, saw its population jump from 59,903 to 88,006.[26] Beaufort County's population growth may have been strangled by deaths from the hurricane early in the decade or perhaps from out-migration of hurricane refugees to Charleston or Savannah, though neither can be said with certainty. The real dive in the sea islands' population, however, did not occur until the decade of 1900 to 1910.[27] For example, in 1900, 8,819 people lived in the St. Helena Township, which encompassed St. Helena Island (8,285 of that population was Black). By the 1910 census, St. Helena Township had lost 25 percent of its people; by 1920, another 10 percent had departed.[28] From 1919 to 1928, St. Helena lost another 17 percent of its population.[29] Once it began, the reduction was severe and ongoing.

Observers noted a similar difficulty in pulling apart the causes of the demographic and agricultural decline in the state to isolate a single source. Mary

Fletcher Stackhouse, a white South Carolina scholar in the 1920s, argued, "Negro tenants are leaving the state in no small number to seek employment elsewhere[,] this emigration due chiefly to their inability to get credit because of the weevil invasion and declining value of the crop and land."[30] African Americans were not fleeing because the weevil destroyed their crops but because of banks, creditors, and landlords who withheld credit in reaction to fear over the weevil—a form of labor control that backfired against them when African Americans chose to leave rather than live under an oppressive system.[31]

Edward T. Hendrie Shaffer, in a widely-read 1923 article for the *Atlantic Monthly* titled "The New South—the Negro Migration," similarly noted that the boll weevil's destruction occurred amid the "intricate social and political problems" of the Jim Crow South.[32] In South Carolina over the course of six months in 1923, over 50,000 African Americans left the state.[33] Shaffer, like the white politicians who complained of the state's Black majority, saw the out-migration of African Americans as an opportunity, as long as they moved away from the South entirely: "Nearly all the exodus, however, is from the purely agricultural districts which have always teemed with an overwhelming mass of negro labor. . . . If this mass will of its own accord transfer itself to other sections, then, many feel, one pressing phase of the boll weevil problem will have been well solved."[34] Even Shaffer's language dehumanized the state's Black majority, suggesting they were as detrimental to the state as the invasion of the hated boll weevil.

It would be better, Shaffer argued, for white South Carolinians to let African Americans migrate, for "there is no place for such a class either in the new scientific warfare required to grow cotton under boll weevil conditions or in the intensive struggle which it takes to bring to success a modern diversified farm."[35] "There exists a universal belief among negroes that their race can enjoy greater liberty in certain places of existence in Northern states and that there it is less subject to danger from the dreaded Ku Klux and other forms of mob violence," Shaffer noted, and so should "his exodus continue to the point where his numbers show a marked decrease, the result may be a beneficial influx of white immigration into the Southern states." Progressive farming practices could, after all, work only when African Americans were gone, because they had "shown neither willingness nor aptitude" to engage in "those new methods of scientific and intensive farming" that would be the only hope for South Carolina's future agricultural success. (The Black farmers across the Lowcountry who had long adopted diversified farming methods would have been surprised to hear this.) While Shaffer did not refer to the Lowcountry in particular, this attitude closely mirrored the language that white residents of the Lowcountry deployed to describe Black land use patterns in the sea islands.

Shaffer's article attracted a great deal of attention in South Carolina and angered members of the established white elite, including Ambrose Gonzales, because Shaffer refused to ignore white supremacy and its economic expressions. (By then, Gonzales was the lone editor of *The State*. This was because in 1903, Ben Tillman's nephew James Tillman, infuriated by the newspaper's coverage of his gubernatorial campaign, assassinated Gonzales's brother Narciso in cold blood and in broad daylight on the streets of Columbia.) What angered Gonzales, though, was not Shaffer's eagerness for a Black exodus, a favored point of his own, but Shaffer's criticism of the antebellum economy's inefficient use of enslaved labor and monoculture. Gonzales, who must have never had a good faith conversation with a Black freedman or woman, countered, "I never heard a freedman complain of having been cruelly punished while a slave."[36]

While many white elites in South Carolina resented Shaffer's characterization of the South during slavery, none had a rebuttal for Shaffer's key point: that the issues of the environment and Jim Crow were interwoven. Even James Henry Rice, a naturalist and white farmer who adulated the Lowcountry's history of rice slavery, acknowledged in a 1923 letter to a friend in Columbia that "the negro is emigrating. . . . Nothing is surer than that the negro population will leave the state. Ten years will about wind up the negro population."[37] (He was right about the Black majority, though not the African American population overall.) William Henry Frierson, a South Carolina lawyer born in 1854, penned an article, "The Negro Exodus," in which he considered the Great Migration the most acceptable solution for what he saw as the great ill of emancipation. "Born of sectional hate, fanned into a flame by fanatical zeal, [emancipation] was violent and abortive," he wrote. It "thrust on the South the gravest race problem the world has ever known."[38] He saw three historic solutions to such a problem: extermination, colonization, and amalgamation, none of which could properly handle the current predicament. "What are we going to do with the negro?" he asked—and answered, "Better let him go North. . . . It begins to look to me that destiny has stamped upon this race, now so numerous in a country in which they do not properly belong, the solution which will come from a wide distribution."[39] Frierson, Rice, and other white commentators tied the Great Migration not as much to a specific turning point but a complex combination of factors over which they frequently argued.

While white South Carolinians scrapped over the cause of the out-migration and the prospects of South Carolina's floundering economy, Black sea islanders continued to move away from the region. Kiser's interviews provide invaluable insight into the reasons why Black sea islanders left.[40] While his perspective on the factors that pushed African Americans out of what he described as "an isolated, all-black society" did not account for the numerous ways in which political, economic, and social ills intersected in the lives of Black sea islanders, the interviews showcase why they migrated away from St. Helena Island. Kiser had a dim view as to what constituted racism, claiming that "racial hatred has had very little to do with migration from St. Helena," while ignoring account after account that described the material and economic practices of Jim Crow. But the mechanisms of white supremacy were nonetheless present in the interviews and hampered the lives of Black sea islanders in tangible and painful ways. Almost all described depressed wages, exploitative lending practices, exhausting labor, and racist attitudes from white South Carolinians—and some also mentioned the 1893 hurricane.

After the 1893 hurricane, Kiser reported that St. Helena islanders observed that "many youngsters left in order to help their families" as they had trouble making a wage on the sea islands.[41] The hurricane, Kiser asserted, had triggered an early form of emigration from the sea islands that grew in the years to come. This state of matters was worsened by the limited land available for inheritance by the 1890s, as the "gross acreage of tillable land remained practically constant and there had been few annexations to the small tracts originally bought by ex-slaves. Each child of the family as co-heir could not expect to inherit enough land for a farm of his own."[42] These young people leaned into the flexibility of their community's economic practices, which did not necessitate the constant presence of every family member, at the expense of the family economy's primary function, which was to nurture a home hearth for Black sea islanders. Kiser found that Black sea islanders described the hurricane as reducing them to a condition of real need, with a psychological effect that Kiser described as "discouragement."[43] The hurricane revealed to many Black sea islanders, especially young people, that the situation there would not likely improve over the long term.

Black women in particular characterized the hurricane as a critical turning point in their lives and in their decision to migrate, especially because the poverty that came after the storm threatened single African American women, whose opportunities for employment were limited. Hannah Pinkney, who left the island in 1896, cataloged a wearying list of tragedies. Her husband died and left her with three young children to support just before the hurricane destroyed her whole crop of cotton, corn, and sweet potatoes.[44] Though "the ensuing winter was the hardest she has ever experienced," she was able to lean on a network of family and friends to survive and plant the next spring—only to find the land too wet to yield.[45] She tried the next year but failed again. With no other viable options, she moved to Savannah with her family, unable to maintain a farm on her own under the circumstances of bad weather and sodden earth. Families left their homesteads because they had no choice.

Two other women's stories also serve as a reminder that the hurricane caused lasting illnesses, killing unknowable numbers of people many years after the storm hit the coast. Essie Roberts recounted how she and her family barely survived the storm and suffered for years after.[46] Her husband, who had thrown his weight against a wall to hold the house in place against a tree that prevented it from sweeping out to sea, had his skull cracked open by a falling beam. That injury changed everything for the family. Unable to work because the injury instigated a period of "declining health," Roberts worked as a maid in town for one of the Red Cross volunteers to support her family for a time, but her husband's illness required her return. He lingered in poor health for years until he died in 1901, when Roberts took a permanent job as a maid for the Christensen family during their yearly summer trips to Brookline, Massachusetts. Roberts's cousin, in whose care she had placed her children, eventually sent them to her up north once she was settled.

Viola Ware, a young African American woman who lived with her father and mother when the hurricane hit, described how the hurricane exacerbated her father's asthma.[47] The chronic illness worsened to the point that he was unable to work, and her mother also grew ill. Ware served as their caretaker until 1899, when a cousin in New York invited her to visit. Ware "thought it might be a chance to help out at home if I could get away and make some money. Truth, though, I didn't think I was leavin' for good when I left."[48] Except, of course, Ware did leave for good, like many other Black sea islanders around the same time. She recalled that "the situation on [St. Helena] Island that caused me to leave more than anything else was the 1893 storm."[49]

Many other St. Helena islanders did not directly mention the hurricane in their reasons for departure, but they did describe conditions of economic downturn, exploitation, and stagnation that the hurricane had substantively worsened. One man, who left St. Helena in 1904 as a twenty-year-old, recalled a paucity of jobs and white merchants' abuse of credit: "Work was scarce—couldn' make a devilish lot. What you make you couldn't hardly get it off. . . . Grocery man took it all."[50] Another young woman, Rebecca, also noted the harmful practices of white merchants on St. Helena:

> Folks ain't got no means of makin' no money on the Island. 'Nother thing is this: When people start out to make a crop, lot of 'em don' have enough money to buy seed. Well, they borrow money from one of the merchants on the Island. These merchants furnish 'em little stuff to eat till crop is made. You know, corn meal, flour, molasses. Might have to sell 'em a few little tools too. Well, all of this don't amount to much expense but still sometime people don't make enough to pay for it if they don' make nothin' off the crop. What the merchant do then? Well, before he loans any money, he takes a mortgage. If he just loans a little for seed, he takes a chattel mortgage. Well, some of 'em is kind enough to let the debt run over another year, but most of 'em take your animal. Then what you gonna do?[51]

Sometimes taking control of the property title was unnecessary. Removing the primary means by which a landowner could work the earth was sufficient. A young man and World War I veteran named Joe who migrated in 1909 said the same: "You have had a bad year. You owe for your seed or fertilizer or rations for the year. You can't pay it if you don't have it. Then what are you going to do? Well, they will take your cow or your pig or your ox or your mule or your cart. What are you going to do next year if you got no mule? What are you going to eat if they take your cow or pig? That's the time when people get disgusted and discouraged and have to move."[52] He also felt that the few whites who had arrived on St. Helena Island in recent years had an eye toward exploitation, more so than the whites who had lived there since the Civil War and Reconstruction. He spoke kindly of Mr. MacDonald of the Corner Store, noting that he "would always carry you over if you couldn't pay him at the end of the year." But the new storeowners had fewer scruples and were not as inclined to be generous. Within the context of an island, from which it was difficult to reach the mainland for supplies and where it was easy to obtain a monopoly over local markets, a small number of whites could exercise an outsized influence over the lives of Black sea islanders. After so many bad years on the sea islands after the 1893 hurricane, such seasons must have come with regular frequency to the region.

Other Black sea islanders found the endless hard labor wearing, especially within the context of economic, environmental, and political precarity that the hurricane had heightened. Rebecca said that she left "simply to better my condition."[53] One woman, who migrated in 1913, declared, "I never did, even when I was a child, have no intention of farming all my life. I never worked much on a farm. Father never did compel me to go out and dig in the ground."[54] The labor would yield little, she knew: "I could look around and see how things was going. In summer while you was working your back off in the cotton field, prices would be tree high. Then when you finally pick your cotton and carry it to the store, prices would be down so low that after you sold it, you wouldn't get enough to run you through the winter."[55] A man who migrated in 1917 when he was seventeen had made a similar calculation: "I could work and dig all year on the Island and best I could do would be to make $100 and take a chance on making nothin'. . . . So I decided I'd be a fool to stay there n' dig all my life at that rate. Payday, little as it was, would come once a season."[56] All of these factors backed Black sea islanders into a corner.

Young sea islanders also found life on St. Helena confining in comparison to the entertainments and prospects that cities could offer. Another man, who also moved in 1917 but at the age of thirty-one, saw out-migration as inevitable: "The people will continue leaving the South and will continue coming North just as long as they have better opportunities in the North. It's getting too easy now for people to travel to be willing to tolerate hardships which they can remedy by picking up and travelling a few hundred miles. And it's just due to the narrowness of the southern people that they don't provide a better means for the people to make a living."[57] One man also attributed migration in part to a rebellious, youthful resentment of a static, unrewarding setting: "Young people grow up now and say, 'I want to get away from here. No diggin' in the soil for me. Let other man do the diggin'. I'm through with farmin'. I would be a fool to move back. Have to be out in the hot sun diggin' in the soil. I can't see it.'"[58] The constant cycle of hard labor, debt, and little cash was grinding. Black sea islanders had difficulty mustering reasons to stay because they faced increased exploitation at the hands of white supremacists and diminished economic returns, with little relief.

Some articulated a vision of an oppressive South ruled by hostile whites. A woman who moved to Harlem in 1910 when she was twenty-one told Kiser, "All southern people are against the colored. Been tryin' to hold them down. Want to keep 'em diggin' all their life."[59] One young man presciently predicted that "thirty years from now, the Island will be owned by the white people. . . . It will no longer be an Island owned by colored people," in part because of the soaring taxation rate in the region: "Look what taxes are doing. When I was down there our taxes was around four dollars. Now they jumped to fifteen."[60] A woman who left in 1895 as a fifteen-year-old saw the situation as intractably marked by ill treatment at the hands of whites: "I have a kind of respect for the Island, but since I moved away and had my eyes opened I've just got disgusted with the merciless treatment the folks get down there. They work hard, and get nothing for it. . . . I could never see why the white people of the South want to keep their feet on the necks of the blacks. They want the Negroes to stay down there, but they want them to stay down. They won't give a good man a chance."

Hard work and high prices increasingly marked Black life on the sea islands and, for those who felt compelled to leave, obscured the benefits to living within the rural communities of the Lowcountry. Black sea islanders, especially of younger generations, found themselves more and more reluctant to do back-breaking manual labor as economic opportunities contracted, political rights diminished, and environmental perils multiplied. Because of the hurricane, "successive years of poor crops and the decline of phosphate mining," Kiser attested, "effected a gradual change of attitudes," pushing Black sea islanders from their homes and their hard-won land.[61] Black sea islanders became refugees from their own homes, driven out by an array of environmental, economic, cultural, and political forces.

With the sea islands quieter, their population small and their economic future bleak, white coastal residents turned to the power of history to deal with the profound changes that the region was undergoing. White landowners, the descendants of the enslavers, grasped at fantasies of masterly control of the Lowcountry landscape and Black labor alike. In the early 1920s, James Henry Rice waxed in florid prose about what the Lowcountry's history of slavery meant to wealthy white South Carolinians:

> In full flower it was the wonder and envy of mankind; even in its ruins it appeals alike to the cynic and the historian. Youth and beauty, splendor and power, combined with great wealth and lives of chivalric lustre, have always interested the human race and it is likely they always will. The rice planter is no more, but what he gave to civilization in grace and beauty of living, with so many elements that go to form the man of full stature and the splendid ruin of his past, these things will not perish, but will live as long as the coastal region harks back to imperishable glory.[62]

Rice, as described earlier, found the 1893 hurricane culpable for the decline of the Lowcountry.[63] The maintenance of Lowcountry cultivation had once meant the maintenance of white supremacy. Though white South Carolinians could not force the continuation of Lowcountry agriculture as they had under enslavement or even before the Great Migration, they could and did develop a narrative of Lowcountry history that assuaged their racist anxieties over slavery's end and the exodus of Black workers from the region. They fashioned histories that featured the ingenuity of white "planters" in the agricultural innovation of rice cultivation and the beauty of the Lowcountry under enslavement.[64] These stories, a Lost Cause of the Lowcountry, gained popularity in the 1920s, as both agriculture and the Black population in the region collapsed from the long, slow grind of environmental change that the hurricane exacerbated.[65]

These reimaginings of the history of Lowcountry rice cultivation muted the harsh conditions of labor in the region's fields. This was a deliberate act. Elizabeth Allston Pringle, known as the "woman rice planter," was deeply sympathetic to Rice's ideas. White Lowcountry residents in the early twentieth century added their own interpretation to the Lost Cause mythology, in which white southerners enshrined the nobility of the Confederacy and slavery to justify their postwar creation of the racist regimes of Jim Crow. They connected environmental control of the Lowcountry with white supremacy. In the region, the emergence of Jim Crow was as much an environmental phenomenon as a political one. White landowners' drive to destroy the political gains of Reconstruction was matched by their desire to control both Black labor and coastal landscapes. But as hurricanes wore away the commercial production of rice and African Americans migrated away from the Lowcountry, the white elite turned their attention from the material to the memorial, forging a narrative about the beauty of Lowcountry agriculture that cast a glossy veil of moonlight and magnolias over the brutalities of enslavement in the past and the difficulties of labor in the present.

Before Pringle had to relinquish rice as a crop in the 1910s, she attempted to control Black labor for growing the crop, showed a keen disdain for the African Americans who worked her land, and displayed a sense of determination to preserve aspects of the white supremacist regime under which she had grown up. In practically the same breath as her complaints about Black workers as "idle" and "shambling," she described the Lowcountry landscape in romantic terms.[66] Her cousin, Alice Ravenel Huger Smith, was famous for her watercolors depicting the Lowcountry in soft pastel colors, in which Black workers serve as picturesque accessories to the salt marshes and forests of the coast.[67] Pringle frequently described scenes as though she were passing them along to Smith for artistic inspiration. The autumn rice harvest, she wrote, "is the gayest week of the year."[68] She smiled approvingly at the "real harvest weather—crisp, cool, clear; and the bowed heads of the golden grain glow in the sun." Against the backdrop of this scene of vivid blue and gold, "men, women, and children all carry what look like immense loads," Pringle wrote, "on their heads, apparently without effort." In Pringle's illusory Lowcountry, even child labor was easy, natural, an intrinsic and harmonious part of the rice-gentled landscape.

This was a profound cognitive dissonance given both her own rancor toward Black workers, most of all when they asserted themselves, and the arduous reality of labor in a rice field. But in the Lost Cause of the Lowcountry—which

Pringle reified in her writing and Smith in her insipid paintings—Black workers labored without thought or care, rice thrived, and the racial hierarchy that wealthy white women benefitted from and perpetuated could remain stable. There was no disconnecting any of these elements. They were part of a cohesive white supremacist mythology that centered on the falsehood of the simple, undemanding cultivation of the fecund Lowcountry environment through rice and passive Black labor.

Rice, again indulging his purple prose, likewise declared that at rice labor camps, "people dwelt in dreamland, with asphodel, roses and all manner of delights."[69] The enslaver was not a vindictive wielder of the whip but "an intellectual man, fitted by reading and by travel to entertain men of varied tastes." Ambrose Gonzales, a close friend of Rice's, took a similar tone. He mourned the emptying of the Lowcountry and the ruination of its labor camps, musing in his newspaper *The State*:

> The great canals of the tidal region, whose waters once flooded broad fields of golden grain, have long been "gathered back into the lonely arms of the forest." The stately mansions, whose gracious hospitality these fields sometime supported, crumbled into dust and ashes during the "March to the Sea," and are now but ivied ruins! Of quondam slave and slaveholder, few remain; but, between these, as with their descendants to a lesser degree, there is kindlier feeling than exists between blacks and whites anywhere else today. . . . Slaveholders were never slave-drivers—[but] men and women whose lives were ruled by the self-imposed maxim, Noblesse Oblige![70]

This selective memory of the Lowcountry cast the region as refined, with a cultivated, elegant conjoining of environment and economy, rather than a vast slave labor camp that kept hundreds of thousands of African Americans captive, killed infants and children at horrific rates, and broke the bodies of young Black men and women. In this perverse, shared fantasy, gorgeous gardens surrounded the lovely big house, and the golden, bowed heads of ripe rice lined orderly fields, which were tended by enslaved African Americans whose humanity blurred into something insignificant and generic. Indeed, Alice Ravenel Huger Smith often literally erased the individuality of African Americans, rendering their faces indistinct brush strokes of brown paint in watercolor after watercolor.

This false impression of mastery garnered power from the Lowcountry's reduction in circumstances. White men and women from the Lowcountry shook their heads at the contraction of cultivated acreage and the departure of thousands of African Americans, who fled because of diminished economic opportunities, destructive hurricanes, and the acceleration of Jim Crow. These changes meant the decline in white fortunes and in a social position that they and their ancestors had jealously guarded through violence and paranoia for centuries. To compensate, white Lowcountry residents forged a narrative of the Lowcountry that exalted the so-called planter class and the bitter fruits of enslaved labor. They told themselves and anyone who would listen pretty stories to drown out the cacophony of evidence that they and their ancestors were wrong and that they had been cruel.

Despite the dominance of white supremacist ideologies that painted glossy pictures of Lowcountry agriculture under enslavement, Black South Carolinians living through Jim Crow also articulated their own histories and experiences of labor that pushed against white landowners' characterizations. African Americans constructed their own narratives that make it clear that work in Lowcountry fields was difficult and dirty and that they often did it only because they were made to do so, like the many St. Helena islanders who pointed to the grinding nature of farm labor as a motivation for migrating away from what had been home. No lovely pastoral tableaus appear in these narratives. Instead, Black South Carolinians' histories focus on the harshness of labor in the region, the desperate need of white landowners for Black labor, and how African Americans strove to make postwar systems of farm labor work for them, as best was possible under the circumstances under Jim Crow.

These histories would have been unwelcome to white South Carolinians, intent as whites were on a dual project of depicting the slavery of the antebellum South as a stable system of economic prosperity and of subduing Black resistance to Jim Crow. Mack Taylor, a freedman living in Winnsboro in the 1930s, demonstrated that Black South Carolinians had acute memories of the stringencies of enslavement, arguing that his ancestors "was fetched here 'ginst our taste. Us fell de forests for corn, wheat, oats, and cotton; drained de swamps for rice; built de dirt roads and de railroads."[71] Gabe Lance of Sandy Island gave a similar history, recounting, "All dem rice-field been nothing but swamp. Slavery people cut canal and dig ditch and cut down woods—and dig ditch through the raw woods. All been clear up for plant rice by slavery people."[72] White South Carolinians' Lost Cause of the Lowcountry twisted slavery into something unrecognizable to Black South Carolinians and used that fallacious narrative to feed the ideology of Jim Crow as a necessary hierarchy ordering the labor and political control of Black southerners. Black South Carolinians in the meantime kept histories alive that made vividly clear what labor in the Lowcountry truly meant to them.

The physical remnants of commercial rice production remain, etched deep into the riverside mud by centuries of African-descended laborers, enslaved and free, carving dikes, ditches, and dams out of the earth. Some fields are grown over with grass, into which ducks and waterfowl nestle and roost—others are flooded with murky water or studded with invasive water hyacinth. From above, the straight lines of the rice fields appear at odds with the sinuous, muscular curves of the rivers. From the ground, the fields look like sodden grasslands or the floodplain forests that they originally were, leggy cypresses and tupelos reaching for the heights of their ancestors. Hurricanes may have flung open the gates, caved in the trunks, washed out the dams, and filled ditches with thick silt. But the white supremacy that ordered African Americans back to the ruined fields, demanded control over Black labor, and resented Black expertise on rice broke the back of the "rice kingdom" that its logic had created in the first place.

Decades after the hurricane of 1893 swept the sea islands, South Carolinians grappled with its consequences. Sometimes its legacies—of chronic illness, economic precarity, or soured soil—indelibly marked the lives of African Americans. Sometimes its traces were more faint and tangential, in its erosion of industries and agriculture already on the brink of decline. And sometimes the hurricane became part of an accretion of oppression, yet another catastrophe among others that made getting by harder for coastal African Americans. They responded in many ways. Some tried to manage the damage incurred, exploring cooperative solutions in conjunction with federal assistance. Others found their livelihoods too diminished for staying to be viable. White South Carolinians regarded the departure of African Americans and the reduced power of those who remained as a victory, though at the same time a reduced labor force and decades of environmental catastrophe had consequences for their economic well-being—whether they admitted it or not.

Children of the Civil War, the white elite may have thought that they understood what loss and change felt like. But as African Americans fled the Lowcountry, riding the rails and steamships for Harlem and Philadelphia, white landowners began to feel something sinking, a seismic shift, as the terrain they once knew rearranged itself under their feet. Without dominion over labor camps full of Black workers, who were they? Rather than grapple seriously with that question—rather than examine how their experiences of power had warped not only their souls but their whole society—they dug their heels into the porous soil of the Lowcountry; retrenched into their townhouses in Charleston; sunk their claws into local, county, and state institutions; and crafted misty, cobwebby myths about the golden life of what they called the rice kingdom. But a Lowcountry leeched of its Black majority was becoming an empty place.

If the hurricane had not struck the Lowcountry of South Carolina, would coastal African Americans' flexible economy have crashed? Would the commercial production of rice have shriveled and phosphate crumbled? Would African Americans have begun to move away from the sea islands in the first few decades of the twentieth century? Would elite whites have constructed racist mythologies of antebellum life in the Lowcountry? It is easy enough, looking at similar demographic, economic, and social processes elsewhere in the South, to say that yes, these things would have come to pass, some way or another. And yet that is a deeply unsatisfactory counterfactual that does not acknowledge either the material or the memorial damage that the hurricane inflicted. For, of course, the hurricane did strike the coast, hurting rice and phosphate irreparably and, with those industries, African Americans' economic practices. It did sicken, kill, and impoverish thousands, and leave the survivors with wounds both physical and psychological for the rest of their lives.

The storm's long-term effects took many forms. For some, the hurricane, a jarring event, was an easy signpost that Black and white South Carolinians alike could point to when trying to make meaning out of the trajectory of their lives and that of their communities. For others, it was the axis on which their fortunes turned, a moment that wrecked their families and the livelihoods on which they had come to depend, forcing them to adapt and to rebuild their notions of what survival meant. And for some, the hurricane faded in and out of their conscious understanding of what shaped their lives—capturing meaning at certain moments, losing significance at others. As Lowcountry residents wrestled with the sweeping environmental, political, economic, and demographic transformation overtaking the region in the late nineteenth and early twentieth centuries, the hurricane played a role not only in the material changes of the region but in the many conflicting narratives that coastal South Carolinians cultivated for themselves about those changes.

The Great Sea Island Storm of 1893 struck at a moment of extreme political and economic precarity, damaging cornerstones of African Americans' livelihoods. The hurricane gradually—though not inevitably—coalesced with the pressures of white supremacy to force an out-migration of Black sea islanders. Those who stayed would continue to face hurricanes, white supremacists, and, eventually, corporate developers, which would all challenge their livelihoods and their independence. The sea islands, slivers of sandy earth that swell above the marshes and ocean that surround them, do not seem sturdy enough to shield Black sea islanders' cultural traditions and political legacies from the bruising blows of nature and politics. But for Black sea islanders then and their descendants today, the idea of picking up and leaving never was and never will be simple.

Epilogue

Quash Stevens, after the Storm

Quash Stevens found his world fractured. The hurricane swept past Kiawah Island, rendering it "Destersory."[1] He and his wife Julia had suffered through a terrifying night as the hurricane raged around their house and knocked down one of its chimneys, perhaps damaging the diploma from the Hampton Institute in Virginia that Julia had proudly hung on the wall to commemorate her 1871 graduation.[2] Quash wrote to his would-be sister-in-law Adele, then his primary correspondent among the white Vanderhorsts, in Charleston of the destruction. The Vanderhorst "big house" from the early 1800s had lost half its roof. Barely a drop of fresh water was to be found on the island for three days, forcing Quash and his family to cook with salt water until rainfall saved them. Every family had lost their food stores. Eleven houses were flattened, as was every chimney on the island. Miles of fencing had been swept away by the flood. On Seabrook Island, across the inlet, a man lost his wife and five children.[3] The cotton crop was ruined, with only a little left that was "yealer," Quash dismissing it, "Doint Tink it Worth Pickin."[4] There was almost nowhere to store cotton anyway, since nearly all the barns on the island were gone. Fortunately, at least, the stock had survived the night. It was difficult to discern a way forward on Kiawah.

After three weeks of hunger, thirst, despair, and deafening silence from Charleston, Quash wrote reproachfully to Adele: "I have not Horde from you since the greaite stome. I rote to you Buth got no replie."[5] Perhaps Quash thought that Adele would send help, but Charleston's relief committee came to the island's aid first, when some of the men from Kiawah traveled into the city to fruitlessly find work. Adele knew they were already suffering. Less than two weeks prior, Quash and his family had been struck down by chills and a fever—maybe malaria, since Quash called it "the Woirset attack I Had for 20 Years."[6] With the hurricane came deteriorating conditions: "the Place smeile offowl and I Fear We Will Have graite siken," and besides that, Quash wrote, "I Still Have Sickns."[7] "I see Nothing Left for Ouest Buith starveishen or Leave Some other Parte of the Countery," Quash lamented, his reproof to Adele for abandoning him and Kiawah in their hour of greatest need.

Quash had always lived in this in-between space, surviving on the marginal benefits that the Vanderhorsts dispensed and for which he vied. Quash, born in May of either 1840 or 1843 on the Round O rice labor camp on the Ashepoo River, was never acknowledged as the enslaved son of his powerful and wealthy enslaver Elias Vanderhorst, and the Vanderhorsts kept him enslaved until the Thirteenth Amendment forced their hand.[8] While he had another first name, James, that appeared occasionally in records, he preferred to go by Quash, a day name of West African origins meaning "Sunday."[9] Perhaps his enslaved mother, herself from South Carolina, sought to honor the traditions of parents or grandparents who had been born in West Africa.[10] Quash somehow survived to adulthood, as the fetid, mosquito-infested rice fields of the Lowcountry were death traps for enslaved children.[11] When he reached adulthood, Quash served his white half brother Lewis Vanderhorst during the Civil War after Lewis enlisted in the Charleston Light Dragoons.[12] The company comprised well-to-do white men who brought their enslaved servants to war to provide them with a semblance of their life as privileged sons of the Lowcountry aristocracy. Elias's wife, Anne, then deeded Quash to another one of his half brothers, Arnoldus IV, in August 1864.[13] When the Thirteenth Amendment was ratified in 1865, we do not know how Quash met emancipation. He had spent the war being traded among his erstwhile half-brothers as a piece of family property. He did seem to have seen his connection to the Vanderhorsts as the most likely option for making a living, especially since he married shortly after gaining his freedom and now had a family to consider.

Shortly after the war ended, Quash asked Arnoldus IV and his wife Adele whether he should return to Round O or go to Kiawah. "Write and let me know I think we can do well on ether one of those places raising Stock," Quash wrote in his steady cursive and phonetic spelling, which hinted at what was likely a Gullah dialect.[14] Quash's father, Elias, wondered whether emancipation would break not only Quash's bondage but also his ties to the family. He wrote to Arnoldus IV, "I hope Quash remains faithful," still focused on what Quash could do for *them*, not in what they owed to *him*, how Quash might have envisioned his freedom, or what it might mean to Quash to keep that faith.[15]

Either to keep Quash close to the family, of but not in it, or to offload a challenging property, the Vanderhorsts sent Quash to Kiawah in 1866.[16] Maybe, when he arrived on Kiawah after the Civil War with his new wife by his side, he breathed in the ocean breezes deeply, a freshness rare in the rice lands of his childhood, and hoped that their future children would grow up strong and healthy. There he stayed for decades as the self-declared "planter" of Kiawah: this was a title usually reserved for rich white men, suggesting that he imbued his life there with a stature not afforded him either by his white family members or white southern society.[17] He and his wife, who must have married as soon as the war ended, had at least four children—Eliza, William, Annie, and Laura—born between 1865 and 1873.[18] At some point between the birth of Laura and 1880, his wife passed away. In 1881, Quash married Julia Gibbs, a thirty-five-year-old schoolteacher from Charleston.[19] Julia either bore no children or none who lived for long. She may have suffered from miscarriages or illness at childbirth, as suggested in an 1883 letter in which Quash wrote to Adele, "My Wife Is still sick and yet over Her Troble."[20] Kiawah was not an easy place to live for Quash, his family, or the other few dozen Black tenants there with them, as he may have hoped it would be.

But Kiawah did represent a possibility for Quash. The island stretches from the Stono to the Kiawah River, an endless sandy beach facing the ocean to the east, and broad flats of softly shaded gold, brown, and green salt marshes to the west. On its south end, a teardrop of land falls into the Kiawah River's inlet, where dolphins nose their way up to hunt in the rich estuarine waters that surround the island. To the north, ridges of ancient sand dunes forested with laurel and live oak, loblolly pine, and palmetto spread like fingers pressed into the pluff mud of the salt marsh. The nine-mile-long island, barely a mile and a half wide, had once been a refuge for the Kiawah people. There, they distanced themselves from the English invaders who arrived on Albemarle Point in 1670 before establishing their permanent settlement of Charles Town on the peninsula opposite the point in 1680.[21]

As European diseases, warfare, and dispossession laid waste to Indigenous peoples in the Lowcountry, white colonizers set cattle loose to graze on stolen, emptied Indigenous land, including Kiawah, for little cost and effort.[22] By the American War for Independence, enslaved laborers on Kiawah grew and processed indigo; produced provisions, including tanyas, a starchy West African root similar to taro; and herded cattle.[23] From the late eighteenth until the twentieth century, a mere third of the island was cleared for crops.[24] The rest remained wooded with maritime forests, red bay, wax myrtle, magnolia, and holly tangling in a dense understory below the low boughs of pine, live oak, and palmetto.

Amid the war, the northeastern half of the island came into the possession of Elizabeth and Arnoldus Vanderhorst.[25] Arnoldus died in 1815 and deeded the property to his children, including Elias, Quash's father.[26] Arnoldus had himself fathered at least one son with an enslaved woman, Hagar Richardson, and made his white children's inheritance contingent on manumitting Hagar, her children, and her father.[27] Elias complied. From his father, he received half of Kiawah and the rapacious belief that a rich white man could take whatever he desired from his human property. Kiawah, though, was not the boon that Elias may have hoped it would be. He struggled to turn a profit from the island's operations. "Everything goes wrong here," he bemoaned in 1840, just before Quash's birth.[28] "Perhaps it will drop right into the sea one of these days," his wife, Ann, prophesized of the "forlorn island."[29]

Kiawah, almost completely under Vanderhorst ownership by the late nineteenth century, was the perfect isolated property to lease to Quash. Elias died in 1874, leaving Kiawah to his son Arnoldus IV, Quash's half brother.[30] Arnoldus died in a hunting accident on Kiawah in 1881, the fault of a shotgun with a malfunctioning trigger that Quash had warned him about—and whom Arnoldus had fatefully ignored.[31] Kiawah then became the property not of Quash, who had successfully managed Kiawah's forty to fifty Black tenant farmers, the production of cotton crops, and stock raising for fifteen years, but Arnoldus IV's widow Adele and their disagreeable son Arnoldus V. Once again, Quash was passed over.

The Vanderhorst family always kept Quash at arm's length. Quash penned notes to his family, asking them to write him more frequently, to communicate with him more clearly, and to allow him to see them more often. Of the hundreds of letters he must have written to the Vanderhorsts, twenty-eight survive.[32] Quash was frank, at times plaintive, both about troublesome conditions on the island and his desire to be treated with more respect. "I Havenot Hord From you Since I sine you in the City," he reproached "Mis Adall" in 1887; "I Weaint To the City on Tuesasday . . . and Wais soperise Not to Fine you theair if I hade Knowen That you Wais going Soo soone I wood Come Doine at owence," he chided her in 1893.[33] He provided for them—sent them the island's cotton, butter, cream, cows, lambs, blackberries picked by his children, even cash—and they responded infrequently enough that he was left asking for explanations.[34]

Family was a troubled concept between Quash and the Vanderhorsts, and the Vanderhorsts deliberately kept Quash insecure and off-kilter. Or, maybe just as painfully, they simply didn't think about him, except when necessary. Quash may have guessed at the Vanderhorsts' indifference, whether intentionally feigned or an inherent apathy. His frequent admonishments suggest that he did and that he wanted them to recognize their callousness, to see themselves for who they were, and to realize how they ought to treat him. The hurricane only increased this friction between Quash and the Vanderhorsts, especially because Adele failed to respond with appropriate urgency to the desperate conditions on Kiawah after the storm.

Even with the derangement of the hurricane and mounting proof of the Vanderhorsts' indifference, Quash did not yet abandon Kiawah. But as the years passed, Quash did not gain any optimism about the future of either

Kiawah or South Carolina. A year later, agriculture and politics both looked grim. The crops, cotton, and corn "is Loste" after unending rainfall that left "much f the Lands" "ondr Waatr."[35] He complained to Adele that he had to ride fifteen miles to vote, only to be blocked by hostile poll managers. He despised Tillman, writing that with Tillman's political ascendance, "the Hole state is in a Bad Fiex . . . No Man Life is safe Now." A year after that, Quash was desperate for information, fearing that Adele and her son Arnoldus V were planning to sell Kiawah out from under him. "I Ham Sorey that I did Not Know soner That you Wais in [Charleston] as I Wood Like to Have Talk With you," Quash wrote, assuring Adele that "I Will Doo all that I can to make [Kiawah] Pay you as Well as myself."[36] He asked Adele to give him first preference if she was indeed planning to sell and wrote beseechingly, "Now Mis Adall Pleais rite and Lete me Know so my mind Can Bee at Araist." Though white visitors to the island may have playfully (and patronizingly) referred to Quash as "the Cassique of Kiawah," it must have been increasingly obvious to Quash that his work on Kiawah was not valued by the Vanderhorsts and that he was as unprotected from dispossession by the people who should have claimed him as family as he was from the hurricanes rolling in off the ocean.[37]

Quash's days on Kiawah were numbered. Arnoldus V, stubborn and abrasive with little regard for a man who under different circumstances would have been his uncle, chafed at Quash's level of control over Kiawah and began to encroach on Quash's management.[38] The paternalist condescension of the older Vanderhorsts gave way to single-minded segregation in Arnoldus, who wanted to excise the Stevenses from the Vanderhorsts. Quash was running out of patience for Arnoldus. He threatened to leave in 1900, but Adele convinced him to stay another year.[39] He reluctantly signed the one-year lease.[40] That year was not a smooth one. Arnoldus picked a fight with Quash, hiring a new manager for the stock from among the remaining Kiawah islanders and threatening legal action over cattle that Quash had sold.[41] Quash worried over money, a vicious drought, sicknesses that laid his family low, and how he would pay rent if another hurricane struck.[42]

Just months before the lease was set to end in January 1901, Quash composed a letter to Arnoldus that revealed his wounded feelings and his sense of pride in his comprehensive knowledge of Kiawah. Quash reminded Arnoldus that he had informed him of his intentions to leave Kiawah some time ago, for "I saw no way [to stay] under present condition." "I have been managing [the island] for 34 years," Quash wrote, a simple statement weighted with decades of labor, sacrifices, and loss unrecognized.[43] In a final appeal to Adele, Quash attempted to communicate both his side of the story and the irreparable damage to the bonds between him and the Vanderhorsts. He wanted to compel her to understand what he had done for her and that he did not make the decision to depart lightly. "I Have Bieain in your sorves For Neaily Fortey years and Have Doin the Beaist I Know How," he wrote.[44] His letter reflected a man with long and deep patience, at the very end of what he could tolerate.

Whether Adele responded was not recorded. Arnoldus would not compromise. The Vanderhorsts did not care. Quash left Kiawah. As soon as the Kiawah lease expired, Quash and his son William bought Seven Oaks, a large farm on the far eastern edge of Johns Island, for $3,000.[45] Quash's immediate reason for leaving is evident—Arnoldus V's meddling with Kiawah, which Quash took as a personal insult—but the full complexity of his feelings is unknown to us.

Something in Quash had shifted. Certainly the fight with Arnoldus eroded any last wisps of confidence Quash had in his business relationship with the Vanderhorsts. Perhaps William, a graduate of Claflin University in Orangeburg, South Carolina, and by then a thirty-two-year-old blacksmith, helped convince his father to let go of Kiawah and the Vanderhorsts.[46] William was newly married, and it may be that Quash wished to give his son a fresh start away from the control of the Vanderhorsts. Maybe tragedy had poisoned Kiawah. By 1900, Quash had lost his first wife and fathered several children, but it is not clear how many lived to adulthood, and he and Julia suffered from recurrent health problems.[47] Any hopes for a healthful climate on a sea-facing island to nurture well-being in their sons and daughters had been dashed.[48] It is also possible that Quash had relinquished any aspirations to being treated by the Vanderhorsts with respect, much less as a branch of the family.[49]

Whatever the reason, in 1901, Quash, Julia, William, and William's wife, Lilla, a twenty-nine-year-old schoolteacher, left their family settlement on Kiawah and moved to their new farm on the Stono River.[50] He had little further communication with the Vanderhorsts, at least that the Vanderhorsts retained. If Quash ever wrote to Adele again, the letters did not survive.[51] In his last letters to Arnoldus, he asked for a fair price for his cows on Kiawah. "I have offered you the cattle as low as I Possibly can take," he wrote, "Pleais Lete me Have the money Due me as I aim in much Neede of it."[52] The Stevenses would have needed cash to operate their new farm. Johns Island was, at least, familiar territory, just across the Kiawah River from their old home. Julia had already established a reputation there as an educator. Along with the young ones of Kiawah, she had taught children on Johns Island to read and led Sunday school classes there for many years.[53] The farm was farther inland, less susceptible to a hurricane's storm surge, with more access to fresh water than Kiawah. It was a couple miles from the bridge over the Stono River connecting Johns Island to James Island, shortening the trips that the family took into Charleston, where they had cultivated friendships and community ties.[54]

There, the Stevenses rebuilt their lives. Quash spent a decade at Seven Oaks while William and Lilla began a family, Lilla birthing four boys who all had the chance to meet their grandfather. Maybe during those years of Quash's life, he felt free, easier, and happier, more so than he ever had. Perhaps he struggled to rid himself of resentment he may have felt toward the Vanderhorsts, to cut those blood ties and their hold on him for good. Maybe some days he succeeded, and some he did not. His family with him at Seven Oaks—his wife, son, daughter-in-law, and grandsons—must have been a balm for whatever sorrow or anger may have gripped his heart as he reflected on his six decades at Round O and on Kiawah. Maybe he told himself that Seven Oaks and the Stevenses were his present, not Kiawah and the Vanderhorsts. But little record of those years remains in a physical archive. He reappears into our view when he and William sold Seven Oaks for $3,500 in 1910, likely due to Quash's declining health, though it seems the family stayed on the property.[55]

Quash died of heart failure at Seven Oaks a few months later on March 20.[56] He was buried in the graveyard of the Centenary Methodist Episcopal Church in downtown Charleston, a historic African American church established in 1866.[57] William continued to farm on Johns Island until his own untimely death, sometime before 1920.[58] Julia subsequently moved back to her hometown of Charleston to continue teaching.[59] Lilla and her four sons, James Quash Jr., William Jr., John, and Harold Arnoldus, traveled to her parents' comfortable, middle-class home in Columbia.[60] At least two of Quash's grandsons, William and Harold, eventually journeyed north, hoping, like so many other Black southerners, to escape the relentless violence of the South and to follow career opportunities.[61] Julia, though, stayed in Charleston, the city of her birth, where she passed away in 1933. She was buried at the Centenary graveyard too, perhaps near her husband Quash.[62]

Without Quash, Kiawah's numbers dwindled. The Vanderhorsts had made the island an unpleasant place to live. Through inattention and malice, they chased away Quash and his family, who had poured so much of their lives into the community there. Many Black families followed. Having pushed out Quash, Arnoldus V ran through schemes to keep Kiawah in what he considered the legitimate family line, hawking it as a winter hunting resort for rich white northerners, ordering the island's Black residents to harvest palmetto fronds for Easter services in Charleston, advertising for overseers who were all ultimately unsatisfactory, and renting large pieces of land for pasture to whomever he could snag.[63] He wished to maintain the island as a heritage property, and he saw it a tangible embodiment of his family's now-diminished wealth and power.[64] Adele, it seems, had washed her hands of her son Arnoldus V and Kiawah's management and passed away in 1915.[65] Arnoldus V died in 1943, unsurprisingly with no children and no heirs.[66]

Because of Arnoldus V's ineptitude, animosity toward Quash, and self-defeating obnoxiousness, Kiawah passed out of the Vanderhorst family. The island fell into the lap of Adele's last living executor of her will, William Weston. He paid Charlie Scott, the last Black resident from the old days, $100 per annum to be the island's caretaker. Scott lived until his death in the settlement on Rhett's Bluff that Quash and his family had inhabited for decades.[67] The island, hollowed out, was sold by Weston in 1950 to the Royal Lumber Company. The company's owner C. C. Royal had the land logged and established a modest beachside development, with houses constructed using salvaged bricks from abandoned buildings on Kiawah.[68] In 1974, Royal's heirs sold Kiawah to Coastal Shores, Inc., a subsidiary of the Kuwait Investment Corporation, for $17 million.[69] The corporation bulldozed almost every vestige of the island's agrarian history to build sprawling golf courses and ritzy beach houses; shipped off the surviving, half-wild marsh tackies to a slaughterhouse in Walterboro; and privatized the island, except for a public beach on the sandy south end, too unstable for constructing luxury homes.

The corporation did save the Vanderhorst "big house" from demolition. (Quash had carefully maintained the house through the late 1800s, even keeping the dining room "so arranged as to suggest preparation for an evening meal," and into which he let the other tenant farmers of Kiawah when the hurricane blew down their houses.[70]) The property, renovated in the 1990s, went on the market and sold in 2021 for $20.5 million.[71] Houses on the island, exclusive and expensive, sell for ever-higher prices—for now. So goes the history of Kiawah and many other sea islands: the islands lapsed into a cycle of economic attrition, environmental erosion, political oppression under white supremacy, and demographic contraction, which spat them into a new trajectory of corporate development, commercial privatization, and tourist reclamation under racial capitalism. The Lowcountry has been reconstituted as a new New South, scrubbed of its history for easy consumption by and immense profitability for a favored and lily-white few.

In many ways, the Great Sea Island Storm prepared the sea islands for their conquest by corporate developers sixty years later. As a significant turning point in Lowcountry history, the hurricane signaled a slow demise of the coast's agrarian institutions, for better and worse, depending on the institution and the beneficiary. As African Americans moved away, as fields lay fallow, and as landowners griped over property taxes on land that no longer yielded, the Lowcountry underwent a profound change. Wealthy northern industrialists and bankers consolidated huge tracts of coastal land for winter retreats in the 1920s, until they lost their fortunes or use for the land. Corporations then purchased vast swaths of the Lowcountry, swallowing entire islands with the permission of a laissez-faire state government that had no interest in protecting the Black landowners who were left. They furthered the damage to Black autonomy on the sea islands incurred by the hurricane and Jim Crow in the late nineteenth and early twentieth centuries.

Development and an ostentatious fascination with the Lowcountry's history would ideally be two forces at odds: but in the Lowcountry, development and tourism have historically been designed to mutually reinforce each other and to exclude the Gullah-Geechee. Today, that tide shows signs of turning. The Lowcountry's history is undergoing an encouraging renaissance of public interest. The Reconstruction Era National Historical Park in Beaufort County was founded as a monument in 2017 and a historical park in 2019 and is located among one of the few Gullah-Geechee communities that has retained some of their land. The International African American Museum is scheduled to open in Charleston in 2022. But, just as the historian Melissa Cooper has pointed out in the debates over the Gullah Geechee Cultural Heritage Corridor, officially designated in 2006 by the U.S. Congress, many Black sea islanders wondered what a heritage corridor could do to stem the pressures of rising taxes and developers hungry for land.[72] As Black sea islanders from the 1890s to the present recognized, little is of more importance to their communities than hardier protections for Black landownership, not even superseded by an institutionalized appreciation for their culture and history.

There must be a balance between celebration of heritage and material protection from dispossession. Downtown Charleston is already lost. Its workers' homes and alley-way apartments are spruced and priced up beyond all recognition, with Black Charlestonians pushed north of the low-lying Charleston Neck, once the target of the Sanitation and Drainage Commission's pet projects.[73] Some islands—Kiawah, Seabrook, much of Hilton Head—have been cordoned off by developers, with effects nearly as detrimental for Black sea islanders as the storm surge that overtook them in 1893. Islands that are not predominantly privatized—James and Johns, among others—are rapidly gentrifying, with property taxes and prices out of reach for descendants of Black sea islanders who might want to return to family land. St. Helena is the predominant holdout, with some protective covenants for Black landowners. What will be preserved in the coming years, and for whom? And of what is left from the looting of racial capitalism, what will the climate crisis spare? Black sea islanders, as always, mount lines of defense against dispossession, from their land and from their history. The latter, at least, is attracting more and more attention. The future of the former remains to be seen.

The same capitalist greed that transformed the Lowcountry in the twentieth century, writ globally, may drown the region over the next couple hundred years. The ocean's salt waves could breech the Charleston Battery for good, the way that hurricanes and king tides already do. The line between sea and horizon may draw closer to shore, and tides will creep higher until the tips of cordgrass dip beneath the water's surface. Someday the plaque describing the cascading effects of the 1893 hurricane on the Beaufort waterfront might be permanently underwater. The Atlantic could submerge the lush golf courses of Kiawah and lap at the doors of the wealthy white landowners who bought their multimillion-dollar homes with no real thought for the history that lay under their feet. Their insurance might save them, or they might sell before the island disappears. Or maybe not. As Quash did on that dark night in August 1893, they may look out from their homes one day and see only water. As Ann Vanderhorst prophesized, Kiawah, alongside the other islands lacing the edges of the South, could wash into the sea. Barrier islands are by nature flexible slices of land designed to absorb storm surges and to shift with the ocean's ebb and flow: they are slight, impermanent, and susceptible to sea-level rise.

The hurricane was a scrying mirror held up to the Lowcountry, reflecting a bleak, tragically possible future: a region that, during the storm, was inundated by a swollen sea and might be again by waters pushed higher by global warming. Indeed, the hurricane even swept away the last remaining sand bar of Edingsville Beach, once a resort community on Edisto Island lined with sixty fine houses belonging to the white elite.[74] But the overweening greed of developers meant that they laid eyes on pieces of land, half submerged in the ocean during hurricanes and high tide, and saw a commodity. That greed blinded them to the old patterns of coastal ecosystems and the imminent future of coastal flooding. Black sea islanders past and present held onto the coast because of their ties to the region and because of the yoked political and economic oppression under Jim Crow that prevented them from owning land in other parts of the South. Developers have dug in because they already see the islands as a sunk cost. The white elite of the Lowcountry once did the same. Just as the hurricane revealed the futility of the white elite's deep-seated need to control the Lowcountry landscape and their cruelties in manufacturing and profiting from Black suffering, so too may the climate crisis be both a product of the capitalist drive that built the tourist economy of the Lowcountry and that regime's downfall.

A future with a drowned Lowcountry is painful to contemplate. But we cannot look away. As much as the combined forces of nature, white supremacy, and capitalism have taken from the Lowcountry, there is still something left that must be saved, above all for the Gullah-Geechee. The region's grim outlook should not inculcate inaction: that would only guarantee such a future. From the vantage of the hurricane, we can observe an ocean-washed Lowcountry. We see the opportunistic maneuverings of the white elite to consolidate their rule, the legacies of slavery, the mass death that visited Black households, uneven victories, and accelerated processes of attrition and Black migration. Yet from that same viewpoint, we have also borne witness to the generations of African Americans who, to the present, never allowed the likelihood of suffering and oppression to forestall their determination to fight for their loved ones, their communities, and a better world. We must heed not only the roar of the rising ocean, ever louder across the decades, but also the voices from the Lowcountry thronging for justice, for a living, for community, for a home. In reckoning with the Great Sea Island Storm of 1893's effects on the Lowcountry, we must reckon with the Lowcountry today. There, the past is always present, and hurricane season will always return.

Notes

Preface

1. This list was compiled by the archivist Grace Cordial of the Beaufort County Library, from the Coroner's Inquisition Records Storm of 1893, for which I am very grateful. "Storm of 1893 Death List," Hurricane of 1893 vertical file.

2. Tragically, of course, the 1890 census was destroyed in a fire in 1921. The information about the following people is thus largely drawn from the 1880 census, and I try to make it clear that these are possible trajectories for them and that they may have been in different living situations in 1893 than they were in 1880. For the Hunts: "United States Census, 1880," database with images, *FamilySearch* (https://familysearch.org/ark:/61903/1:1:M691-WGG), Abby Hunt, Sheldon, Beaufort, South Carolina, United States; citing enumeration district ED 48, sheet 222D, NARA microfilm publication T9 (Washington, D.C.: National Archives and Records Administration, n.d.), FHL microfilm 1,255,221.

3. "United States Census, 1880," database with images, *FamilySearch* (https://familysearch.org/ark:/61903/1:1:M691-HQY), Phillip Brisbane in household of Daphne Brisbane, Saint Helena, Beaufort, South Carolina, United States; citing enumeration district ED 49, sheet 246D, NARA microfilm publication T9 (Washington, D.C.: National Archives and Records Administration, n.d.), FHL microfilm 1,255,221.

4. "United States Census, 1880," database with images, *FamilySearch* (https://familysearch.org/ark:/61903/1:1:M691-6Y5), Renty Capers in household of Josiah Capers, Saint Helena Island, Beaufort, South Carolina, United States; citing enumeration district ED 50, sheet 256A, NARA microfilm publication T9 (Washington, D.C.: National Archives and Records Administration, n.d.), FHL microfilm 1,255,221.

5. *Backwater Blues* is a monograph on the 1927 Mississippi River floods that makes a similar argument through analyzing the blues recordings that Black musicians created after the floods: the floods, as Mizelle argues, "heightened" African Americans' "long tradition of intellectual narration and articulation of environmental landscapes." Mizelle, *Backwater Blues*, 15.

Introduction

1. Documentation that Quash lived on that thumb of land, now known as Rhett's Bluff, provided in Natalie Adams et al., "Chapter 5: The History of Kiawah Island," in Michael Trinkley, ed., *The History and Archaeology of Kiawah Island*, 103.

2. Quash Stevens to Adele Vanderhorst, August 29, 1893, 12/213/15, Vanderhorst Family Papers. I am leaving Quash's spelling as is, for it captures something of what was likely his Gullah dialect, and he can be perfectly well understood. I first encountered Quash's letters while looking through the Vanderhorst Family Papers at the South Carolina Historical Society. After a quick search, it was clear that, while Quash's letters have not been

analyzed much by historians, they are pretty well documented on SCIWAY (website), "Quash Stevens Letters," accessed Feb. 9, 2022, https://www.sciway.net/hist/chicora/quash.html; and through the Chicora Foundation's excellent survey of Kiawah mentioned in note 1, in appendix 1, "Letters from Quash Stevens, 1868–1893," 429–35. The Chicora Foundation also released a booklet of Quash's letters, *"Your Servant, Quash": Letters of a South Carolina Freedman*. However, those collections only include letters from 12/213/15 in the papers, whereas more letters from Quash appear in folders 16 and 17, mostly from 1900 cataloguing his final months on Kiawah and his fights with Arnoldus over the property. A letter to Adele from September 3, 1893, signed "J. Q. Stevens," which may be by Quash's wife, Julia, also describes the horrible condition of the island after the storm.

3. A photograph of Quash's son William shows him mounted on what is, due to its diminutive stature, a marsh tacky. Photograph in George C. Rowe, *Negroes of the Sea Islands*, 715.

4. Quash identified himself as a planter in his marriage certificate to Julia Gibbs in 1881. See *Charleston, South Carolina, U.S., Marriage Records, 1877–1887*. He also wrote of problems with drought on Kiawah. See his letters to Adele, July 14, 1883 ("Water is get Hard I Hav To Dig all the Time to get Water For them our Well on the Iland Has Bin Dry and if rain Doint Com Sone I doint Know Wat We Will Doo For the Waoints of Water Theair Has not Bin aney rain Heair 10 Weeaks or more"); Sept. 26, 1887 ("the Cattel Is Not Doing Well" due to drought, "I have Nevr Seen it So Dry Sin on the Iland"). Adams et al. also describe how fresh water was scarce on Kiawah, "Chapter 5: The History of Kiawah Island," 12.

5. Adams et al., 56.

6. This is the accepted provenance of Quash; the Chicora Foundation's booklet of his letters discusses it on page 14.

7. First, I must credit Scott Gabriel Knowles's concept of the slow disaster, which is regularly invoked throughout the book. See Knowles, *Disaster Experts*; Knowles, "Slow Disaster in the Anthropocene"; and his website Slow Disaster, accessed Feb. 9, 2022, https://slowdisaster.com/, which explores the idea through blog posts. I must also call attention to Horowitz and Remes, *Critical Disaster Studies*, the introductory text to the burgeoning field of critical disaster studies, which contests the meanings of disaster and recovery to reveal how power shapes and structures those meanings. For disaster histories that have been useful to my understanding of how to write about them (see hurricane histories later), see Barry, *Rising Tide*; Church, *Paradise Destroyed*; Davis, *Gulf*; Donegan, *Seasons of Misery*; Dyl, *Seismic City*; Ermus, *Environmental Disaster*; Morris, *Big Muddy*; Remes, *Disaster Citizenship*; Steinberg, *Acts of God*; Valencius, *Lost History*; Erikson, *Everything in Its Path*; and Kierner, *Inventing Disaster*.

8. He describes "the disaster of slavery" as requiring the assimilation of formerly enslaved people into a new democracy that would "preserv[e] the ideals of popular government" and "overthrow . . . a slave economy." However, "it was this price which in the end America refused to pay and today suffers for that refusal." A familiar tale. Du Bois, *Black Reconstruction in America*, 325.

9. Not enough has been written about Jim Crow in the Lowcountry, though that is beginning to change. Three books look at Charleston: Kytle and Roberts, *Denmark Vesey's Garden*; Williams and Hoffius, *Upheaval in Charleston*; and Yuhl, *Golden Haze of Memory*. This is not an academic monograph, but for understanding Jim Crow in Charleston, one must read Garvin Fields, *Lemon Swamp*. One vital book happens to cover Jim Crow in Beaufort County, among other eras, and is the kind of meticulously detailed local study

that is invaluable to any researcher interested in a specific place: Rowland and Wise, *Bridging*. For two overviews of Jim Crow's early years in South Carolina, which both touch on the Lowcountry, see Perman, *Struggle for Mastery*; and Kantrowitz, *Ben Tillman*.

10. The hurricane, as such a watershed moment, has garnered mention in a few statewide and local studies and the occasional academic treatment, mentioned in books and as the subject of two articles. Fraser, *Lowcountry Hurricanes*; and Rowland and Wise, *Bridging*, gives brief but extremely informative breakdowns of the storm. Stewart touches on the hurricane's impact on rice in "*What Nature Suffers*," which was perhaps the first Lowcountry environmental history. Moser Jones, *American Red Cross*, discusses Clara Barton's role in the American Red Cross recovery effort. McKinley briefly mentions phosphate and the hurricane in *Stinking Stones and Rocks*. Ras Michael Brown refers to the hurricane as part of cultural memory in *African-Atlantic Cultures*. Bland, "'A Grim Memorial,'" also looks at the hurricane through the lens of public discourse and is an excellent scholarly work on the storm and its aftermath.

11. I draw from studies of hurricanes, as well as the growing field of disaster studies. On hurricanes, one must begin with Schwartz, *Sea of Storms*, though the 1893 storm is not included; Horowitz's magisterial *Katrina*; Mulcahy, *Hurricanes and Society*; Johnson, *Climate and Catastrophe*; Pérez, *Winds of Change*; Rohland, *Changes in the Air*; and Smith, *Camille, 1969*. Horowitz's *Katrina* was absolutely vital as I prepared the manuscript; his introduction is necessary reading for any disaster historian.

12. A paraphrasing of Phillips's opening lines of *Life and Labor in the Old South*.

13. This may call to mind the same type of obfuscation that occurs with a focus on Charleston architecture.

14. First, on southern beaches' transformations in the twentieth century, see Kahrl, *Land Was Ours*. My brief overview of Lowcountry history owes debts to many historians, who have helped craft fuller, more accurate narratives of the region's history over the last few decades. See Wood, *Black Majority*; Dusinberre, *Them Dark Days*; Littlefield, *Rice and Slaves*; Joyner, *Down by the Riverside*; Coclanis, *Shadow of a Dream*; Jones, *Born a Child*; Olwell, *Masters, Slaves, and Subjects*; Carney, *Black Rice*; Edelson, *Plantation Enterprise*; Fields-Black, *Deep Roots*; Smith, *Carolina's Golden Fields*; Sutter and Pressly, *Coastal Nature, Coastal Culture*; Swanson, *Remaking Wormsloe Plantation*; Foner, "The Emancipated Worker," in *Nothing but Freedom*; Porcher, *Market Preparation of Carolina Rice*; Strickland, "'No More Mud Work,'" *Southern Enigma*.

15. Let us not get too caught up in the nature/human divide, as per ensuing debates over Cronon, "The Trouble with Wilderness; or, getting back to the wrong nature," *Uncommon Ground*, but find a medium between the power of nature and the power of humanity. The South brings some meaningful nuance to the conversation.

16. For more on how white southerners in particular understood the swampy southern landscape at large, see Johnson, *River of Dark Dreams*; Kirby, *Mockingbird Song*; Miller, *Dark Eden*; Rosengarten, *Tombee*; Wilson, *Shadow and Shelter*; and Vileisis, *Discovering the Unknown Landscape*. For more on malaria, see Humphries, *Malaria*.

17. The literature on Indigenous South Carolina includes Dubcovsky, *Informed Power*; Tortora, *Carolina in Crisis*; Feeser, *Red, White, and Black*; La Vere, *Tuscarora War*; and Ramsey, *Yamasee War*.

18. For one of the earliest texts on this topic that takes Gullah culture seriously rather than as an antique curiosity or devolved English, see Turner, *Africanisms*; he was

thorough in his cataloguing of the Gullah language, demonstrates its grammatical and linguistic coherence, and rejected the racist term "baby talk" to describe it. Other texts on Gullah culture include Manigault-Bryant, *Talking to the Dead*; Montgomery, *Crucible of Carolina*; Creel, *"A Peculiar People"*; Cross, *Gullah Culture in America*; Jones-Jackson, *When Roots Die* (published posthumously); Pollitzer, *Gullah People*; and Baird and Twining, *Sea Island Roots*. A more recent scholarly text that explores the history of Gullah studies is Cooper, *Making Gullah*. Two of the most famous Gullah historians (Gullah in what they study and Gullah themselves) are Emory Shaw Campbell and Marquetta L. Goodwine. Campbell's recent book, cowritten with Thomas C. Barnwell Jr. and Carolyn Grant, *Gullah Days*, is excellent. Goodwine's *The Legacy of Ibo Landing* is as well.

19. For more on the Stono Rebellion, of course, see Wood, *Black Majority*. The Port Royal Experiment is best explored in Rose's iconic *Rehearsal for Reconstruction*. And for more on Black residents of the Lowcountry in the couple decades after the Civil War, see Marszalek, *Black Congressmen*; Miller, *Gullah Statesman*; Dorsey, "'Great Cry,'" in *African American Life*, though that whole book is excellent generally; Schwalm, *Hard Fight for We*; and Saville, *Work of Reconstruction*.

20. With a nod here to Joyner, *Down by the Riverside*, who wrote that "no history, properly understood, is of merely local significance" (xvii).

21. Stewart, *"What Nature Suffers,"* 3.

22. That Reconstruction lasted in some places into the 1890s is not a new idea. See Foner, *Reconstruction*.

23. This is not a complete list, but these are key books that bring together the history of the pre–World War II Jim Crow South with at least some environmental or labor history: Sutter and Manganiello, *Environmental History*; Aiken, *Cotton Plantation South*; Bryan, *Price of Permanence*; Hunter, *To 'Joy My Freedom*; Kelley, *Hammer and Hoe*; Johnson and McDaniel, "Turpentine Negro," in *"To Love the Wind"*; and Kirby, *Rural Worlds Lost*. There are also a few great essays on the topic in Arnesen, *Black Worker*, such as Kelly, "Industrial Sentinels Confront the 'Rabid Faction'"; and Woodruff, "The Organizing Tradition." The same can be said of the recent volume, Hild and Merritt, *Reconsidering Southern Labor History*, which is an important collection overall.

24. Where else to start with the new South historiography but Ayers, *Promise of the New South*, which emphasizes how the roil and toil of life in the era formed a violent, uneasy region. And without Woodward, *Origins of the New South*, who would the new South historians argue with?

25. For historians on continuity over change, see Wayne, *Reshaping of Plantation Society*; Higgs, *Competition over Coercion*; and Blackmon, *Slavery by Another Name*.

26. Mauldin, *Unredeemed Land*, 158.

27. "Things get better and they get worse, often at the same time." Horowitz, *Katrina*, 14.

Chapter One

1. Quoted in Niels Christensen, "The Sea Islands and Negro Supremacy," 1889, 2, box 7, folder 654, Christensen Family Papers, South Caroliniana Library.

2. Biographical information from Salley, "Notes and Queries," in the *South Carolina Historical and Genealogical Magazine*, 103–5.

3. Two scholars have written about the concept of the "margin" or "edges" in ways that informed my thinking here. "Islands, Edges, and Globe: The Environmental History of the Georgia Coast" explores the idea of the Georgia Lowcountry as an "edge" representing "ecotones, political boundaries, or economic exchanges" that are fruitful for thinking through the region's history. Stewart, *Coastal Nature, Coastal Culture*, 45. Enslavers "encountered the spatially embedded character of their power . . . as the fearful edge of their own apprehension. At the margins of fields . . . it was clear that the Cotton Kingdom was less an accomplished fact than an ongoing project." Johnson, *River of Dark Dreams*, 243

4. An excellent discussion of Black sea islanders' Emancipation Day celebrations can be found in Barnwell, Grant, and Campbell, *Gullah Days*, ch. 4, "Emancipation."

5. See Rose, *Rehearsal for Reconstruction*.

6. Rose, 5.

7. Rose described the day as bright and still. Rose, 11.

8. Barnwell, Grant, and Campbell, *Gullah Days*, 27.

9. Foner, *Nothing but Freedom*, 82.

10. Rose, *Rehearsal for Reconstruction*, 15.

11. Barnwell, Grant, and Campbell, *Gullah Days*, 23.

12. Strickland, *Countryside*, 153.

13. Aziz Rana (and other historians) points out that many Republicans, including Lincoln, did not believe that wage labor was truly freedom either. See Rana, *Two Faces*.

14. Foner on the Combahee rebellion can be found in Foner, *Nothing but Freedom*, 91–96; and on a rebellion on Langdon Cheves's land, see Strickland, *Countryside*, 152. Schwalm also discusses U.S. military interruption of Black claims to Lowcountry land. *Hard Fight for We*, 158–59.

15. For more on the economy, see Schwalm, which she describes as a "family economy"; and Foner, 79.

16. The estimate of less than 5 percent comes from Oubre, *Forty Acres*, 195. Though the Freedmen's Bureau did, with limited success, facilitate the purchase of federally confiscated lands at low prices by formerly enslaved people, it did not have sufficient support or institutional heft to engage in any widespread project of land redistribution. Oubre does point out that landownership rates grew over the decades, until by 1900 when 25 percent of Black farmers in the South owned land; but this was the result of Black-led efforts, not because of any actions taken by the federal government (196). This is still much lower than the landownership rate within Beaufort County.

17. Lawrence S. Rowland, "Port Royal Experiment, 1861–1870s," *South Carolina Encyclopedia*, 2016, http://www.scencyclopedia.org/sce/entries/port-royal-experiment/.

18. Rose, *Rehearsal for Reconstruction*, 356–57.

19. Rose, 285; Oubre, *Forty Acres*, 194–95; and Dorsey, "'Great Cry,'" 229–33.

20. Christensen, "Sea Islands and Negro Supremacy," 7.

21. Williams and Hoffius corroborate this image of Charleston in 1886 in *Upheaval in Charleston*, 11–14. See also Kytle and Roberts, *Denmark Vesey's Garden*, for more on Charleston as a contested site of the memory of enslavement.

22. See Wise and Rowland with Spieler, *Rebellion, Reconstruction, and Redemption*, 497–99.

23. One of the clearest contemporary descriptions of the two days' system comes from Hammond, *South Carolina*, 29–31.

24. Notebook of Ben Tillman, 1894, box 1, envelope 9, in series 8 "Diaries and Notebooks, 1861–1905," Benjamin Ryan Tillman Papers, ms. 0080.

25. Genovese discusses this phenomenon in *Roll, Jordan, Roll*, 11–12. While many aspects of this work are now controversial, the simple fact that sea island planters lived on their land for only part of the year is not. Rosengarten in *Tombee*, 90, also discusses how planters favored Beaufort as a location for second and sometimes third homes in the antebellum period.

26. Clyde Vernon Kiser discusses the success of black yeomanry on St. Helena in his study, *Sea Island to City*, 63. Woofter, *Black Yeomanry*, as one might expect, also discusses this subject, though keep in mind that it is an extraordinarily problematic book that elides the complexity of Gullah culture (he often refers to African American sea islanders as "pure African") and diminishes the hardships of slavery (see p. 34, where he says, "The slave's task was not heavy").

27. Anne Simons Dea, *Two Years of Plantation Life*, n.d., Part I-A, 44, Anne Simons Dea Papers, South Caroliniana Library.

28. Brown also refutes the idea of the sea islands as isolated as a major premise of *Afro-Atlantic Cultures*.

29. Dea, *Two Years*, 43.

30. Descriptive details from Laura M. Towne and Ellen Murray, "Report of the Penn Normal and Industrial School, 1898," box 51 "Annual Reports, 1890–1907," folder 444 "Annual Reports, 1908–1912," Penn School Papers, Southern Historical Collection, Wilson Library, University of North Carolina at Chapel Hill.

31. For a quick but respectful rundown of root work, or hoodoo, see Barnwell, Grant, and Campbell, *Gullah Days*, 228–30.

32. Hammond, *South Carolina*, 57.

33. See Harris, *Patroons and Periaguas*, for more on the history of boatmen in early South Carolina.

34. Credit here to Glave and Stoll, "*To Love the Wind*," for their conceptualization of what the lens of African American history does for environmental history, a field that long ignored those contributions. Those two fields, brought together, are vital for understanding the Lowcountry's history fully. Also see Roane, "Plotting the Black Commons," in which Roane, engaging with the literature on Black ecologies, analyzes how African Americans in the antebellum and Reconstruction-era lower Chesapeake Bay created new landscapes of use, work, and pleasure to escape white domination and to envision a future apart from the commodification of both Black labor and the environment.

35. For more on land tenure rates in Beaufort, see Strickland, *Countryside*, 160–62, which includes an incredible chart detailing land tenure rates over several decades in every Lowcountry county in South Carolina.

36. In one of the first scholarly articles on the phosphate industry, Tom W. Schick and Don H. Doyle described this as a "conservative desire to pursue an independent course as self-sufficient yeomen." Schick and Doyle, "South Carolina Phosphate Boom," 31. I do not think I would call it conservative, as their take rings of Genovese, given that it required bucking every economic and political pressure under racial capitalism.

37. This language does, I am aware, evoke the rhetoric of the "moral economy," as Thompson discusses in "The Moral Economy." I am reluctant to turn overmuch to those eighteenth-century concepts, however, because of the tangled nature of racial capitalism

and slavery in shaping the experiences of Black yeoman farmers in this era—forces that shaped the world that English freeholders lived in, of course, but that were not as direct for obvious reasons.

38. For more on the intersection of gender, race, and labor in the Lowcountry amid the Civil War and immediately after, see Saville, *Work of Reconstruction*.

39. Hammond, *South Carolina*, 57.

40. See Strickland, "'No More Mud Work.'"

41. Hammond, *South Carolina*, 469.

42. For more on rice in the late nineteenth century, see Stewart's chapter, "The Limits of the Possible," in "*What Nature Suffers*," 193–97; and Coclanis, *Shadow of a Dream*, 129–50.

43. Hammond, *South Carolina*, 31.

44. Hammond, 28.

45. Hammond, 35.

46. Hammond, 30.

47. Lewis M. Grimball to sister Elizabeth on Sept. 21, 1893, 8, box 3, folder 48, Grimball Family Papers #980.

48. Leigh, *Ten Years*; see p. 25 (on "lazy"); and p. 168 (on Chinese laborers).

49. Pringle, *Woman Rice Planter*, 7.

50. Pringle, 3–5. Hammond calculated that the state's average yield was twenty bushels of rice from each acre, which approximately split the difference between the two. Hammond, *South Carolina*, 57.

51. For more on everyday rebellion in the Lowcountry, see Jones, *Born a Child*.

52. Dea, *Two Years*.

53. Dea, 31.

54. Leigh, *Ten Years*, 27.

55. For a description of the task system, see Joyner, *Down by the Riverside*, 43–45.

56. Leigh, *Ten Years*, 54 (on refusing regular wages); 55 (on the task system); 57 and 124 (on not doing uncompensated work); and 79 (on land purchase).

57. Leigh, 79.

58. See Foner, *Nothing but Freedom*, which describes that strike wave.

59. Uncle Ben Horry, interviewed by Genvieve W. Chandler in Murrells Inlet, Georgetown County, "Volume XIV: South Carolina Narratives," part 2, *Slave Narratives*, 300.

60. From the USDA censuses of South Carolina. The 1850 production number is in the 1870 census, on p. 90. The 1939 number is from the 1940 census in volume 1, part 27, "South Carolina," in table 4. These censuses can be found online at http://agcensus.mannlib.cornell.edu/AgCensus/censusParts.do?year=1940.

61. James Henry Rice, "Georgetown County's Vivid Story," a column excerpted from *The State*, n.d., likely 1920s, James Henry Rice Papers.

62. Rogers, "Phosphate Deposits," 192.

63. McKinley, *Stinking Stones*, 20.

64. McKinley, 22.

65. Rogers, "Phosphate Deposits," 197.

66. Rogers, 206.

67. Rhett, *Handbook*, 125.

68. Rhett, 125.

69. Rhett, 125.

70. Jessie A. Butler interviewed Henry Brown, "South Carolina, Part I," *Slave Narratives*, 119.

71. Chlotilde R. Martin interviewing Sam Polite, "South Carolina, Part 3," *Slave Narratives*, 276.

72. Hammond, *South Carolina*, 31.

73. McKinley, *Stinking Stones*, 37.

74. Schick and Doyle, "South Carolina Phosphate Boom," 20.

75. Rogers, "Phosphate Deposits," 219.

76. McKinley, *Stinking Stones*, 11, 8, and 18.

77. Rhett, *Handbook*, 125.

78. McKinley, *Stinking Stones*, 78.

79. McKinley, 90.

80. McKinley, 80.

81. Oliphant, *Evolution*, 9.

82. McKinley, *Stinking Stones*, 94; also Oliphant, *Evolution*, 9.

83. Oliphant, 9.

84. "Prisoners of the State," Nov. 22, 1891, *The State*, 3.

85. Benjamin Tillman, "Message of Benjamin R. Tillman, Governor, to the General Assembly of South Carolina at the regular session commencing November 17, 1894," (Columbia: Charles A. Calvo Jr., State Printer, 1894), 34, box 1, folder 23, P U Series, Tillman Papers.

86. McKinley, *Stinking Stones*, 99.

87. McKinley, 110.

88. Rogers, "Phosphate Deposits," 217.

89. Schick and Doyle, "South Carolina Phosphate Boom," 18.

90. "Third Annual Report of the Board of Phosphate Commissioners to the General Assembly of South Carolina," *Regular Session of 1893* (Columbia: Charles A. Calvo Jr. State Printer, 1893), 336.

91. McKinley, *Stinking Stones*, 120.

92. McKinley, 120.

93. From *The Rural Carolinian*, 1873, quoted in Shuler and Bailey, *History of the Phosphate*, 22.

94. The 167-pound bag from McKinley, *Stinking Stones*, 136; the rest Shuler and Bailey, *History of the Phosphate*, 22.

95. Indeed, the pollution from fertilizer factories on the Charleston Neck lingered for generations, with the Environmental Protection Agency labeling the former locations of the Wando, Stono, Atlantic, Ashepoo, Pacific Guano, and Etiwan companies as Superfund sites in the late twentieth century. "Addressing a Century of Pollution," *Post and Courier*, May 11, 2014.

96. Phosphate executives to Benjamin Tillman, Oct. 16, 1893, item 35, box 24, folder 15, Governor Tillman Letters, South Carolina Department of Archives and History; Rogers, "Phosphate Deposits," 197.

97. Schick and Doyle, "South Carolina Phosphate Boom," 22.

98. Kantrowitz, *Ben Tillman*, 186.

99. Rogers, "Phosphate Deposits," 218.

100. Rogers, 218.

101. Number as of Nov. 10, 1893, "Report of the Comptroller-General," in *Reports and Resolutions*, 419.

102. One of the best books on the changes in the American South's farms after the Civil War, which links the degradation of living conditions, environmental exhaustion, the rise of poverty, and the oppressions of Jim Crow, is Aiken, *Cotton Plantation*.

Chapter Two

1. Jesunofsky's report "Memoranda" in the Charleston *Year Book—1893*, 261.

2. See Fraser, *Lowcountry Hurricanes*, 53.

3. See Fortenberry, "For Refuge and Resilience."

4. A useful public-facing source synthesizing hurricanes in the Lowcountry is Nic Butler, "Hurricanes in Lowcountry History, Part 2," *Charleston Time Machine*, Charleston County Public Library, accessed Feb. 9, 2022, https://www.ccpl.org/charleston-time-machine/hurricanes-lowcountry-history-part-2.

5. It is also worth noting that the disaster nearest at hand in 1893, however, was not a hurricane but the 1886 earthquake. This powerful quake killed roughly sixty Charlestonians and was felt as far away as Boston, Massachusetts, with aftershocks that frightened anew its survivors for weeks. See Williams and Hoffius, *Upheaval in Charleston*.

6. From the National Oceanic and Atmospheric Administration, "North Atlantic Storms," 207.

7. Alice Louisa Fripp Diary, Aug. 26 and 27, 1893, 46, Fripp Family Papers, 1887–1903, Caroliniana Library.

8. Anne Simons Dea Diary, Aug. 27, 1893, 22, Anne Simons Dea Papers, 1893–1910, Caroliniana Library.

9. Schwartz, *Sea of Storms*, 24.

10. Harris, "Devastation," 240.

11. From the folklorist John Bennett's essay, "A Low Pressure Area," n.d., but likely from the 1910s, 1, Charleston Hurricane, 1893 Vertical File, South Carolina Historical Society.

12. Though the story of the Mosquito Fleet comes from the Bennett essay, there is a contemporary scientific study that found that fish can and do indeed evacuate to deeper water one or two days ahead of tropical disturbances, sensing subtle changes in barometric pressure and water temperature. See Bacheler et al., "Tropical Storms Influence."

13. Bennett, "Low Pressure Area," 3.

14. I refer here to the literature on knowledge systems and rice in West Africa, specifically Wood, *Black Majority*; Carney, *Black Rice*; and Fields-Black, *Deep Roots*.

15. Schwartz, *Sea of Storms*, 143.

16. Dukes, "Red Rockets," 84, Hurricane—1893 Vertical File.

17. Dukes, 87.

18. Work on the Weather Bureau has largely fallen to historians interested in the role of science and technology within the administrative state in the Gilded Age. For a bureaucratic history of the National Weather Service, see Hughes, *Century of Weather Service*; and for the Weather Bureau, see Whitnah, *Weather Bureau*. A fascinating article (which also surveys that literature) on the uses and misuses of the warning systems in place in this era is Pietruska, "'Tornado Is Coming!'"

19. Reprint of Lewis Jesunofsky's daily diary in the days leading up to and during the storm, in Charleston Hurricane, 1893 Vertical File. Biographical information about Jesunofsky's service in Charleston comes from the unpublished but meticulously researched volume by Laylon Wayne Jennings at the Beaufort District Collections. Jennings, "History of Storms," 87.

20. Belvin Horres, "Hurricane of '93 Brought Charleston's Greatest Disaster," newspaper clipping from Aug. 27, 1953, containing a reprint of Jesunofsky's diary from Aug. 20–28, in Charleston Hurricane, 1893 Vertical File.

21. *Year Book—1893*, reprint of Jesunofsky weather journal, 257. What the Weather Bureau would soon learn was that at that moment, not one but four hurricanes were churning the Atlantic—a catastrophe unknown until that year, which would not be repeated for more than a century. Ships faced storm after storm.

22. Horres, Jesunofsky's entry from Aug. 26.

23. "At the Mercy of the Storm," *News and Courier*, Aug. 28, 1893, 1.

24. No title, *News and Courier*, Aug. 28, 1893, 1.

25. No title, *News and Courier*, Aug. 28, 1893, 3.

26. Entry on Aug. 27, 1893, Susan Hazel Rice Diaries, PAD #5, July 15, 1893–Oct. 31, 1893, Beaufort District Collections.

27. C. Mabel Burn account, Hurricane—1893 Vertical File.

28. The congregation of Queen AME was older, dating to 1865, but a new sanctuary opened in September 1892. Campbell, Grant, and Barnwell, *Gullah Days*, 258.

29. Three hundred visitors did escape, returning to Savannah at noon on Sunday; "A Poor Time for a Picnic," *News and Courier*, Aug. 28, 1893, 2. The advantage of a railroad connecting mainland to sea island was still a rare privilege; Port Royal Island just outside Beaufort was the only other, and that was because of the island's U.S. naval station.

30. Alice Louisa Fripp Diary, Aug. 27, 1893, 46.

31. Anne Simons Dea Diary, Aug. 27, 1893, 22.

32. Alice Louisa Fripp Diary, Aug. 27, 1893, 46.

33. Horres, Jesunofsky's entry on Aug. 27, 1893.

34. "A Wild Night at West End," *News and Courier*, Aug. 29, 1893, 8.

35. No title, *News and Courier*, Aug. 28, 1893, 2.

36. "Rescued at Great Risk," *News and Courier*, Aug. 28, 1893, 2.

37. "The Island Heard From," *News and Courier*, Aug. 28, 1893, 2.

38. "A Wild Night at West End," 8.

39. No title, *News and Courier*, Aug. 28, 1893, 1.

40. No title, *News and Courier*, Aug. 28, 1893, 1.

41. "A Wild Night at West End," 8.

42. This is a big claim, but given the anecdotal stories of the surge's height along the coast, combined with Jesunofsky's comparison of the storm surge levels in Charleston to others, it seems to be a fairly accurate statement. See Jesunofsky, "Memoranda," *Year book—1893*, 251.

43. "The State's Survey," *The State*, Sept. 1, 1893, 4.

44. Rachel C. Mather in Katherine M. Jones's primary source collection, *Port Royal*, 320.

45. C. Mabel Burn, interview on Nov. 3, 1959, in Hurricane—1893 Vertical File.

46. "Out of the Depths," *News and Courier*, Sept. 3, 1893, 1.

47. James Henry Rice to Sophia Brunson, Aug. 13, 1929, folder 11 "17 May 1926–17 Aug. 1932," James Henry Rice Papers (1868–1925), Caroliniana Library.

48. Mather, in Jones, *Port Royal*, 320.

49. Weary, in Jones, 323.

50. Maggie Waring, in Jones, 321.

51. Charlotte Edwards, in Jones, 323.

52. Lewis M. Grimball to his sister Elizabeth Munro, Sept. 21, 1893, 4, folder 48, Grimball Family Papers.

53. Kiser, *Sea Island to City*, 98.

54. Laura M. Towne and Ellen Murray, "Report of the Penn Normal and Industrial School, 1898," box 51 "Annual Reports, 1890–1907," folder 444 "Annual Reports, 1908–1912," Penn School Papers.

55. Towne and Murray, "Report."

56. Towne and Murray.

57. "The Storm of 1893," 6, in Hurricane—1893 Vertical File.

58. Fripp Diary, 47, from a letter she wrote to her brother Edgar on Sept. 7, 1893, and copied into her journal.

59. Harris, "Devastation," 241.

60. Harris, 242; also corroborated in Mather, in Jones, *Port Royal*, 321.

61. Mather, in Jones, 321.

62. "The Storm of 1893," 1, in Hurricane—1893 Vertical File.

63. "The Charleston Cyclone," *The State*, Aug. 30, 1893, 5.

64. Towne and Murray, "Report."

65. Towne and Murray.

66. Towne and Murray.

67. Towne and Murray.

68. Mather, in Jones, *Port Royal*, 321.

69. Towne and Murray, "Report."

70. Towne and Murray.

71. "Johnson Atkins sons experiences on the night of the storm, as told by himself to Mr. McDonald, when he came to get rations and nails," 1893, box 55, folder 468, Penn School Papers.

72. "The Hon. William Elliott's Plain Statement," *The State*, Sept. 3, 1893, 1.

73. "The Hon. William Elliott's Plain Statement," 1.

74. This is repeated again and again in different accounts. "The Storm of 1893," 6, Hurricane—1893 Vertical File; also see "The City of Dreadful Night," *News and Courier*, Sept. 2, 1893, 1.

75. "The City of Dreadful Night," 1.

76. "A Terrible Tale from Dewee's Island," *News and Courier*, Sept. 2, 1893, 2.

77. "News from Pon Pon," *News and Courier*, Sept. 2, 1893, 2.

78. Stewart, "*What Nature Suffers*," 149.

79. Fripp Diary, Sept. 7, 1893, 48.

80. "The Death Roll Growing," *News and Courier*, Sept. 2, 1893, 1.

81. No title, *The State*, Aug. 31, 1893, 1.

82. "Coroner Wells's Report," *The State*, Sept. 3, 1893, 1.

83. No title, *News and Courier*, Sept. 2, 1893, 2; the death count is from "A Devastated Island," *News and Courier*, Sept. 2, 1893, 1.

84. "Desolation and Death," *News and Courier*, Sept. 2, 1893, 1.

85. "600 Lives Lost," *The State*, Sept. 1, 1893, 1.

86. "600 Lives Lost," 1.

87. "Desolation and Death," 1.

88. "Desolation and Death," 1. Joel Chandler Harris repeats the tale with accuracy (or, at least, it matches the *News and Courier* version) though without Arthur's name in his essay, Harris, "Devastation," 239.

89. "1500 DEAD," *The State*, Sept. 3, 1893, 1.

90. This story can be found in several places: "Desolation and Death," 1; "Its Harvest Was Death," *New York Times*, Aug. 30, 1893; a 1959 interview with C. Mabel Burn in *A History of Storms on the South Carolina Coast*, 73–74, in the Hurricane—1893 Vertical File. Burn noted that Hazel was the only white person she remembered dying in the hurricane. Susan Hazel Rice, Dr. Hazel's little sister, of course recounts his death in her diary, Susan Hazel Rice Diaries. Her family was informed of his death on August 29; she wrote sadly that he "does not look very natural" with his "poor discolored face" when his body was delivered to Beaufort for burial in the Baptist cemetery. Rachel C. Mather mentions his heroic death in a story she assembled of the hurricane, "Fearful Night of Terrors," which includes accounts from other survivors, one of which also brings up Dr. Hazel, in Jones, *Port Royal*, 320.

91. "The Death Roll Growing," *News and Courier*, Sept. 2, 1893, 1.

92. "The Hon. William Elliott's Plain Statement," *The State*, Sept. 3, 1893, 1.

93. Anne Simons Dea Diary, Sept. 18, 1893, 31.

94. C. Mabel Burn, interview.

95. Their names are included in "Conversation with Joe Rivers," "The Storm of 1893," Hurricane—1893 Vertical File, 5.

96. No title, *News and Courier*, Sept. 2, 1893, 1.

97. Mather, in Jones, *Port Royal*, 321.

98. Arthur Tolliday's account, in Jones, 324.

99. Towne and Murray, "Report."

100. "A Miserable Failure," *News and Courier*, Aug. 29, 1893, 8.

101. African American spirituals, sung amid crisis and in private or church settings, have their own long history. One excellent place to start is Giovanni, *On My Journey Now*. For a scholarly work on their co-option and adaptation to mainstream entertainment, see Graham, *Spirituals*. For a definitive text on African American music, see Epstein, *Sinful Tunes and Spirituals*.

102. Jesunofsky recorded that the winds were from the east at the storm's height, in *Year Book—1893*, 255.

103. Horres describes it that way in "Hurricane of '93 Brought Charleston's Greatest Disaster."

104. No title, *News and Courier*, Aug. 30, 1893, 1.

105. "A Rough Night at the Hospital," *News and Courier*, Aug. 30, 1893, 1.

106. From the Frederick Christensen Diary, Sept. 2, 1893, n.p., Christensen Family Papers, Caroliniana Library.

107. "The Blow at Mount Pleasant," *News and Courier*, Aug. 30, 1893, 2.

108. No title, *New York Times*, Aug. 31, 1.

109. Causeway, "The Situation in Beaufort," *News and Courier*, Sept. 2, 1893, 1; Edisto, "The Death Roll Growing," *News and Courier*, Sept. 2, 1893, 1; Drawbridge, "The Wreck of the New Bridge," *News and Courier*, Aug. 29, 1893, 1.

110. Conversation with Joe Rivers, 2, Hurricane—1893 Vertical File

111. Fripp Diary, Aug. 28, 1893, 46.

112. "Storm-Swept Savannah," *The State*, Aug. 30, 1893, 1.

113. "The Hon. William Elliott's Plain Statement," 1.

114. Jennings, "History of Storms," 32.

115. Dea Diary, Aug. 28, 1893, 22.

116. Jesunofsky, *Year Book—1893*, 255.

117. No title, *The State*, Aug. 29, 1893, 4.

118. "Destruction Wrought," *The State*, Aug. 29, 1893, 8. Also see No title, *The State*, Aug. 29, 1893, 5.

119. No title, *The State*, Aug. 29, 1893, 5.

120. Jesunofsky, *Year Book—1893*, 248.

121. "New-York Out of the World," *New York Times*, Aug. 30, 1893, 1.

122. Jesunofsky, *Year Book—1893*, 249.

123. Genevieve Chandler interviewing an "ex-slave," "Volume XIV, Part 1," *Slave Narratives*, 115.

Chapter Three

1. "A Storm on the Coast," *New York Times*, Sept. 2, 1893.

2. "Nearly 200 Dead," *New York Times*, Aug. 31, 1893, 1,

3. "Cyclone Couplets," *News and Courier*, Aug. 28, 1893, 2.

4. "The Story of the Storm," *News and Courier*, Aug. 29, 1893, 1.

5. "Cyclone Couplets," *News and Courier*, Aug. 28, 1893, 2.

6. "Sweep of the Mighty Wind," *New York Times*, Sept. 2, 1893.

7. "The Story of the Storm," *News and Courier*, Aug. 29, 1893, 1.

8. All quotes from "Rising from the Ruins," *News and Courier*, Aug. 30, 1893, 1.

9. N. Jones, *Born a Child*, 26, elucidates a similar vision of the perceived besiegement that slaveholders felt from all sides and how their defensiveness and insecurity led to an increase in their brutality.

10. This section keeps in mind Olwell's quote about how "violent changes" are accompanied by "a series of equally violent continuities"; he was speaking about South Carolina's colonial-era slave society, but it is applicable to this era too. Olwell, *Masters, Slaves, and Subjects*, 12.

11. With allusion here to Bryan, *Price of Permanence*, in which he argues throughout that white southern wheelers and dealers attempted to wield control over resources and the environment just as they did over African Americans.

12. All information about the street cleaning in this paragraph comes from *Year Book—1893*, 23 and 72.

13. "Rising from the Ruins," *News and Courier*, Aug. 30, 1893, 1.

14. "The Story of the Storm," *News and Courier*, Aug. 29, 1893, 1.

15. *Year Book—1893*, 78.

16. "The Story of the Storm," 1.

17. "The Story of the Storm," 1; and "Getting Along Nicely," *News and Courier*, Aug. 31, 1893, 1.

18. Writing about Charleston's use of convict labor for projects of municipal improvement is still scant. However, the 1892 *Year Book*, the 1893 *Year Book*, and others include detailed reports on the use of convict labor in the city.

19. Oliphant, *Evolution*, 11.

20. *Year Book—1893*, 163.

21. *Year Book—1893*, 157.

22. For a biography of James Henry Hammond, see Faust, *James Henry Hammond*.

23. All of this information is also located within the introduction to the James Henry Hammond Jr. Papers at the Caroliniana Library.

24. Found in "Testimony of Mr. Harry Hammond," *Hearings before the Industrial Commission*, 816, James Henry Hammond Jr. Papers, Caroliniana Library. He was a supervisor for the 1880 census. Shortly thereafter, the state Board of Agriculture commissioned Hammond to compose the 1883 handbook of South Carolina, *South Carolina*.

25. With some reference here to Bryan's conceptualization of "permanence." Bryan, *Price of Permanence*.

26. "A Striking Idea," *The State*, Sept. 11, 1893, 4.

27. Lewis M. Grimball to his sister Elizabeth Munro, Sept. 21, 1893, 4, folder 48, Grimball Family Papers.

28. Reed gave a long account to *The State*. "1500 DEAD," Sept. 3, 1893, 1 and 5. All the details given in the next couple of paragraphs are drawn from this description.

29. "Shrieks," *The State*, Sept. 3, 1893, 5. "Providence," in Letter from George A. Reed, "Sheriff's Office, Beaufort County," to Governor Benjamin Ryan Tillman, n.d., box 27, folder 32 "Letters Received & Sent, Dec. 15, 1893—Jan. 2, 1894," Governor Benjamin Ryan Tillman (1890–94) Papers, South Carolina Department of Archives and History (hereafter SCDAH).

30. "1500 DEAD," 3.

31. "1500 DEAD," 1.

32. Or, at least, there was no coverage of its damage in any account that I have found.

33. From a biography of Abbie Holmes Christensen in Tetzlaff, *Cultivating a New South*, 78–79.

34. From the Frederick Christensen Diary, Sept. 2, 1893, in the Christensen Family Papers.

35. At first, he thought that the hurricane had cost him $8000 to $10,000. By September 5, his estimate had risen to $15,000. Letter from Niels Christensen to Abbie Christensen, Sept. 5, 1893, 1, Christensen Family Papers.

36. "The Charleston Cyclone," *The State*, Aug. 30, 1893, 3; and Rowland and Wise, *Bridging*, 3.

37. Rowland and Wise, 16.

38. "Horrors upon Horrors," *The State*, Sept. 1, 1893, 1.

39. "Sweep of the Mighty Wind," *New York Times*, Sept. 2, 1893.

40. Kiser, *Sea Island to City*, 98.

41. Kiser, 98.

42. "1500 DEAD," 3.

43. "1500 DEAD," 1.

44. "1500 DEAD," 1.

45. "Rough on the Rice Planters," *News and Courier*, Sept. 1, 1893, 2; Clay Hill from "Eighty Inquests on Combahee," *News and Courier*, Sept. 1, 1893, 1.

46. "The Charleston Cyclone," *The State*, Aug. 30, 1893, 1.

47. "Another Account," *News and Courier*, Sept. 1, 1893, 1.

48. "Desolation and Death," *News and Courier*, Sept. 2, 1893, 1.

49. "Out of the Depths," *News and Courier*, Sept. 3, 1893, 1.

50. "The Sea Island Situation," *The State*, Sept. 6, 1893, 1.

51. Jones-Jackson discussed these burial practices in *When Roots Die*, 26. It is also something that can be learned simply by visiting the sea islands, where efforts to preserve these graveyards are ongoing. On Dataw Island, for example, which is owned by a corporation, African Americans have successfully lobbied to preserve a graveyard there; similar efforts are underway on St. Helena Island.

52. "1500 DEAD," 1.

53. "1500 DEAD," 1.

54. "1500 DEAD," 1.

55. From Conversation with Joe Rivers, 7, Hurricane—1893 Vertical File.

56. See "Coroner's Inquisition Records, Storm of 1893," Hurricane—1893 Vertical File.

57. "As to the Rice," *News and Courier*, Aug. 31, 1893, 1.

58. "As to the Rice," 1.

59. "Out of the Depths," 1.

60. "An Official Census," *News and Courier*, Sept. 18, 1893, 2.

61. Mather, *Storm-Swept Coast*, 40.

62. "Seeing Is Believing," *News and Courier*, Sept. 6, 1893, 1.

63. "Out of the Depths," 1.

64. "Seeing Is Believing," 1.

65. "1500 DEAD," 1.

66. Interview with Diana Brown of Edisto Island, n.d., c. 1930, 4–5, box 32, folder 9, Lorenzo Dow Turner Papers, Melville J. Herskovits Library of African Studies.

67. "Seeing Is Believing," 1.

68. "Seeing Is Believing," 1.

69. The obsession with smell calls to mind Kiechle, *Smell Detectives*, in which she discusses how smells informed concepts of public health; and Nash, *Inescapable Ecologies*, especially in her brilliant introduction where she describes the connections between environment and health.

70. Mather, *Storm-Swept Coast*, 38.

71. "The Sea Island Situation," *The State*, Sept. 6, 1893, 1. The focus on the senses in this section owes credit to the overall ideas in Smith, *Sensory History Manifesto*.

72. "The City of Dreadful Night," *News and Courier*, Sept. 1, 1893, 1.

73. "Fearful on the Islands," *The State*, Sept. 14, 1893, 8.

74. "Seeing Is Believing," 1.

75. "The Sea Island Situation," *The State*, Sept. 6, 1893, 1.

76. "A Voyage of Desolation," *News and Courier*, Sept. 18, 1893, 1.

77. "Red Cross in Possession," *News and Courier*, Sept. 19, 1893, 4.

78. Mather, *Storm-Swept Coast*, 56.

79. "Getting Along Nicely," *News and Courier*, Aug. 31, 1.

80. Susan Hazel Rice Diary, Aug. 29, 1893.

81. Susan Hazel Rice Diary, Sept. 15 and 16, 1893.

82. Elkinton, *Selections*, 315, at the Caroliniana Library.

83. Elkinton, *Selections*, 315.

84. Lewis M. Grimball to his sister Elizabeth Munro, Sept. 21, 1893, 4, folder 48, Grimball Family Papers.

85. Lewis Grimball to his sister Elizabeth, Oct. 15, 1894, 2, Grimball Family Papers.

86. Lewis Grimball to Berkeley Grimball, June 2, 1899, 1 and 3, Grimball Family Papers.

87. Berkeley's sisters received a letter of condolence on his death from G. A. Courtenay on July 18, 1899. G. A. Courtenay to the Misses Grimball, July 18, 1899, 1, folder 53, Grimball Family Papers.

88. Elkinton, *Selections*, 309.

89. From Caroline Boineau, "Capturing the History of Edisto Island: Oral History Interview with Alice Stevens," Apr. 12, 1993, Lowcountry Digital Library, Edisto Island Historic Preservation Society. See 13–14 of the transcript at https://lcdl.library.cofc.edu/lcdl/catalog/lcdl:122743.

90. Mather, *Storm-Swept Coast*, 55–56.

91. Mather, 106.

92. Mather, 106.

93. "Johnson Atkins's experiences on the night of the storm, as told by himself to Mr. MacDonald, when he came to get rations and nails," 1893, box 55, folder 468, Penn School Papers.

94. Quoted in G. G. Johnson, *Social History*, 206.

95. "Stories from my father's life: Uncle Abram," July 8, 1918, box 55, folder 469, Penn School Papers.

96. All quotes from Harris, "Devastation," 242.

97. From Silver, *To Live as Free Men*, 00:06:39–00:08:50. As far as I know, this is the only video and sound recording that still exists of an African American relaying their story of the hurricane.

98. This calls to mind Mizelle, *Backwater Blues*, 15: "African Americans have a long tradition of intellectual narration and articulation of environmental landscapes that were only heightened during the 1927 flood . . . historical commentary of the flood was rooted in the idea that race and nature were interrelated burdens; the perils of rain, wind, and water only exacerbated existing vulnerabilities of second-class citizenship."

99. Diana Brown could only credit him for her survival: if it hadn't been for him, she said, "I wouldn't uh been here." Interview with Diana Brown, n.d., c. 1920s, box 32, folder 9, 7, Lorenzo Dow Turner Papers.

Chapter Four

1. Mather, *Storm-Swept Coast*, 37.

2. Mather, 105.

3. A good book on Reconstruction memory in South Carolina is Baker, *What Reconstruction Meant*; another on Civil War memory, particularly outside of the Lowcountry of South Carolina and as it pertains to Tillman, is Poole, *Never Surrender*.

4. One key book on white conservatism in South Carolina after Reconstruction is Holden, *In the Great Maelstrom*; he approaches white supremacy through four white thinkers, academics, and politicians in the state rather than through policies and practices. For more on governance structures in the South at this time, see Ayers, *Promise.*

5. For more on this dynamic, see Karp, *This Great Southern Empire.*

6. "Fearful on the Islands," *The State*, Sept. 14, 1893, 8.

7. "Helping the Helpless," *News and Courier*, Sept. 10, 1893, 8.

8. Quoted in Kantrowitz, *Ben Tillman*, 513.

9. "Relief for the Sufferers; Gov. Tillman sends a special agent down," *The State*, Sept. 3, 1893, 1, references that Tillman discussed the storm "while pacing up and down his office."

10. "The Exodus to Chicago," *The State*, Aug. 14, 1893, 8.

11. Telegraph reprinted in full in "600 LIVES LOST," *The State*, Sept. 1, 1893, 1.

12. "600 LIVES LOST," 1.

13. "600 LIVES LOST," 1.

14. "600 LIVES LOST," 1.

15. "Relief for the Sufferers," 1.

16. "Relief for the Sufferers," 1.

17. "600 LIVES LOST," 1.

18. "Relief for the Sufferers," 1.

19. "A Record of the Dead" and "The Appeal of Robert Smalls," subsections of "600 LIVES LOST," 1 and 5.

20. "The Appeal of Robert Smalls," subsection of "600 LIVES LOST," 1 and 5.

21. Miller's biography of Smalls sums up his career and describes him in similar terms; see E. Miller, *Gullah Statesman*, 245–50.

22. A quote from Lewis Grimball bemoaning that assistance will "make the negro lazy and sassy and less inclined to work than he was before," in Lewis M. Grimball to his sister Elizabeth Munro, Sept. 21, 1893, 3, folder 48, Grimball Family Papers.

23. Ayers discusses the formation of a Black middle class in southern towns and cities in this era in *Promise of the New*, 68–72.

24. "The Appeal of Robert Smalls," subsection of "600 LIVES LOST," 5.

25. "600 VICTIMS," *New York Times*, Sept. 1, 1893.

26. Frederick Christensen Diaries, 73, Christensen Family Papers.

27. Account from "Murder and Arson," *The State*, Sept. 8, 1893, 5.

28. For more on this topic, see Rowland and Wise, *Bridging*, ch. 8, "Beaufort's African American Community at the Turn of the Century."

29. Joseph Barnwell was a descendant of a Port Royal Island planter family. He gave an address in 1880 to the South Carolina Historical Society, of which he was then president, declaiming the "tyranny, the corruption, the imbecility" of Reconstruction to the room, a subversion of "the whole government of the State" and the "historic race which had ruled it for two hundred years." He told his audience, without blushing at the bold irony of the statement, that the formerly enslaved were unfit to govern and that their experiment with democratic participation failed because their "only experience of government had been the slaver." It was no matter to Barnwell who had been the cruel enslaver setting a poor example or consideration for how such a role might have corrupted the enslaver for governance. His beliefs apparently did not shift much over the subsequent years. For the

speech, see Joseph W. Barnwell, "Dual state governments: Carolina in the revolutions of 1719, 1776, and 1876: address before the South Carolina Historical Society," at Charleston, S.C., May 18, 1880, 8, Hathi Trust Digital Library, https://catalog.hathitrust.org/Record/100550768. For biographical details, see the "Finding Aid," Joseph W. Barnwell Papers, South Carolina Historical Society.

30. In "Mayor Ficken's Annual Review," in *Year Book—1893*, 24; the committee's members are also listed in "Charleston's Charity," *The State*, Sept. 3, 1893, 5.

31. Millar and Williams, *Upheaval in Charleston*, 157.

32. Elkinton, *Selections*, 322.

33. "System in Relief Work," *The State*, Sept. 10, 1893, 8.

34. "System in Relief Work," 8.

35. "System in Relief Work," 8.

36. This poster can be found within the Christensen Family Papers, "Post on the Rules for the Government of the Sea Island Relief Committee," folder 621.

37. "Post on the Rules."

38. Schwartz, *Sea of Storms*, 181–82.

39. Schwartz, 189–91.

40. Remes, *Disaster Citizenship*, takes an interesting perspective on the Progressives and mutual aid. He contends that Progressive reformers and bureaucrats fundamentally misunderstood "working-class culture" in their application of disaster assistance, and so those afflicted built their own localized, kinship-based "architecture of mutual aid." Another excellent book on the role of mutual aid societies is Beito, *From Mutual Aid*, which charts the role of organizations, churches, and fraternities across races, ethnicities, and classes in providing assistance that their members construed as basic rights. Immigrants and African Americans, Beito finds, were prone to participating in and forming such organizations as a means of shelter and security from the persecution that they faced from the government, nativists, or racists.

41. No title, *The State*, Sept. 9, 1893, 1; and "For Sea Island Sufferers," *The State*, Sept. 7, 1893, 1.

42. Mayor Ficken's "Annual Review," in *Year Book—1893*, 24.

43. Ginzberg in *Women and the Work of Benevolence* looks at changes in women's work outside the household and contrasts the antebellum with the postbellum era. She argues that while something more like radicalism inflected white, middle-class women's activism before the war, a classist desire to control the poor and immigrants, particularly within urban spaces, guided their activism after. Gilmore's *Gender and Jim Crow* admittedly examines a period after the hurricane, but her insights about the power of Black women to mobilize after the rise of Jim Crow, and the ways in which white women rejected or worked with Black women, provide a look at the ways in which charity work, activism, and race operated in the South.

44. There is substantial documentation of this tradition across the South, from older scholarship to new: see Hoffman's "From Slavery to Self-Reliance"; throughout Hahn, *Nation under Our Feet*; a core argument within Ortiz, *Emancipation Betrayed*.

45. "Columbia to the Rescue," *The State*, Sept. 3, 1893, 1.

46. "All to the Sufferers' Aid," *The State*, Sept. 2, 1893, 8.

47. "Charleston's Charity," *The State*, Sept. 3, 1893, 5.

48. "For the Cyclone Sufferers," *The State*, Sept. 8, 1893, 3.

49. "Contributions Secured in New York," *The State*, Sept. 6, 1893, 1.

50. "Col. Elliott's Mission," *The State*, Sept. 10, 1893, 1.

51. Clipping from the *Brooklyn Eagle*, Sept. 9, 1893, in folder "Newspaper Clippings," Sea Island Relief Operations, Red Cross File, 1863–1957, American National Red Cross, 1878 to 1957, in the Clara Barton Papers, Library of Congress, https://www.loc.gov/item/mss119730621/, image 6.

52. "The Work of Relief," *The State*, Sept. 12, 1893, 1.

53. "Cloth for the Sufferers," *The State*, Sept. 12, 1893, 8.

54. "Even the Children Work," *The State*, Sept. 12, 1893, 8.

55. "A Noble Response," *The State*, Sept. 9, 1893, 8.

56. "A Noble Response," 8.

57. "For Sea Island Sufferers," *The State*, Sept. 7, 1893, 1.

58. From Adams, *Jacksonville Auxiliary Sanitary Association*.

59. "Chastened by the Cyclone," *News and Courier*, Sept. 10, 1893, 5.

60. "The Story of the Storm," *News and Courier*, Aug. 29, 1893, 1.

61. "A Colored Mass Meeting," *News and Courier*, Sept. 8, 1893, 3.

62. "Brave Old Charleston," *The State*, Sept. 4, 1893, 1. More documentation on Black donations can be found in "The Colored Fund," *News and Courier*, Sept. 7, 1893, 3.

63. "The Colored Men Give," *The State*, Sept. 2, 1893, 8.

64. "The Colored Men Give," 8.

65. "Starving!," *The State*, Sept. 8, 1893, 3.

66. Jervey, "Butlers of South Carolina," 303.

67. "The Wards of the Nation," *News and Courier*, Sept. 2, 1893, 3.

68. According to "Brawley, William Huggins, 1841–1916," in the Biographical Directory of the United States Congress, accessed Feb. 21, 2021, http://bioguide.congress.gov/scripts/biodisplay.pl?index=B000775.

69. "Out of the Depths," 1.

70. "The Wards of the Nation," 3.

71. For a biography of George Murray, see Marszalek, *Black Congressman*. All biographical details of Murray come from this biography. For a general text on Black populists in this era, see Ali, *In the Lion's Mouth*.

72. Instead, he completed his college degree at the State Normal Institution in Columbia and worked as a teacher afterward.

73. "Appropriation Asked," *The State*, Sept. 5, 1893, 1.

74. "Appropriation Asked," 1.

75. For a brief contemporary account, see "The Broken Levees—the Lands in the Yazoo Valley Over-flowed," a reprint from the *Times-Democrat* in Greenville, Mississippi, March 30, 1890, 18, *Atlanta Constitution*. An excellent book on the region in this era is Willis, *Forgotten Time*.

76. Cowdrey, *This Land, This South*, 123.

77. "Relief for Brunswick," *The State*, Aug. 26, 1893, 1. The original appeal to the surgeon general is described in "Brunswick's Appeal: Federal Health Authorities to the Rescue," *The State*, Aug. 25, 1893, 1, in which the Marine Hospital Bureau was "expected to furnish food and medicine."

78. "Starvation—Pestilence," *News and Courier*, Sept. 7, 1893, 1.

79. "Starvation—Pestilence," 1.

80. Though he would run against Murray in 1894 and win in an election that Murray, like Miller, contested as fraudulent.

81. "Starvation—Pestilence," 1.

82. An account of this meeting is provided within "Starvation—Pestilence," 1.

83. "Starvation—Pestilence," 1.

84. "The Sea Island Situation," *The State*, Sept. 6, 1893, 1.

85. "Aid for the Sea Islanders; Colored People of Washington Moving Congressman Murray's Plan," *The State*, Sept. 8, 1893, 3.

86. "Government Rations," *The State*, Sept. 8, 1893, 3.

87. "Col. Elliott's Mission," 1.

88. "Col. Elliott's Mission," 1.

89. "Col. Elliott's Mission," 1.

Chapter Five

1. "Relief for the Sufferers," *The State*, Sept. 3, 1893, 1.

2. "The Red Cross Muddle," *The State*, Sept. 12, 1893, 8; and Barton, *Story of the Red Cross*, 78.

3. "The Red Cross Muddle," 8.

4. "The Red Cross Muddle," 8.

5. "The Red Cross Muddle," 8. All of the quotes are from this published exchange.

6. Jones's *The American Red Cross* has been invaluable in understanding Clara Barton's character and the nature of the Red Cross throughout the nineteenth century. For an excellent book on the Red Cross's larger efforts, see Irwin, *Making the World Safe*. And, as mentioned in the introduction, Bland's "'A Grim Memorial'" is an excellent article that also discusses the recovery efforts in the Lowcountry. Furthermore, all the information in this paragraph comes from the newspaper article on the journey, "A Voyage of Desolation," *News and Courier*, Sept. 18, 1893, 1.

7. Clara Barton Diary, Sept. 29, 1893, box 5 "Diaries and Journals," folder "May 1893–May 1894," Clara Barton Papers, Manuscript Division, Library of Congress, Washington, D.C., https://www.loc.gov/item/mss119730041/, image 143 (hereafter Clara Barton Diary, image [number]).

8. There were, of course, efforts of recovery in the wake of, for example, war. The Civil War is the most instructive example, and the U.S. Sanitary Commission during the war provided some interesting precedents for federal intervention in the management of environmental factors: see Giesberg, *Civil War Sisterhood*.

9. For more on progressives and the South, see "The Shield of Segregation" in McGerr, *Fierce Discontent*.

10. Strickland, "Traditional Culture," 153.

11. This of course calls to mind both Du Bois's *Black Reconstruction* and Kelley's *Hammer and Hoe*. What distinguishes this case is the timing: 1893 was three decades removed from the Civil War and for many African Americans was a moment when hope was contracting rather than expanding. The 1920s and 1930s of Kelley's narrative possessed a new globalism to them, as socialist and communist activists turned to the rural South for the first time to reimagine these ideologies for rural African Americans.

12. This account is given in Jones, *American Red Cross*, 11.

13. Clara Barton, "Names of the Soldiers," n.d., box 114, folder "Sea Islands, S.C., Miscellany," Relief Operations, Red Cross Files, Clara Barton Papers, https://www.loc.gov/item/mss119730619/, image 11 (hereafter Red Cross Miscellany, image [number]). The name Middleton is common because it was the name of a wealthy white family who enslaved hundreds of people in South Carolina; today, Middleton Place is a tourist site.

14. See more on Wagner in Egerton, *Thunder at the Gates.*

15. Barton, *Story of the Red Cross*, 79–80 (first quotation on p. 79–80; second on p. 80; third on p. 79). Barton transcribed the veterans' Lowcountry Gullah dialect clumsily.

16. Barton, 80.

17. Barton, 80.

18. Dr. Egan's testimony in Washington, D.C., June 2, 1895, box 116, folder "Speeches and Writings, 1893–June 1895 and undated," Sea Islands Relief Operations, Red Cross Files, Clara Barton Papers, https://www.loc.gov/item/mss119730626/, image 12 (hereafter Red Cross Speeches and Writings, image [number]).

19. Barton quoted in "Report of Shipping Room, Medical, Surgical and Sanitary Departments," 5, box 115, folder "General Reports, Nov. 1893–July 1895 and undated," Sea Islands Relief Operations, Red Cross Files, Clara Barton Papers, https://www.loc.gov/item/mss119730622/, image 64 (hereafter Red Cross General Reports, image [number]).

20. Jones, *American Red Cross*, 73. Drs. Hubbell and Egan, George Pullman, and John McDonald with his wife Ida Battell, a nurse from Milwaukee, constituted the core of this group. McDonald had been aboard the doomed steamer the *Savannah* as it floundered off the coast during the hurricane. Pullman, who had joined Barton the year before as her financial secretary, had powerful relatives: his uncle was the Pullman railroad car magnate. Harriette Reed, a wealthy Bostonian and frequent Red Cross volunteer, traveled to Beaufort periodically to assist. Admiral Beardslee, Mrs. Gardner, and H. L. Bailey, all South Carolina residents, joined Barton's crew and agreed to work as volunteers for the Red Cross.

21. Transition described in an untitled document/memorandum, Oct. 2, 1893, folder 653, Christensen Family Papers.

22. Letter from Niels Christensen to Abbie on Apr. 8, 1894, folder 26, M55, Christensen Family Papers. Niels Christensen expressed frustration that Barton did not take advice from anyone. Christensen's wife, Abbie, was a fascinating figure in her own right, with leftist politics and a distaste for white Democrats in the South so severe that she preferred to live in her home state of Massachusetts than with her husband in Beaufort: see Tetzlaff, *Cultivating a New South.*

23. "Mrs. Reed's Clothing Report," n.d., 5, Red Cross General Reports, image 16.

24. See Jones, *American Red Cross*, for more on the Red Cross's earlier efforts. The Jacksonville yellow fever recovery was particularly fraught for the Red Cross. See "Work of the Red Cross," *The State*, Oct. 18, 1893, 2, for a quick list of these efforts.

25. I do dispute here McGerr's assertion of the radicalism of the progressives; they are only radical if one ignores the socialist and communist labor organizers of the industrialized North and Midwest or the Black workers of the sea island coast, whose visions for a truly egalitarian, just nation and whose methods for getting there far eclipse the modest, paternalistic plans for reform of the progressives.

26. Barton, *Story of the Red Cross*, 79.

27. Barton, 198.

28. In an undated notebook written in Barton's hand, Red Cross Speeches and Writings, image 55.

29. In an undated notebook written in Barton's hand, Red Cross Speeches and Writings, image 55.

30. In an undated notebook written in Barton's hand, Red Cross Speeches and Writings, image 55.

31. Barton, "Sea Island Relief Methods in the Field," Red Cross Speeches and Writings, images 18–21.

32. Barton, "Sea Island Relief Methods in the Field," Red Cross Speeches and Writings, images 18–21.

33. Barton, "Sea Island Relief Methods in the Field," Red Cross Speeches and Writings, images 18–21. The number of sea island refugees crowding Beaufort can be found in Barton, *Story of the Red Cross*, 81.

34. Barton, "Sea Island Relief Methods in the Field," Red Cross Speeches and Writings, images 18–21.

35. Barton, "Sea Island Relief Methods in the Field," Red Cross Speeches and Writings, images 18–21.

36. Clara Barton Diary, Oct. 3, 1893, image 145; entry opens "very rainy—warm."

37. Dr. Egan, undated testimony, 6, Red Cross Speeches and Writings, image 4.

38. Dr. Egan, undated testimony, 6, Red Cross Speeches and Writings, image 18.

39. Dr. Egan, undated testimony, 6, Red Cross Speeches and Writings, image 4.

40. As determined from the Daniel Smith illustrations, which suggest that particular building. Many thanks to Christopher Barr, the interpretive supervisor with the Reconstruction Era National Historical Park, for helping me puzzle that out.

41. From Admiral Beardslee's report in Barton, *Red Cross*, 208.

42. Dr. Egan, undated testimony, Red Cross Speeches and Writings, image 3.

43. Dr. Egan, undated testimony, Red Cross Speeches and Writings, image 3.

44. Clara Barton Diary, Oct. 8, 1893, image 147.

45. A peck is a volume measurement equivalent to two gallons. Barton, "Mid-Field Report," Jan. 7, 1894, folder 621, Christensen Family Papers.

46. Clara Barton to Mr. Charles Hebard of Chestnut Hill, Philadelphia, Pa., Jan. 30, 1894, box 114, folder "Correspondence, Aug. 1893–Feb. 1895 and undated," Sea Islands Relief Operations, the Red Cross Files, image 30 (hereafter Red Cross Correspondence, image [number]).

47. Barton, "Sea Island Relief Methods in the Field," Red Cross Speeches and Writings images 18–21.

48. Clara Barton Diary, Oct. 14, 1893, image 151.

49. Dr. Egan, undated testimony, Red Cross Speeches and Writings, image 10.

50. Dr. Egan, undated testimony, Red Cross Speeches and Writings, image 13.

51. Dr. Egan's account in Barton, *Red Cross*, 227.

52. As recounted in Clara Barton's diary in late October [undated, but that is when it appears in her diary sequentially], image 151.

53. Clara Barton Diary, Nov. 27, 1893, image 173.

54. Clara Barton Diary, Nov. 27, 1893, image 173.

55. See Schwalm, *Hard Fight for We*, for more on this dynamic; it is the central contention of her book.

56. See Schwalm, 206.

57. Barton mentions that "forty women were hired to come over from the islands and cut potatoes." Barton, *Story of the Red Cross*, 85.

58. Gardner's report in Barton, *Red Cross*, 260.

59. Barton, 258.

60. Barton, 258.

61. Mary Jenkins wrote in February 1894, "We the people pray Day and night that the god of heaven will keep you." In another undated letter, she told Barton that "the good what you have done for us we aren't able to give you the thanks but we pray night an day that the god of heaven will bless you." Red Cross Correspondence, image 34.

62. In Barton, *Red Cross*, Mrs. McDonald's report, 220.

63. In the "Report of Shipping Room, Medical, Surgical and Sanitary Department," undated, Red Cross General Reports, image 63.

64. Mr. C. C. Richardson to Clara Barton, n.d., Red Cross Miscellany, image 22.

65. Mr. C. C. Richardson to Clara Barton, n.d., Red Cross Miscellany, image 22.

66. Barton, *Red Cross*, 260.

67. Jacqueline Jones has written about call-and-response songs among Black women in "My Mother." Lauri Ramey has also done so in *Slave Songs*.

68. Fraser, *Lowcountry Hurricanes*, 184–87.

69. Mather, *Storm-Swept Coast*, 80.

70. "Appealing to Miss Barton," *The State*, Sept. 20, 1893, 8.

71. In Barton's *Midfield Report*, published widely in newspapers on Jan. 7, 1894; it can be read in folder 621, Christensen Family Papers.

72. In "ON THE VERGE OF STARVATION," *Columbia Register*, May 25, 1894, 1.

73. Barton, *Midfield Report*.

74. Clara Barton Diary, image 153.

75. Moser, *American Red Cross*, 32.

76. Moser, 41.

77. Most disaster historians tend to point to the 1950 Disaster Relief Act as the institutionalization of disaster relief in federal policy.

78. "The Sea Island Situation," *The State*, Nov. 12, 1893, 1.

79. Ellen Murray to Ben Tillman, Sept. 27, 1893, item 3, folder 33, box 23, Ben Tillman letters received, Governor Benjamin R. Tillman Papers.

80. This quote and all others in the paragraph are from "Message of Benjamin R. Tillman, Governor. To the General Assembly of South Carolina, at its regular session commencing November 26th, 1893" (Columbia: Charles A. Calvo Jr., State Printer, 1893), Hurricane—1893 Vertical File.

81. In Barton, *Midfield Report*.

82. "Sea Island Sufferers," *The State*, Nov. 29, 1893, 1.

83. "Sea Island Sufferers," 1.

84. "Sea Island Sufferers," 1.

85. Dr. Hubbell, untitled report, Red Cross Speeches and Writings, image 23.

86. Dr. Egan's undated testimony, Red Cross Speeches and Writings, image 7.

87. Dr. Egan's undated testimony, Red Cross Speeches and Writings, image 7.

88. In a report from Hilton Head Island, dated Nov. 3, 1893, Red Cross General Reports, image 2. It appears to be an early report written by a Red Cross representative, perhaps Barton herself.

89. "Labor on Port Royal Island at Gray's Hill," handwritten note by Dr. Hubbell, n.d., box 115, folder "Labor Reports, A–L," Sea Islands Relief Operations, Red Cross Files, https://www.loc.gov/item/mss119730623, image 54 (hereafter Red Cross Labor Reports, image [number]).

90. Barton, "Methods of Sea Island Relief," Red Cross Speeches and Writings, image 19.

91. Dr. Egan's report in Barton, *Red Cross*, 224.

92. Dr. Egan's report in Barton, 224.

93. "Sea Islands Report: Hilton Head," no author, Nov. 3, 1893, 2, Red Cross General Reports, image 2.

94. D. E. Washington, "Labor on St. Helena Island at Chaplin & Fripp," n.d., Red Cross Labor Reports, image 112.

95. P. W. Washington, "Labor on Port Royal Island at Retreat," box 115, folder "Labor reports, M–Y," Sea Islands Relief Operations, Red Cross Files, image 114 (hereafter Red Cross Labor Reports, image [number]).

96. This is Barton's estimate during an "exit interview" in "The Work of the Red Cross," *News and Courier*, Jun. 23, 1894.

97. "The Work of the Red Cross."

98. Wesley L. Jackson to Clara Barton, n.d., ca. 1894, Red Cross Labor Reports, image 60.

99. Friday Smalls to Clara Barton, May 5, 1894, Red Cross Labor Reports, image 107.

100. No name to Clara Barton, Mar. 2, 1894, Red Cross Labor Reports, images 24–25.

Chapter Six

1. "Starving!," *The State*, Sept. 8, 1893, 3.

2. Schwartz, *Sea of Storms*, 200, refers to this kind of rhetoric as a form of "theatre."

3. For some examples of this language, see an article in the *Columbia Register* on May 25, 1894, entitled "CURSING THE RED CROSS! Bitter Feeling against It in Beaufort County."

4. This may call to mind a similar thread that runs through Steinberg's *Acts of God*, in which he emphasizes that a difficulty of recovery efforts is determining where the destruction of a natural disaster ends and the deprivations of society begin.

5. For more on Black disenfranchisement in South Carolina, see Perman, *Struggle for Mastery*.

6. "A Noble Response," *The State*, Sept. 9, 1893, 8.

7. "Distress on Jenkins Island," *The State*, Sep. 11, 1893, 5.

8. "The Red Cross Muddle," *The State*, Sept. 12, 1893, 1.

9. Jones, *American Red Cross*, 28–29.

10. Jones, 28–29.

11. Barton had requested assistance from the secretary of the treasury, who granted her use of a couple revenue cutters. The letter of thanks she penned describing this was reprinted in "Sea Island Sufferers," *The State*, Nov. 29, 1893, 1.

12. Barton, "Mid-Field Report," Jan. 7, 1894, folder 621, Christensen Family Papers.

13. Thomas Martin to J. E. Hemphill, Nov. 7, 1893, box 114, folder "Correspondence, Aug. 1893–Feb. 1895 and undated," Sea Islands Relief Operations, Red Cross Files, images 7–9 (hereafter Red Cross Correspondence, image [number]). The Red Cross representative's name is spelled at different times both as "McDonald," "Mcdonald," and "MacDonald," sometimes both within official reports. McDonald is used here.

14. John McDonald to J. C. Hemphill, Nov. 18, 1893, Red Cross Correspondence, images 20–23.

15. Mather, *Storm-Swept Coast*, 75.

16. Clara Barton to Ben Tillman, Dec. 1893, Governor Benjamin R. Tillman Papers, South Carolina Department of Archives and History.

17. Clara Barton Diary, Jan. 9, 1894.

18. Niels Christensen to wife, Abbie Christensen, Mar. 1894, folder 26, Christensen Family Papers.

19. "To the People of the Islands (To be read by their Pastors)," Mar. 23, 1894, 6, Red Cross Speeches and Writings, image 46.

20. Hubbell to Barton, n.d., Red Cross Labor Reports, image 113.

21. "D. C. Washington to his parishioners," Mar. 23, 1894, 6, Red Cross Speeches and Writings, image 48.

22. "Response of Mr. Pullman," Red Cross Speeches and Writings, image 49.

23. "Response of Tillinghast," Red Cross Speeches and Writings, image 50.

24. Red Cross Labor Reports, image 39. William Grant is listed as the boss of Pocotaligo on the "Main Land," in an undated report.

25. This trial is transcribed in its entirety in the *Index to the Miscellaneous Documents*, 651. It is a fascinating trial well worth reading.

26. Eugene Gregorie to Colonel William Elliott, Mar. 29, 1894, Red Cross Correspondence, image 59.

27. Petitions included consecutively within box 115, folder "Resolutions and Statements, Sept. 1893–June 1894 and undated," Sea Islands Relief Operations, Red Cross Files, https://www.loc.gov/item/mss119730625/, images 2–26 (hereafter Red Cross Resolutions, image [number]).

28. Quoted in Ayers, *Promise of the New South*, 257.

29. Foner, *Nothing but Freedom*, 91–107, provides an excellent rundown of the strikes, their repression, and the lingering consequences.

30. Here, I allude to Hahn, *Nation Under Our Feet*.

31. Untitled petition, n.d., Red Cross Resolutions, image 7.

32. Untitled petition 1 to George Pullman, Red Cross General Field Secretary, Mar. 28, 1894, Red Cross Resolutions, image 11.

33. Untitled petition to George Pullman, Apr. 3, 1894, Red Cross Resolutions, image 15.

34. Untitled petition 2 to George Pullman, Mar. 28, 1894, Red Cross Resolutions, image 12.

35. Untitled statement notarized by William Grant, Apr. 2, 1894, Red Cross Resolutions, image 14.

36. Untitled petition 2 to George Pullman, Mar. 28, 1894, Red Cross Resolutions, image 12.

37. Untitled petition to Clara Barton, Apr. 3, 1894, Red Cross Resolutions, image 17.

38. Untitled petition to Clara Barton, Apr. 3, 1894, Red Cross Resolutions, image 18.

39. Untitled petition to Clara Barton, Apr. 3, 1894, Red Cross Resolutions, image 18.

40. Untitled petition to Clara Barton, Mar. 24, 1894, 3, Red Cross Resolutions, image 10.

41. Untitled petition to Dr. Hubbell, Jun. 22, 1894, 1, Red Cross Resolutions, image 22.

42. Undated letter, William Grant to Clara Barton, ca. June 1894, Red Cross Resolutions, image 20.

43. An estimate from Barton, *Midfield Report*.

44. "Cursing the Red Cross," *Columbia Register*, May 25, 1894, 1.

45. "The Governor's Appeal," *Watchman and Southron*, May 30, 1894, 1.

46. "Give, and Quickly!," *The State*, May 18, 1893, 4.

47. Mrs. McCants to Governor Tillman, May 28, 1894, folder 3 (first quotation); N. S. Rune to Governor Tillman, June 5, 1894, folder 12 (second quotation); E. P. Shedel to Governor Tillman, June 4, 1894 (third quotation); Jacob Lomox to Governor Tillman, May 31, 1894, folder 8 (fourth quotation); J. P. Pickett to Governor Tillman, June 4, 1894, folder 11 (fifth quotation); all in box 35, Governor Benjamin R. Tillman Papers.

48. "Receipts and Expenditures," *The State*, June 17, 1894, 5.

49. Letter from the Sea Island Relief Committee to Benjamin Ryan Tillman, June 1, 1894, folder 9, box 35 "Governor Benjamin Ryan Tillman Letters Received and Sent, May 28–July 7, 1894," Governor Benjamin R. Tillman Papers.

50. "The Red Cross Vindicated," *News and Courier*, June 1, 1894, 1, box 115, folder "Newspaper clippings, 1893–1894," Sea Islands Relief Operations, Red Cross Files, image 30 (hereafter Red Cross Newspaper Clippings, image [number]). Barton wrote the letter to Barnwell on May 27, 1894, according to the newspaper.

51. "Gratitude of Bluffton," *The State*, June 16, 1894.

52. Barton, *Red Cross*, 268.

53. Dr. Magruder of the U.S. Marine Hospital Service, Red Cross General Reports, images 67–69.

54. "Memoranda of Sea Islands Relief Methods," n.d., Red Cross Speeches and Writings, images 33–34, quotation on 34.

55. H. L. Bailey to Clara Barton, June 1, 1894, Red Cross General Reports, image 24.

56. Dr. Hubbell's report in Barton, *Red Cross*, 237.

57. Dr. Hubbell's report in Barton, 242.

58. Dr. Hubbell's report in Barton, 242.

59. "Happy . . .": a quote from Mrs. Harriette L. Reed, an associate of the Red Cross's, included in Barton, *Red Cross*, 264. "Grateful . . .": in Barton, *Story of the Red Cross*, 79.

60. "Helping a Stricken People," *New York Times*, Apr. 16, 1894.

61. Testimony of Dr. E. Winfield Egan, June 2, 1895, Red Cross Speeches and Writings, images 1–13, quotations on 13.

62. Dr. Hubbell's report in Barton, *Red Cross*, 242.

63. "A People's Gratitude: The Vanderbilts Render Honor unto Clara Barton." A reprint from the Charleston *News and Courier*, July 1, 1894. From Red Cross Miscellany, images 2–4.

64. Kaufman quoted in "A People's Gratitude," image 3.

65. No title, *New York Times*, May 31, 1894.

66. Niels Christensen to Abbie, June 10, 1894, folder 26, Christensen Family Papers.

67. From Committees to Barton, Feb. 26, 1894, image 34; Mar. 20, 1894, image 47; Mar. 30, 1894, image 2; and June 7, 1894, image 82, for examples of the thanks given, Red Cross Correspondence.

68. S. J. Pinckney to Barton, Mar. 20, 1894, Red Cross Correspondence, image 47.

69. M. C. Black to Barton, June 2, 1894, Red Cross Correspondence, image 74.

70. Women's Aid to Barton, June 5, 1894, Red Cross Correspondence, image 75.

71. Residents of Bonnie Hall to Clara Barton, June 7, 1894, Red Cross Correspondence, image 82.

72. "Commending the Red Cross," June 8, 1894, newspaper not named, Washington, D.C., Red Cross Newspaper Clippings, image 30.

73. Barton, *Red Cross*, 198.

74. Untitled, June 12, 1894, *Evening Telegram*, Red Cross Newspaper Clippings, image 51.

75. "The Work of the Red Cross," *News and Courier*, Jun. 23, 1894, Red Cross Newspaper Clippings, image 41.

76. "The Work of the Red Cross."

77. "The Work of the Red Cross."

78. "Berating Miss Barton," *News and Courier*, Jul. 6, 1894.

79. A bulletin by Clara Barton, Feb. 26, 1895, 2, Red Cross Correspondence, image 104.

80. Letter from J. B. Hubbell to Mr. Wistar, reprinted in an untitled article from *American Friend*, Aug. 5, 1894, Red Cross Newspaper Clippings, image 53.

81. Letter from Michael Smalls and others to Clara Barton, July 2, 1894, reprinted in an untitled article from *American Friend*, Aug. 5, 1894, Red Cross Newspaper Clippings, image 53.

82. Numerous historians have written about this topic. Kelley's *Hammer and Hoe* is most explicit in vocalizing the simultaneous power and fragility of these alliances. Foner's *Reconstruction* also points toward those alliances, as do Ayers's *Promise of the New South*, Miller in *Gullah Statesman*, Rosengarten's *All God's Dangers*, and Gilmore's *Defying Dixie*.

Chapter Seven

1. "The Work of the Red Cross," *News and Courier*, June 25, 1894, Red Cross Newspaper Clippings, image 39.

2. Kiser, *Sea Island to City*, 104.

3. See Du Bois, *Black Reconstruction*, ch. 9, "The Price of Disaster," 325.

4. Elkinton, *Selections*, 304.

5. The phrase "negro domination" appears throughout speeches of the day. See one example in McCabe to C. A. Woods, June 28, 1895, folder 5 "1 Aug 1893–28 June 1895," Montgomery Family Papers, South Caroliniana Library.

6. This was not new. For more on how enslaved African Americans were made to "build" early republic South Carolina, see Quintana, *Making a Slave State*.

7. Gilmore, *Gender and Jim Crow*, 3.

8. Tillman quoted in "Tillman Seems Uneasy," *The State*, Oct. 30, 1894, 1.

9. Some historiography on the early years of Jim Crow, some of which incorporate an environmental history angle, includes, on the role of environmental planning generally in the New South and Jim Crow, Bryan, *Price of Permanence*; Woodward, *Strange Career*; Ring, *Problem South*; Mauldin, *Unredeemed Land*.

10. David Duncan Wallace describes all of this—as a negative. See Wallace, "South Carolina Constitution," 22–25.

11. South Carolina's Eight Box Law of 1882 mandated that during an election, every state office should have a separate, labelled ballot box; any ballot placed in the incorrect box would be invalidated. This system was essentially a literacy test, but it also was an easy system for corrupt precinct managers to mislabel boxes, misidentify boxes to voters who asked for clarifications, and participate other forms of deliberate obfuscation. The law also instituted strict voter registration requirements.

12. For more on fusionist tickets, see Marszalek, *Black Congressman*; Perman, *Struggle for Mastery*; and Wise and Rowland, *Rebellion, Reconstruction, and Redemption*.

13. The best discussion of the 1868 convention is still Du Bois, *Black Reconstruction*, ch. 10, "The Black Proletariat in South Carolina."

14. Quoted in Perman, *Struggle for Mastery*, 104.

15. Calling upon Kantrowitz's line from *Ben Tillman*, 2, in which he writes that white supremacy was "more than a slogan and less than a fact," instead "a social argument and a political program."

16. *Journal of the Constitutional Convention*, 2, a speech delivered on Sept. 10, 1895.

17. State of South Carolina Executive Chamber, Feb. 18, 1895, folder 1 "11MSS24 May 1845—November 1895," Tillman Papers.

18. "Tillman Seems Uneasy," 1.

19. "Tillman Seems Uneasy," 1.

20. For more on these conventions, see Perman, *Struggle for Mastery*, 102; "What the Negroes Want," *The State*, Jan. 25, 1895, 8; "Negro Preachers Combine," *The State*, Feb. 15, 1895, 2.

21. Wallace, "South Carolina Constitution," 29.

22. "Coming to the Convention," *The State*, Aug. 26, 1895, 8.

23. "Coming to the Convention," 8.

24. Here, it is useful to turn to studies on Civil War memory and books on white supremacy in South Carolina: Kytle and Roberts, *Denmark Vesey's Garden*; Baker, *What Reconstruction Meant*; Poole, *Never Surrender*; Blight, *Race and Reunion*; and Brundage, *Southern Past*.

25. Again, I think of Hahn, *Nation Under Our Feet*; Schwalm, *Hard Fight for We*; and John Scott Strickland's two essays as key works that address the idea of Black labor in the South justifying citizenship.

26. "Speeches at the Constitutional Convention by General Robert Smalls," 9.

27. "Speeches at the Constitutional Convention by General Robert Smalls," 20.

28. "Speeches by the Negroes in the Constitutional Convention," in *The Suffrage*, 6.

29. Constitution, article 2, Right of Suffrage, section 4c, in *Journal*, the appendix.

30. *Journal*, 469.

31. *Journal*, 724.

32. *Journal*, 727.

33. *Journal*, 730.

34. E. Miller, *Gullah Statesman*, 212.

35. E. Miller, 213.

36. Rowland and Wise, *Bridging*, 83.

37. The first law prohibiting emigrant agents from working in South Carolina without first obtaining a state license seemed to be on the books in 1891; see "Emigrant Agents to Pay Licenses," *The State*, Dec. 23, 1891, 1. The law was reaffirmed frequently, including one

in 1907 that increased the fine's floor from $500 to $1000; see "Emigrant Agents," *The State*, Feb. 22, 1907, 3. One famous case occurred in Greenwood, a small town far inland; see "Emigrant Agent Arrested; Crowds of Negro Women Leaving Greenwood for New York," *The State*, Mar. 15, 1903, 9. References to Florida, where emigrant agents brought Black South Carolinians to turpentine and phosphate camps, appear in "Bound for Florida," *The State*, July 15, 1899, 2; and "For Violating Emigrant Law," *The State*, Jan. 22, 1904, 3.

38. "A Striking Idea," *The State*, Sept. 11, 1893, 4.

39. *Journal*, 731.

40. Language of "redemption" used in "Greenwood's Negro Exodus," *The State*, Dec. 21, 1898, 4.

41. Tindall, *South Carolina Negroes*, 113.

42. Tindall, 113; quote about the amendments from the article "CONTRACT LABOR LAW UNCONSTITUTIONAL," *The State*, May 24, 1907, 1.

43. Congressional Record, Bill S.1149, Nov. 3, 1894, in Red Cross Newspaper Clippings.

44. "After Clara Barton," *The State*, July 23, 1894, 6.

45. "After Clara Barton," 6.

46. "Negro Emigration," *The State*, Aug. 5, 1895, 4.

47. "The Immovable Negro," *The State*, June 15, 1895, 4.

48. "The Immovable Negro," 4.

49. "The Immovable Negro," 4.

50. "Negro Emigration," 4.

51. "Negro Emigration," 4.

52. "Greenwood's Negro Exodus," *The State*, Dec. 21, 1898, 4.

53. "The Negro-Ridden Planter," *The State*, Jan. 6, 1899, 4; and "The Negro and Cotton," *The State*, Jan. 9, 1899, 4.

54. "Greenwood's Negro Exodus," 4.

55. "The Negro and Cotton," 4.

56. Kantrowitz discusses this push briefly, *Ben Tillman*, 303–4.

57. The Clemson archives contains small, worn, leather-bound notebooks that Tillman kept in his jacket pocket. One is filled with newspaper clippings on the state's demographic trends. See notebook 4, envelope 9, in Diaries and Notebooks, 1861–1905, series 8, Tillman Papers.

58. Benjamin R. Tillman, "The Negro Problem and Immigration," delivered by invitation before the South Carolina House of Representatives, Jan. 24, 1908 (Columbia: Gonzales and Bryan, State Printers, 1908), 14, box "Resolutions—Speeches Tillman, 1912–1918," folder 238 "Speeches—Tillman, 1903–1909," Tillman Papers.

59. Notebook of Ben Tillman, 1894, envelope 8, in Diaries and Notebooks, 1861–1905, series 8, Tillman Papers.

60. Notebook of Ben Tillman, n.p.

61. Notebook of Ben Tillman, n.p.

62. Tillman, "Negro Problem and Immigration," 8.

63. Tillman, 4.

64. Tillman, 11.

65. Tillman, 13.

66. Tillman, 13.

67. Tillman, 13.

68. Tillman, 13.

69. Tillman, 14.

70. Tillman, 18.

71. Tillman, 23.

72. See Prince, *Wetlands*.

73. "Work on Drainage in Low-Country," *The State*, Jan. 25, 1907, 8.

74. Vileisis, *Discovering the Unknown Landscape*. It also worth noting that drainage was a popular antebellum enterprise in the Great Dismal Swamp. See Royster, *Fabulous History*.

75. On maroons in the Great Dismal Swamp: Sayers, *Desolate Place*.

76. See Wilson, *Shadow and Shelter*.

77. See Joyner, *Down by the Riverside*, 37.

78. Perhaps the most classic study of a city and its drainage infrastructure is Colten's invaluable *Unnatural Metropolis*, on the history of the built environment of New Orleans.

79. See the excellent and detailed book on Charleston drainage and landfill Butler, *Lowcountry at High Tide*.

80. "Charleston News Gathered in a Day," *The State*, June 8, 1911, 6.

81. Oddly, no definitive study has been written just on convict labor in South Carolina, though of course a number of excellent books address it in other parts of the South, and Kamerling's recent study compares the prison system in South Carolina to that of Illinois from the 1860s to the 1880s. There is an extensive literature that examines the racist abuse of incarceration throughout the twentieth and into the twenty-first centuries, such as Alexander, *New Jim Crow*. For books on convict labor in the South specifically, see Blackmon, *Slavery by Another Name*; LeFlouria, *Chained in Silence*; Lichtenstein, *Twice the Work*; Mancini, *One Dies, Get Another*; Shapiro, *New South Rebellion*; Kamerling, *Capital and Convict*.

82. From a speech at the South Carolina State House before the General Assembly where Cosgrove addressed the assembly, published in *The State*, Feb. 2, 1907, 8.

83. *Year Book—1903*, 92. Butler, *Lowcountry at High Tide*, discusses Cosgrove and the SDCC, 139, saying that his project laid "the groundwork for growth in the North Area."

84. "Mr. Cosgrove on Drainage," *The State*, Feb. 6, 1907, 8.

85. "For the Welfare of South Carolina," *The State*, Nov. 23, 1903, 8.

86. *Year Book—1903*, 96–97.

87. *Year Book—1903*, 97.

88. "A Striking Idea," *The State*, Sept. 11, 1893, 4.

89. The "immigration movement" was a major issue generally, receiving attention in the *Handbook* (1907), ch. 12, "Commerce, Transportation, and Immigration and Emigration." On p. 511, the *Handbook*'s authors connect the current immigration movement with the desire of colonial and antebellum whites to increase the white population of the state, as a method of rebalancing the state's demographic from what they perceived as a "dangerous" Black majority.

90. Tillman, "Negro Problem and Immigration," 11.

91. Tillman, 16.

92. Tillman, 16.

93. Tillman, 16.

94. This kind of dispossession to institute "proper use" as designated by a government body has some historical parallels; I think of the dispossession of rural peoples, often

American Indian, from National Park lands. For a couple books on the subject, see Spence, *Dispossessing the Wilderness*; and Jacoby, *Crimes against Nature*.

95. "Baldwin's Speech," *The State*, Sept. 15, 1905, 3.

96. Theodore Jervey, pamphlet draft for "Migration of the Negroes," c. 1895, box 28–266, folder 1, Theodore Jervey Papers.

97. *South Carolina General Assembly Reports and Resolutions*, vol. 2, 93.

98. *Year Book—1900*, 148.

99. *Year Book—1900*, 150.

100. *Year Book—1900*, 148.

101. *Year Book—1900*, 150. The three other most common ways in which people arrested were dealt with was by being dismissed: 865, with Black men making up 428; white men, 277; Black women, 128; and white women, 32. Second was sent to the jail: 514, with Black women the largest group receiving that punishment at 293. Third, 364 people were, curiously, "Sent to Hospital," with 218 Black men, 80 white women, 52 Black women, and 14 white women receiving that sentence. No women were sentenced to chain gang labor within the Charleston system, as far as I can tell, but many Black women across the South were forced into convict labor. See LeFlouria, *Chained in Silence*; and Haley, *No Mercy Here*.

102. "South Carolina News and Gossip," *The State*, Sept. 13, 1898, 2.

103. *Year Book—1903*, 93.

104. American Forestry Association, *Forestry, Irrigation, and Conservation*, 99.

105. American Forestry Association, 99.

106. *Year Book—1903*, 96.

107. *Year Book—1903*, 94.

108. *Year Book—1903*, 97, 101–2. The sentencing comes from *Year Book of the City of Charleston, S.C.*, from 1916, 489. One was killed and one escaped, in circumstances that were not reported; and eleven completed their sentences during that time. The reference to the depth of the water is from *The State*, Feb. 2, 1907, 8.

109. *Year Book—1903*, 97.

110. Indeed, the pollution from fertilizer factories on the Charleston Neck lingered for generations, with the Environmental Protection Agency labeling the former locations of the Wando, Stono, Atlantic, Ashepoo, Pacific Guano, and Etiwan companies as Superfund sites. Read more about it at "Addressing a Century of Pollution," *Post and Courier*, May 11, 2014.

111. From South Carolina General Assembly, *Reports and Resolutions*, vol. 2, 93.

112. *Year Book—1903*, 97.

113. *Year Book—1903*, 95, 100.

114. *Year Book—1903*, 96.

115. National Conservation Commission, *Report*, 366.

116. "Cosgrove to the General Assembly," *The State*, Feb. 2, 1907, 8.

117. *Year Book—1903*, 103.

118. National Conservation Commission, *Report*, 366.

119. South Carolina General Assembly, *South Carolina General Assembly Reports and Resolutions*, reports on the SDC's ongoing oversight, on p. 140.

120. A note from T. W. Limehouse, the captain of the convict camp, in *Report of the Sanitary and Drainage Commission for Charleston County, South Carolina*, 42.

121. *Year Book—1892*, 167.

122. See the Charleston *Year Books*, which contain a section "Convict Management" from 1892 to 1922. After that time, the city released yearbooks more sporadically and halted production for a few years during the Great Depression. The *Yearbook, 1932–1935* indicates that the city government opened a prison farm north of Charleston for those convicted of misdemeanors but does not indicate that they were being forced into chain gang labor within the city. See p. 22 for a photograph of the prison farm; the caption states that it was "part of the City's program to provide sanitary and healthful surroundings for the local prisoners serving sentences for misdemeanors."

123. *Year Book—1903*, 107.

124. "For the Welfare of South Carolina," *The State*, Nov. 23, 1903, 8.

125. Oliphant, *Evolution*, 11.

126. Bethea, *Code of Laws*, 2:656.

127. Law quoted in "Leasing of Convicts," *The State*, Jan. 9, 1902, 6. Since 1885, the state had allowed the use of prisoners from the state penitentiary on county chain gangs.

128. "Drainage Unconstitutional," *The State*, June 29, 1908, 4.

129. "To Improve Lands by Drainage Law," *The State*, Apr. 13, 1911, 12, notes the passage of the bill. A. G. Smith, a U.S. Office of Farm Management rep, spoke at the University of South Carolina on drainage, reported in "Makes an Address on Drainage," *The State*, Feb. 20, 1912, 9, though he reported it as three. In *South Carolina General Assembly Reports and Resolutions*, vol. 2, 1012, it was described as two, so the number seems to have varied.

130. On "the father of drainage": "Outline Drainage Work in His Annual Report," *The State*, Jan. 12, 1911, 6.

131. "Outline Drainage Work in His Annual Report," 6.

132. *Reports and Resolutions*, 1912, 872–873.

133. *Reports and Resolutions*, 1912, 873.

134. "Drainage with Tiling in the Lower Part of the State," *The State*, June 19, 1910, 11.

135. "To Improve Lands by Drainage Law," *The State*, Apr. 13, 1911, 12.

136. *Reports and Resolutions*, 1913, vol. 2, 1012.

137. "Outline Drainage Work in His Annual Report," 6.

138. *South Carolina General Assembly Reports and Resolutions*, 1921, vol. 1, 155.

139. "Drainage Saves Lands from Loss," *The State*, Oct. 14, 1906, 2.

140. First, a majority of resident landowners had to present the county clerk of the court with a signed petition signaling their intent to form a drainage district; then, the clerk of court would appoint a board of supervisors to report on the practicality of the proposal; third, hearings had to follow surveys, plans, and cost estimates; fourth, the clerk of the court would officially declare the district established and commissioners appointed; and finally, the state would issue bonds for twenty-five years at a 6 percent interest rate, with construction work through contract. In *South Carolina General Assembly Reports and Resolutions*, 1921, vol. 1, 872.

141. The *South Carolina General Assembly Reports and Resolutions* issued yearly updates on the state and progress of chain gangs, which were operated by the counties but over which the state government had oversight. The conditions for county chain gangs were usually atrocious, with only the SDC receiving a "decent" rating from the state's 1,000-point evaluation system. The state oversight committee frequently railed against the use of state convicts by the counties, arguing that the conditions were too frequently unsanitary and unhealthy

and that the prisoners were too likely to suffer abuse. See, for example, *Reports and Resolutions*, 1920, vol. 2, 93. The prevalence of chain gang forced labor in South Carolina and the use of forced labor in creating a certain sort of landscape strikes me as a Jim Crow update of Johnson, *River of Dark Dreams*, ch. 8, "The Carceral Landscape."

142. For more on the Great Migration generally, see Wilkerson, *Warmth of Other Suns*. For more on the Great Migration from the Lowcountry specifically, see Kiser, *Sea Island to City*.

Chapter Eight

1. The Weather Bureau's perceived failures were highlighted in the local press: see "Weather Bureau Criticised," *News and Courier*, Aug. 31, 1911, 3.

2. A useful overview of the 1911 hurricane appears in Fraser, *Lowcountry Hurricanes*, 210–18.

3. Rossa B. Cooley, "Report of the Principal," *Annual Report of the Penn Normal, Industrial and Agricultural School of St. Helena Island, Beaufort County, South Carolina, Fiftieth Year, 1911–1912*, 13, box 51, folder 446, Penn School Papers.

4. "Hum of Business Goes Merrily On," *News and Courier*, Aug. 31, 1911, 1.

5. "Can Take Care of Themselves," *News and Courier*, Sept. 9, 1911, 4.

6. "Can Take Care of Themselves," 4.

7. Doyle, *New Men, New Cities*, 172.

8. Doyle, 172.

9. Du Bois, *Black Reconstruction*, 700.

10. Fraser, *Lowcountry Hurricanes*, 193.

11. Fraser, 201. See also "Cyclone Sweeps South Coast," *The State*, Oct. 3, 1898, 1.

12. Rowland and Wise, *Bridging*, 184; Kiser, *Sea Island to City*, 105.

13. Kiser, 103.

14. Bryan, *Price of Permanence*, 88–92, provides an excellent rundown of the late years of phosphate.

15. "Rising from the Ruins," *News and Courier*, Aug. 30, 1893, 1.

16. "Out of the Depths," *News and Courier*, Aug. 30, 1893, 1.

17. "Out of the Depths," 1.

18. "Sweep of the Mighty Wind," *New York Times*, Sept. 2, 1893, 1.

19. "Third Annual Report of the Board of Phosphate Commissioners," *Reports and Resolutions*, 328.

20. Multiple sources confirm Tillman's dislike of phosphate: see Kantrowitz, *Ben Tillman*, 186–89; McKinley, *Stinking Stones*, 163; and Kiser, *Sea Island to City*, 100.

21. "25%" from Number as of Nov. 10, 1893, "Report of the Comptroller-General," *Reports and Resolutions*, 419; the $75,000 decision from "Third Annual Report of the Board of Phosphate Commissioners," 325.

22. Rogers, "Phosphate Deposits of South Carolina," 217.

23. Phosphate executives to Benjamin Tillman, Oct. 16, 1893, item 35, folder 15, box 24, Governor Tillman letters.

24. "Third Annual Report of the Board of Phosphate Commissioners," *Reports and Resolutions*, vol. 1, 328.

25. "Annual Report of the State Phosphate Inspector," *Reports and Resolutions*, vol. 1, 709.

26. "In Our Phosphate Fields," *The State*, May 27, 1894, 5.

27. "Annual Report of the State Phosphate Inspector," 717.

28. "Phosphate in Tennessee," *The State*, Apr. 19, 1894, 1.

29. "Competition from Africa," *The State*, Sept. 22, 1894, 1.

30. "Poor South Carolina," *The State*, Dec. 29, 1894, 1.

31. Strangely, there is not much published on the Florida phosphate industry. One reference is Blakey, *Florida Phosphate Industry*.

32. Rogers, "Phosphate Deposits of South Carolina," 218.

33. McKinley, *Stinking Stones*, 100.

34. Tindall, *South Carolina Negroes*, 103.

35. Kiser, *Sea Island to City*, 100.

36. "Getting Along Nicely," *News and Courier*, Aug. 31, 1893, 1.

37. "Rough on Rice Planters," *News and Courier*, Sept. 1, 1893, 2.

38. "Rough on Rice Planters," 2.

39. From Langdon Cheves to Sophia, Sept. 13, 1893, 2, 12/94/5, Langdon Cheves Papers.

40. From Langdon Cheves to Sophia, n.d., late Sept. 1893, 12/94/5, Langdon Cheves Papers.

41. Lewis M. Grimball to Elizabeth, Sept. 21, 1893, 8, Grimball Family Papers.

42. "Getting Along Nicely," *News and Courier*, Aug. 31, 1893, 1.

43. No title, *News and Courier*, Aug. 28, 1893, 2.

44. Mr. Sanders to Berkeley Grimball, Aug. 29, 1893, 3, Grimball Family Papers.

45. Quoted in Stewart, "*What Nature Suffers*," 236.

46. Pringle, *Woman Rice Planter*, 236.

47. From "Georgetown County's Vivid Story," *The State*, n.d., folder 16, James Henry Rice Papers.

48. Rice, from an op-ed in *The State*, "Planting of Rice in South Carolina: Sketch of Its Origin and the Splendid Civilization to which It Gave Rise—the Day of Its Decline and Causes," n.d., folder 16, James Henry Rice Papers.

49. Rice, in "Planting of Rice in South Carolina: Sketch of its Origin and the Splendid Civilization to which it Gave Rise—The Day of its Decline and Causes," the section entitled "THE STORM OF 1893," *The State*, n.d., c. 1920s, folder 16, James Henry Rice Papers.

50. *Handbook* (1907), 305.

51. Rice, "Planting of Rice in South Carolina."

52. Pringle, *Woman Rice Planter*, 446.

53. Hammond, *South Carolina*, 57.

54. See statistics on truck farming in *Handbook* (1907), 291, which discusses it as still limited to 30,000 acres in Charleston, Colleton, Beaufort, Horry, and Berkeley Counties. The *Handbook* looks at Port Royal Island and its vicinity in 1906 as an example, listing potatoes (by far the most valuable crop, bringing in $180,761), peas (a far second at just $16,255), cucumbers, lettuce, and radishes as the top five most valuable truck-farming crops. Cabbages, potatoes, asparagus, peas, and strawberries seemed to be quite popular in all truck-farming counties. The *Handbook* also contains a chapter on forestry, pp. 538–552. "Yellow" pine (long-leaf pine) and bald cypress were the two most favored trees in the logging industry; both shortleaf and loblolly captured significant parts of the market. In 1905, 9,656 people worked in South Carolina's logging industry (544). There is not much literature on the logging industry in the Lowcountry, though McAlister, *Lumber Boom*, provides some information, almost entirely on Georgetown County.

55. Indeed, South Carolina hit a high point of cotton production, driven largely by a boom inland, in 1904, during which farmworkers picked 1.2 million bales of cotton. *Handbook* (1907), 269. Another uptick until the boll weevil reached a high in 1918. Two essential (and quite different) books on the South Carolina textile industry are Carlton, *Mill and Town*; and Simon, *Fabric of Defeat*.

56. Keyserling, Herbert, 1915–2000, "Jewish Heritage Collection: Oral history interview with Herbert Keyserling," Lowcountry Digital Library, College of Charleston Libraries, 1995-07-181995-07-19, p. 21 of the transcript (hereafter Keyserling interview, [page number]).

57. Information from the National Register of Historic Places Registration Form for "The Corner Store and Office," U.S. Department of the Interior, Oct. 6, 1988, 3–4, for the description of the historic store.

58. Keyserling interview, 22.

59. Keyserling interview, 23.

60. Keyserling interview, 54.

61. For more on how federal programs through the USDA enacted discriminatory policies that led to the 93 percent drop of African American farmers between 1941 to 1974, see Daniel, *Dispossession*.

62. Prof. Ira Williams, State Agent, to Dr. S. A. Knapp, Special Agent in Charge of the USDA program, May 1, 1909, reel 1 mfm 271, "Annual narrative & statistical reports from state offices and county agents, South Carolina, 1909–1944," Penn School Papers, South Caroliniana Library.

63. Williams to Knapp, May 1, 1909, and Ira Williams, "Annual Report," 1911, reel 1.

64. Williams, "Annual Report," 1909, reel 1 mfm 271, "Annual narrative & statistical reports."

65. Williams, "Annual Report," 1911, page not listed.

66. Williams, "Annual Report," 1911, page not listed.

67. Williams, "Annual Report," 1911. For a monograph on the school's history, see Burton and Cross, *Penn Center*.

68. Letter to the editor from Rossa B. Cooley, October 1911, 2, mfm reel 1, Penn School Papers, Southern Historical Collection at the University of North Carolina Library, South Caroliniana Library.

69. Williams, "Annual Report," 1911, 12.

70. Williams, "Annual Report," 1911, 13.

71. J. E. Blanton, County Agent, "Report of the work of the county agent, calendar year 1916," and Frank McKinley, County Agent, "Report of the work of the county agent, calendar year 1916," for the Cooperative Education Work in Agriculture and Home Economics, for the USDA, reel 1, mfm 271, "Annual narrative & statistical reports."

72. Blanton's 1916 report places its starting date at 1912 on the page labeled "ORGANIZATION," "Annual narrative & statistical reports."

73. In L. Hollingsworth Wood to Rossa B. Cooley, Oct. 31, 1911, mfm reel 1, Penn School Papers; the $2,000 amount referenced in Francis Cope Jr. to Mr. J. E. Blanton, Mar. 11, 1912, mfm reel 1, Penn School Papers.

74. Rossa Cooley, "Report given by Miss Cooley at the Board of Trustees' Meeting," Jan. 14, 1913, 2, mfm reel 1, Penn School Papers. The one dollar fee is mentioned in the

L. Hollingsworth Wood letter but confirmed in a contract signed by J. C. Smalls, Jul. 17, 1916, mfm reel 2, Penn School Papers.

75. Cooley, "Report given," 2.

76. For more scholarship on Black farmers after the Civil War and into the twentieth century, see Petty, *Standing Their Ground*; and for more on cooperative work among Black farmers in the twentieth century South, see White, *Freedom Farmers.*

77. See the excellent map in Stackhouse, *Study of the Boll Weevil*, 7.

78. Kiser, *Sea Island to City*, 110.

79. Riggs, "Boll Weevil."

80. Riggs, 11, 9.

81. Riggs, 15.

82. Riggs, 15.

83. Riggs, 16.

84. See Giesen, *Boll Weevil Blues*, who makes this argument: the boll weevil was damaging, to be sure, but southerners also used the weevil's intrusion as a cover for more complex social, political, economic, and ecological problems.

85. See the Penn School Papers, 1915–1923, on reel 2 at the South Caroliniana Library, which contains applications to the society scattered throughout the microfilm. I calculated the averages from those papers.

86. C. L. Baxter, "Annual Report, 1922," on roll 9, 4, "Annual narrative & statistical reports," mentions the abandonment of cotton and the eagerness of farmers for other crops.

Chapter Nine

1. All background information on landownership and the nature of the post–Civil War Ossabaw Island from Dorsey, "'Great Cry,'" 224–52.

2. Dorsey, 244.

3. Dorsey, 228.

4. See Fetig, "Pin Point," 138–39.

5. The Pin Point Heritage Museum is located in the old cannery building.

6. Nixon, *Slow Violence*, 2.

7. Kiser, *Sea Island to City*, 104. Kiser was a sociologist—and a eugenicist. His interviews are sound, but he as a scholar must be regarded with full knowledge of who he was. For more on the Great Migration generally, see Wilkerson, *Warmth of Other Suns*; Lemann, *Promised Land*, which focuses more on the later decades of the migration; Berlin, *Making of African America*; Gregory, *Southern Diaspora*; Chatelain, *South Side Girls*; and, in one of the few studies that seriously considers environmental history and the Great Migration, charting how Black Chicagoans from the South interacted with leisure and nature in a new urban setting, McCammack, *Landscapes of Hope*.

8. Aside from Kiser, there is not much literature at all on the Great Migration and the sea islands, though Cooper, *Making Gullah*, has some information.

9. Quoted in Frederick Christensen Diary, Oct. 30, 1896, 156, Christensen Family Papers.

10. Quoted in Rowland and Wise, *Bridging*, 157.

11. Finnegan, "Lynching." For another book on lynchings in South Carolina, see Baker, *This Mob*.

12. In Marshall Stein, 1935–, "Jewish Heritage Collection: Oral history interview with Marshall Stein," Lowcountry Digital Library, College of Charleston Libraries, 2/19/2015, p. 10 of the interview transcript.

13. Keyserling interview, 52.

14. Keyserling interview, 49.

15. Interview with Harriet Birnbaum Ullman,1927–, and Albert Jacob Ullman, 1923–2008, "Jewish Heritage Collection: Oral history interview with Harriet Birnbaum Ullman and Albert Jacob Ullman," Lowcountry Digital Library, College of Charleston Libraries, 5/24/1998, p. 38 of the transcript.

16. Smalls was described in this way by his grandson Willie, quoted in Miller, *Gullah Statesman*, 244.

17. "The Recent Calamity of the Sea Islands," *The Friend*, Nov. 19, 1898, 149.

18. "Report of the Penn Normal and Industrial School, 1898," (St. Helena: Penn Normal and Industrial School Press, 1898), n.p., folder 444, box 51, Penn School Papers.

19. "Report of the Penn Normal and Industrial School, 1898."

20. Mather, *Storm-Swept Coast*, 55.

21. Kiser, *Sea Island to City*, 118.

22. *Handbook* (1907), 525.

23. *Handbook* (1907), 525.

24. *Handbook* (1907), 525.

25. *Handbook* (1907), 573.

26. *Handbook* (1907), 573.

27. Rowland and Wise also include an extremely thorough and useful accounting of the demographics and statistics of this era in *Bridging*, ch. 9.

28. Kiser, *Sea Island to City*, 104, 107.

29. Kiser, 108.

30. Stackhouse, *Study of the Boll Weevil*, 27.

31. For more, see Giesen, "The Truth about the Boll Weevil."

32. Edward Shaffer, "The New South—the Negro Migration," *The Atlantic Monthly*, Sept. 1922, Oversize Scrapbook, 1894–1945, Edward Terry Hendrie Papers.

33. Shaffer, "The New South."

34. Shaffer, "The New South."

35. Shaffer, "The New South."

36. Ambrose Gonzales, "South Carolina Masters: Humanity of Slave Holders," clip in the Oversize Scrapbook, 1894–1945, Edward Terry Hendrie Papers. Newspaper editors across the state carped at Shaffer and quickly turned to ad hominem attacks. Gonzales doubted that Shaffer could sufficiently understand the racial dynamics in the state, for while he was born in Colleton County, South Carolina, he had committed the grave sin of Yankee parentage, with a father from New Jersey and a mother from New York. Other newspapers piled on. The *Greenwood Index-Journal* concurred that "slavery was a great evil, an economic as well as a moral evil, but it meant for the slaves and their descendants the grandest opportunity to escape from a wild and savage life that has ever come to man of any color in any time. . . . No better friend of the colored man in America can be found today than the former slave owner or his children." A white farmer, writing in to a newspaper, chafed at Shaffer's call to end the "local outbursts of race hatred, of ignorance or of blind

intolerance" against African Americans, so that the white South could progress, grousing that the North was as rife with injustice as the South.

37. James Henry Rice to R. M. Kennedy, Apr. 20, 1923, folder 8, James Henry Rice Papers.

38. William Henry Frierson, "The Negro Exodus," n.d., folder 2 "21 Nov. 1932 to 16 Dec. 1984 & no date," William Henry Frierson Papers, 1854–1932.

39. Frierson, "Negro Exodus."

40. Kiser, *Sea Island to City*, 107.

41. Kiser, 118.

42. Kiser, 118.

43. Kiser, 100.

44. Kiser, 119.

45. Kiser, 119.

46. Kiser, 99.

47. Kiser, 99.

48. Kiser, 100.

49. Kiser, 99.

50. Kiser, 120.

51. Kiser, 120.

52. Kiser, 121.

53. Kiser, 120.

54. Kiser, 124.

55. Kiser, 126.

56. Kiser, 126.

57. Kiser, 127.

58. Kiser, 131.

59. Kiser, 136.

60. Kiser, 135.

61. Kiser, 104.

62. James Henry Rice, "Planting of Rice in South Carolina: Sketch of its Origin and the Splendid Civilization to which it Gave Rise—The Day of its Decline and Causes," n.d., James Henry Rice Papers. However, I can deduce the early 1920s as the date range because that is the period in which Rice wrote his column for *The State*.

63. James Henry Rice, "Georgetown's Vivid Story," n.d., c. 1920s, James Henry Rice Papers.

64. Carney touches on this dynamic, in which whites constructed a history of rice that obscured all African contributions to its rise in the Lowcountry. Carney, *Black Rice*, 160–61.

65. They came to pair this rhetoric, which I discuss here in their writings, with preservation projects and tourism advertisements, as brilliantly described in Yuhl, *Golden Haze of Memory*.

66. Pringle, *Woman Rice Planter*, 7.

67. For an excellent exploration of Smith's paintings, see Kirby, "Unenslaved through Art."

68. Pringle, 35.

69. James Henry Rice, "Planting of Rice in South Carolina," James Henry Rice Papers.

70. Clipped in the E. T. H. Shaffer Oversize Scrapbook, Edward Hendrie Shaffer Papers.

71. Mack Taylor interviewed W. W. Dixon, "Vol. XIV: South Carolina, Part 4," *Slave Narratives*, 157.

72. Genevieve W. Chandler interviewing Gabe Lance, "Vol. XIV: South Carolina, Part 3," *Slave Narratives*, 92.

Epilogue

1. Quash Stevens to Adele Vanderhorst, Aug. 29, 1893. The following description of the destruction on the island is from the same letter or from his letter to Adele on Sept. 13, 1893, 12/213/15, Vanderhorst Family Papers.

2. The accounts that have discussed Quash have not been able to identify much, if any, information about Quash's first wife. I have found a considerable amount on his second wife though. Her name was Julia W. Gibbs, she was born in Charleston in 1850, and she was of the first graduating class from the Hampton Institute. In 1893, just before the hurricane, a representative from the Hampton Institute tracked her down and interviewed her: "He described the house as very neat and tasteful, one of the pleasantest in that vicinity. The wife and home-maker spoke with enthusiasm of Hampton, and he saw her diploma hanging on the wall. . . . She was delighted to hear from Hampton; says she calls it 'the cradle that rocked me.' She had taught at Columbia and Holly Hill, S.C., and is now teaching at her home on John's [*sic*] Island. Says she doesn't know how many she has taught, but 'some have become teachers, some dressmakers, some have been to college and become doctors.' She is now desirous to establish a permanent school on the island. . . . She has taught in the Reformed Episcopal Sunday-school ever since graduating. She says: 'the Negroes are improving in every conceivable way here. We are poor and we know it—ignorant, but whose fault? Not ours. The pupils of the many colored Institutes are grateful to the Northern friends for their never tiring and willing hands to help their hungering 'brothers in black.'" Hampton Normal School Press, *22 Years' of Work*, 23–24. The 1870 census also lists Julia as a student at the institute; see "United States Census, 1870," database with images, *FamilySearch*, Jan. 2, 2021, https://www.familysearch.org/ark:/61903/1:1:MFLR-R3Y, Julia Gibbs in entry for Rebecca Bacon, 1870.

3. J. Q. Stevens to Adele, Sept. 3, 1893, Vanderhorst Family Papers.

4. Quash to Adele, Aug. 29, 1893, Vanderhorst Family Papers.

5. Quash to Adele, Sept. 13, 1893, Vanderhorst Family Papers.

6. Quash to Adele, Aug. 17, 1893, Vanderhorst Family Papers.

7. Quash to Adele, Sept. 13, 1893, Vanderhorst Family Papers.

8. Hacker and Trinkley, "Chapter 5," *History and Archaeology*, 86; or Chicora Foundation, "*Your Servant Quash*," 4.

9. His name James appears as an inheritance—his grandson was named James Quash Jr., and a great-grandson had the same name—and in reference to his wife, Julia, in Hampton Normal School Press, *22 Years' of Work*, 23, where it describes how Julia Gibbs married "Mr. J. Q. Stevens." For more on Quash as a West African name, see Chicora Foundation, "*Your Servant, Quash*," 4; and Inscoe, "Carolina Slave Names," 534–35, which mentions the name Quash, its definition, and its role as a "day name," "used in the tribal custom of naming a child for the day of the week on which he or she was born."

10. She was likely born in South Carolina, based on information Quash gave in the 1880 census.

11. See Dusinberre's tables on the 90 percent mortality rates for children up to age sixteen years at one rice plantation in South Carolina. Dusinberre, *Them Dark Days*, 53.

12. Emerson discusses Quash's service to his brother in *Sons of Privilege*, 27–29. Quash worked as a servant and waiter and was also sent on errands to Round O to fetch "pork, eggs, chickens, and other good things."

13. Adams et al., *History and Archaeology*, 96.

14. Quash Stevens to Arnoldus Vanderhorst, Feb. 18, from Georgetown, n.d., but likely in 1865 or 1866, 12/213/15, Vanderhorst Family Papers.

15. Quoted in Adams et al., *History and Archaeology*, 86.

16. In 1900, Quash wrote that he had been managing Kiawah for thirty-four years. From Quash Stevens to Arnoldus Vanderhorst, Nov. 15, 1900, 3, 12/215/16, Vanderhorst Family Papers.

17. That is what he wrote as his occupation in his marriage certificate to Julia in 1881. See Charleston, South Carolina, U.S., Marriage Records, 1877–1887 [online database] (Provo, Ut,: Ancestry.com, 2007).

18. Quash's children's names are listed in the 1880 census, which also includes Quash's status as a widower by that date: "United States Census, 1880," database with images, *FamilySearch*, accessed Feb. 20, 2021, https://familysearch.org/ark:/61903/1:1:M6S3-N9C, Quash Staveans, Keawah Island, Charleston, South Carolina, United States; citing enumeration district ED 93, sheet 446D, NARA microfilm publication T9 (Washington, D.C.: National Archives and Records Administration, n.d.), FHL microfilm 1,255,224.

19. Unfortunately, as I said, I have been unable to find anything about Quash's first wife, who was the mother of his children. The 1870 census taker did not seem to make it to Kiawah. Julia and Quash's marriage license is from Mar. 24, 1881, and the two were married in Charleston. Quash's last name is misspelled as "Stebins," and Kiawah is spelled as "Kewaw." Both Julia W. Gibbs and Quash were listed as "light brown," and Quash listed his occupation as "Planter." Quash was forty and Julia, thirty-five. See Charleston, South Carolina, U.S., Marriage Records, 1877–1887 [online database] (Provo, Ut.: Ancestry.com, 2007). The 1900 census adds both information and some more confusion. It lists Quash as having been married for thirty-five years (in all fairness, in the 1880 census as a widower, he listed himself as both widowed and still married, so he did not see his first wife's death as somehow erasing those bonds of marriage). From that and the date of birth of their oldest daughter Eliza, we can infer that he and his first wife were likely married in 1865. The 1900 census also lists Julia as having borne either six or two children (the two numbers are written over each other), with what may be a "two" in the column for "still living." It is not, however, clear; it could be a reference to Quash's children with his first wife, given the imprecision over Quash's marital status in the same census entry. So it is difficult to take that census entry without a grain of salt.

20. Quash Stevens to Adele Vanderhorst, May 23, 1883, 12/213/15, Vanderhorst Family Papers.

21. For an overview of that history, see Hacker and Trinkley, "Chapter 5," *History and Archaeology*, 49–50. Hacker and Trinkley point out that while Kiawah Island bears their name, the island was "a rather late location . . . a refuge as they attempted to avoid direct

confrontation with the English who quickly occupied their prime lands in the vicinity of Charleston" (50).

22. Hacker and Trinkley argue that Kiawah was, for its early decades of English occupation, "used solely as range for cattle, a common practice in the early colony." Hacker and Trinkley, 53.

23. Hacker and Trinkley, 58.

24. See the reference to a third being cleared in Rowe, "Negroes of the Sea Islands," 714.

25. Hacker and Trinkley, "Chapter 5," 56.

26. Hacker and Trinkley, 64.

27. Hacker and Trinkley discuss Hagar and her three children, Sarah, Eliza, and Peter; Peter was Arnoldus's son. Elias and John "were made trustees of Hagar and she was given $2,000 for her care and the care of her children until they reached the age of 28." At age twenty-eight, Peter "legally took Vanderhorst as his surname, with Elias serving as witness to the document." The Vanderhorsts were, apparently, not thrilled: "There is some evidence that Arnoldus' manumittion [*sic*] was not looked upon favorably by the family." Hacker and Trinkley, 65. (They cite South Carolina Historical Society 12/195/27–30.)

28. Hacker and Trinkley, 66.

29. Hacker and Trinkley, 67.

30. Hacker and Trinkley, 91.

31. Hacker and Trinkley, 94.

32. As stated above, these letters appear in 12/213/15-17 of the Vanderhorst Family Papers.

33. Quash Stevens to Adele Vanderhorst, Sept. 26, 1887, and July 24, 1893, 12/213/15, Vanderhorst Family Papers.

34. His letters often ended with a description of what he was sending to Adele and Arnoldus. The specific reference to blackberries that his children picked is from Quash to Adele, May 23, 1883, 12/213/15, Vanderhorst Family Papers.

35. Quash Stevens to Adele Vanderhorst, Aug. 22, 1894, 12/213/16, Vanderhorst Family Papers.

36. Quash Stevens to Adele Vanderhorst, Oct. 14, 1895, 12/213/16, Vanderhorst Family Papers.

37. Quoted in Hacker and Trinkley, "Chapter 5," 97, from a letter that Julia A. Blake, a relative of Adele's, wrote to Mrs. Cheves Smyth in 1900, describing a visit to Kiawah in which Quash acted as their host.

38. Hacker and Trinkley, 96, 386, describe him that way. Arnoldus's own brother Elias had trouble getting along with him. Elias wrote to Arnoldus in 1916 saying, "You are a difficult man to help. You seem suspicious of motives, even when all the facts are in your possession. In addition to which it must go your way or not at all. . . . You occupy an impossible position to hang on to." Quoted in Hacker and Trinkley, 104–5.

39. The break is described in Hacker and Trinkley, 97.

40. That lease can be found in 12/213/16, Vanderhorst Family Papers.

41. On November 19, 1900, Quash wrote to Adele that the manager had treated him "So Owfull that I cane note Stain it." He also aired his grievances to Arnoldus in a letter from November 15, 1900. Arnoldus wrote to Quash on November 10, 1900, saying, "I shall be compelled to appoint a thirty party to look after my interests," and noting that Simon

Boggs, who lived near Quash on the island, "is in my employ and any interference with his duties cannot be tolerated. . . . I wish it known that I shall not hesitate to take steps to legally punish any person guilty of a possible breach." All letters in 12/213/16, Vanderhorst Family Papers.

42. On drought, Quash to Arnoldus, July 25, 1900; on sickness, Quash to Arnoldus, July 25, 1900; Quash to Arnoldus, Tuesday 20, 1900 (no month is listed, though in 1900, Tuesdays fell on the 20th in February, March, and November); on hurricane: Quash to Arnoldus, "If a storm coms How Will I pay you," Feb. 23, 1900. All letters in 12/213/16, Vanderhorst Family Papers.

43. Quash Stevens to Arnoldus Vanderhorst, Nov. 15, 1900, 12/213/16, Vanderhorst Family Papers.

44. Quash Stevens to Adele Vanderhorst, Nov. 19, 1900, 12/213/16, Vanderhorst Family Papers.

45. See heated exchanges between the two throughout 12/213/16, Vanderhorst Family Papers. Also Hacker and Trinkley, "Chapter 5," 97; the price and size of Seven Oaks are described in the Chicora Foundation (website), "A Short History of Kiawah Island and Quash Stevens," accessed Feb. 20, 2021, https://www.sciway.net/hist/chicora/quash-1.html.

46. William's occupation and education are described in Rowe, *Southern Workman*, 715. There is also a wonderful photograph in there of William and his wife Julia standing outside of their home on Kiawah. Its resolution was insufficient for inclusion in the book, unfortunately. William is astride a marsh tacky, and Julia is standing on the steps of the house, with another horse hitched up to a carriage. The house, which is quite beautiful, is raised a few feet off the ground with a porch along the front, big glass windows opened to catch a breeze, a salt marsh and the Kiawah River or one of its tributaries in the background.

47. On March 27, 1900, Quash wrote to Arnoldus, "I Ham Not Well," 12/213/16; later that year on July 25, Quash wrote again, "I am down in bed with the fever"; in an undated letter from that same year, he wrote that his wife was down with the fever. He did at least have one of his daughters alive in 1900, though he did not say which, and it may have been his daughter-in-law—she was, however, sick sometime in 1900, which he noted in an undated letter from 1900 to Arnoldus, 12/213/16, Vanderhorst Family Papers.

48. The 1900 census explicitly includes an entry for how many children were born and how many survive; see "United States Census, 1900," database with images, *FamilySearch*, accessed Apr. 23, 2021, https://familysearch.org/ark:/61903/1:1:M3RF-HLK, Q Steven, John's Island Township (south part), Charleston, South Carolina, United States; citing enumeration district (ED) 124, sheet 18A, family 403, NARA microfilm publication T623 (Washington, D.C.: National Archives and Records Administration, 1972), FHL microfilm 1,241,521.

49. Though racial categories in the census are always to be taken with a grain of salt, it is notable that in every census in which Quash, Julia, and William had previously appeared, they had always been classified as "mulatto." But in 1900, immediately before their departure from Kiawah, they were listed "B" for "Black." That could easily have been a product of the hardening lines of racial difference under the newly arisen Jim Crow or due to the proclivities and prejudices of the census taker, W. W. Hanahan. It could also have been the Stevenses' preference.

50. William and Lilla's youngest son, Harold Arnoldus, born 1907 and presumably named after Quash's brother Arnoldus rather than his obnoxious nephew, went on to have an illustrious career as a lawyer and judge in New York. A useful, short biography of him can be found here, and it mentions Lilla's training as a teacher: Historical Society of the New York Courts, "Harold Arnoldus Stevens," a reprint from *The Judges of the New York Court of Appeals: A Biographical History*, ed. Hon. Albert M. Rosenblatt (New York: Fordham University Press, 2007), https://history.nycourts.gov/biography/harold-arnoldus-stevens/. Harold's wife, Ella Clyde Myers, was also from Columbia; they met while attending Benedict College. For some reason, a photograph of Ella in her graduation attire is in the W. E. B. Du Bois Papers. Ella Clyde Myers, 1929, W. E. B. Du Bois Papers, MS 312, Special Collections and University Archives, University of Massachusetts Amherst Libraries, http://credo.library.umass.edu/view/full/mums312-i0059.

51. The last letter from Quash is from 1902, and it was to Arnoldus over some minor business matter connected to Kiawah. Quash Stevens to Arnoldus Vanderhorst, Apr. 14, 1902, 12/213/16.

52. The price for the cattle was fourteen dollars a head. Quash Stevens to Arnoldus Vanderhorst, Feb. 14, 1902; the quote about the need for money from Quash to Arnoldus, Apr. 14, 1902, 12/213/16, Vanderhorst Family Papers.

53. See Rowe, *Southern Workman*, 715, in which he writes: "The children and nearly all of the adults can read and write. This is due to the energy of Mrs. Stevens—formerly Mrs. Julia W. Gibbs, member of the first class of Hampton graduates, 1871—who took into her married life the noble purpose to be helpful to her people, so characteristic of Hampton graduates, and has practically demonstrated what persistent effort will do. The influence of this good woman reaches beyond the bounds of her island home, and finds expression in established Sunday-school work in neglected districts on Seabrook Island and on John's Island."

54. Julia was active as a teacher and in building African American institutions in Charleston. She held a fundraiser for the only Black-operated hospital in Charleston, the Hospital and Training School for Nurses on Cannon Street. See *Hospital Herald* 1, no. 5 (April 1899): 12, accessed Feb. 21, 2021, http://digital.library.musc.edu/cdm/ref/collection/hherald/id/75.

55. See Chicora Foundation, "Short History."

56. See Quash's death certificate, which lists his place of death as his house on Johns Island and his date of death as March 20; the cause was heart failure. See South Carolina Department of Archives and History, Columbia, S.C., South Carolina Death Records, 1900–1924, Death County or Certificate Range: Charleston; South Carolina, U.S., Death Records, 1821–1969 [online database] (Lehi, Ut.: Ancestry.com, 2008).

57. See his death certificate. It is likely that the graveyard that Centenary was using at the time was later converted to private property and has been built over with a private home.

58. I am not certain exactly when William died. However, by 1920, Julia had moved to Charleston to work as a public school teacher and lived and owned her habitation at 51 Cooper Street, as per the 1920 census. "United States Census, 1920," database with images, *FamilySearch*, accessed Feb. 3, 2021, https://www.familysearch.org/ark:/61903/1:1:M6CN-BK5, Julia Stevens, 1920. And, based on Harold Arnoldus's recollection of moving to Columbia after his father's death and before he started high school, it would make sense that William's death was sometime before or around 1920.

59. Perhaps Julia encountered a young Septima Poinsette Clark, the famed civil rights leader who began her own teaching career on Johns Island in 1916 and whose father had also been an enslaved servant for the Light Dragoons alongside Quash. Emerson, *Sons of Privilege*, 108.

60. In 1932, the radical and lifelong activist Modjeska Monteith Simkins moved in right next door to the Johnsons, Lilla's parents. The names of the four boys are in the 1910 census; see "United States Census, 1910," database with images, *FamilySearch*, accessed Apr. 25, 2021, https://familysearch.org/ark:/61903/1:1:M5D9-NYX, Julia W. Stevens in household of William F. Stevens, Johns Island, Charleston, South Carolina, United States; citing enumeration district (ED) ED 74, sheet 13B, family 35, NARA microfilm publication T624 (Washington D.C.: National Archives and Records Administration, 1982), roll 1453, FHL microfilm 1,375,466. Harold Arnoldus's short online biography "Harold Arnoldus Stevens" is the source for their move to Columbia. His maternal grandfather and grandmother were important influences on him, impressing upon him a love of education and learning. Harold Arnoldus's grandfather Rev. James H. Johnson was interviewed by the WPA in 1936, and that document lists his house at 2029 Marion Street in Columbia; see *Federal Writers' Project: Slave Narrative Project, Vol. 14, South Carolina, Part 3, Jackson-Quattlebaum*, 1936, Manuscript/Mixed Material, accessed Apr. 25, 2021, https://www.loc.gov/item/mesn143/. In 1930, the census lists Rev. Johnson and his wife as living in the same house, with his twenty-nine-year-old grandson James Quash right around the corner renting 1316 Elmwood Avenue and working as a U.S. mailman, married to Gertrude, with sons William B., age five, and James Quash Jr., age two. See "United States Census, 1930," database with images, *FamilySearch*, accessed Apr. 25, 2021, https://familysearch.org/ark:/61903/1:1:SPZH-BY1, James Q. Stevens, Columbia, Richland, South Carolina, United States; citing enumeration district (ED) ED 14, sheet 1A, line 13, family 5, NARA microfilm publication T626 (Washington D.C.: National Archives and Records Administration, 2002), roll 2210, FHL microfilm 2,341,944. Rev. Johnson's house was immediately next door to 2025 Marion Street—the house into which Modjeska Monteith Simkins and her family moved in 1932. It is very likely that Rev. Johnson and the Stevens family became acquainted with her during the 1930s. In any case, the two families appear in the 1940 census at 2025 and 2029 Marion Street. As a final note, Lilla Stevens, Quash's daughter-in-law, died of cerebral apoplexy in 1931 in Bennettsville, South Carolina, listed as a schoolteacher, having remarried the Reverend John D. Whitaker; see South Carolina Department of Archives and History, Columbia, S.C., South Carolina Death Records, 1925–1949, Death County or Certificate Range: Marlboro, South Carolina, U.S., Death Records, 1821–1969 [online database] (Lehi, Ut.: Ancestry.com, 2008).

61. In Harold Arnoldus's online biography, he describes the 1926 Lowman lynchings as an event that shook him to his core and inspired him to become a lawyer. Harold Arnoldus loved and was drawn to New York City, where he moved after graduating from Boston College School of Law in 1936 and passing the Massachusetts, New York, and South Carolina bar exams. Though he grew more moderate as he grew older, as a young lawyer, he deliberately pursued cases on labor law, often at the intersection of labor and civil rights—the kinds of cases that Modjeska Monteith Simkins might have supported. His papers can also be found at the Schomburg Center for Research in Black Culture, New York Public Library.

62. See Julia's death certificate, which lists her death as on May 3, 1933, at age sixty-six at the Hospital and Training School at 135 Cannon Street, the African American hospital in Charleston that she had supported. She was listed as the widow of Quash Stevens and a schoolteacher who last worked in 1930. She listed her mother as Martha and her father simply by his last name "Gibbs," and both their and her birthplace as Charleston. She died of "chronic parenchymatous nephritis," with "contributory causes" of "Pellagra—relalive hortic and mitral insufficiency." It seems likely that the graveyard where they were buried is one of the graveyards in downtown Charleston since repurposed for other use. See South Carolina Department of Archives and History, Columbia, S.C., South Carolina Death Records, 1925–1949, Death County or Certificate Range: Charleston, South Carolina, U.S., Death Records, 1821–1969 [database online] (Lehi, Ut.: Ancestry.com, 2008).

63. For more on Arnoldus V's various schemes, see Hacker and Trinkley, "Chapter 5," 100–106.

64. Arnoldus wrote of his desire to maintain Kiawah to Elias, in 1916: "The purpose to maintain the family status as far as possible necessarily involves the preservation of the family setting as expressed in tangible things. I regard the heritage as a distinct asset to myself and all members of the family." Quoted in Hacker and Trinkley, 104.

65. Hacker and Trinkley, 104.

66. Hacker and Trinkley, 108.

67. Hacker and Trinkley, 108.

68. Hacker and Trinkley, 111.

69. Hacker and Trinkley, 112.

70. For Quash's reference to using the mansion for shelter, see Quash Stevens to Adele Vanderhorst, Aug. 29, 1893; for the description of the house's dining room kept in order, see Rowe, "Negroes of the Sea Islands," 714.

71. See Warren Wise, "Kiawah Mansion fetches $20.5 million," *Post and Courier*, June 25, 2021, https://www.postandcourier.com/business/real_estate/kiawah-mansion-fetches-20-5m-setting-a-record-for-a-charleston-area-home-sale/article_ab3953de-cd5b-11eb-9b75-2b804d094cc7.html.

72. Cooper's discussion of the heritage corridor can be found in *Making Gullah*, 197–206.

73. See Coclanis's choice words on Charleston in *Shadow of a Dream*, 159–60: "Indeed, despite the beauty of the lower peninsula and despite Restoration Charleston's 'insidious charm,' the area somehow seems spiritless and lifeless . . . devoid of social richness, at times, it seems, of any humanity at all. More like a museum than a field of human interaction, Restoration Charleston suggests atrophy rather than affirmation. . . . 'Quaint Old Charleston,' which strives always to appear tragic in order to beguile, will never evoke true tragedy but a pitiable *tragedy manqué*."

74. As described in "Memoirs of Eberson Murray," from "Stories of the Past," collected by Chalmers S. Murray, Project #1655 of the WPA. The hurricane of 1885 ravaged the community, which abandoned it largely after that point; and the 1893 storm washed away the final remnants.

Bibliography

Archives

Beaufort District Collection at the Beaufort County Library

Hurricane—1893 Vertical File

"Message of Benjamin R. Tillman, Governor. To the General Assembly of South Carolina at the Regular Session Commencing November 28, 1893." Columbia, S.C.: Charles A. Calvo Jr., State Printer, 1893, 43–47.

"Ballad: The Storm of '93, St. Helena Island." A Song by Rivers Lawton Farn, July 1972.

Storm of 1893 Death List, from the Beaufort County Coroner's Inquisition Records, from Aug. 28–30, 1893. Compiled by Grace Morris Cordial.

"The Storm of 1893." Conversation with Joe Rivers by Mary J. and W. O. Wall Jr., Oct. 18, 1970.

Susan Hazel Rice Diaries, PAD #5, July 15–Oct. 31, 1893.

Robert J. Dukes Jr., "Red Rockets to Geosynchronous Satellites; Flags to Television; The Hurricane Warning System in Charleston, Past and Present," unpublished manuscript.

"Phosphatic Deposits of South Carolina." Map of parts of Charleston, Colleton, and Beaufort Counties, S.C., compiled from various sources by Simons & Howe, Civil Engineers, 1879.

Jennings, Laylon Wayne. "A History of Storms on the South Carolina Coast," unpublished manuscript.

Clemson University Special Collections

Benjamin Ryan Tillman Papers

Library of Congress

Clara Barton Papers, 1805–1958, Red Cross File, 1863–1957, American National Red Cross, 1878–1957, Sea Islands, S.C.

Federal Writers' Project. "Volume XIV: South Carolina Narratives." In *Slave Narratives: A Folk History of Slavery in the United States from Interviews with Former Slaves* (Washington, D.C., 1941).

Lowcountry Digital Library

From the Beaufort District Collection at the Beaufort County Library:
Beaufort Hurricane of 1893 Photograph Collection

Phosphate, Farms, and Family—the Donner Collection
Mather, Rachel Crane. *The Storm Swept Coast of South Carolina*. Woonsocket, R.I.: Charles E. Cook, 1894.
From the Special Collections at the College of Charleston:
The Jewish Heritage Collection Oral Histories

Melville J. Herskovitzs Library of African Studies at Northwestern University

Lorenzo Dow Turner Papers, box 30, folders 7–9.

South Carolina Historical Society

Charleston Hurricane, 1893 Vertical File
Langdon Cheves Papers
John Bennett Papers
Charleston *Year Book*, multiple years;
Year Book—1892. City of Charleston, So. Ca. Charleston, S.C.: Walker, Evans & Cogswell, 1893.
Year Book—1893. City of Charleston, So. Ca. Charleston, S.C.: Walker, Evans, & Cogswell, 1894.
Year Book City of Charleston, S.C. 1900. Charleston, S.C.: Walker, Evans, & Cogswell, 1901.
Year Book City of Charleston, S.C. 1903. Charleston, S.C.: The Daggett Printing Company, 1904.
Vanderhorst Family Papers
Adams, Charles S., ed. *Report of the Sanitary and Drainage Commission for Charleston County, South Carolina*, 1927, no printing information listed, call no. 628 Sanitary, Pamphlet.
Wallace, David Duncan. "The South Carolina Constitution of 1895," Bureau of Publications, USC, Columbia SC, Call # 347.75 1927 Wallace. Delivered as a lecture series at Furman, June 22–25, 1926.

South Carolina Department of Archives and History

Annual Reports of the Department of Agriculture, multiple years
Governor Benjamin Ryan Tillman (1890–94), Letters Received and Sent
Annual Reports of the South Carolina General Assembly, multiple years

South Caroliniana Library

Christensen Family Papers
William Henry Frierson Papers, 1854–1932
Anne Simons Deas Diary
Fripp Family Papers
Benjamin Ryan Tillman Papers, May 24, 1845–Aug. 8, 1965
James Henry Rice Papers, 1868–1935

Edward Terry Hendrie Papers, 1880–1945
Music and Musicians in South Carolina Vertical File, box 3
1886 Earthquake and 1893 Hurricane, Charleston Views
August Kohn Photographs, 1895 Constitutional Convention
Montgomery Family Papers, June 26, 1873–Dec. 4, 1955, and n.d.
Harry Hammond Papers
Penn School Microfilm
Annual narrative and statistical reports from state offices and county agents, South Carolina, 1909–1944, USDA, microfilm
News and Courier records, microfilm
Riggs, Walter. *Boll Weevil: Report of the South Carolina Boll Weevil Commission*. Richard I. Manning, Chairman. Washington, D.C.: Government Printing Office, 1921.
Tindall, George B. "The Campaign for the Disenfranchisement of Negroes in South Carolina." *Journal of Southern History* 15, no. 2 (May 1949).
The Suffrage. "Speeches by the Negroes in the Constitutional Convention," compiled by Mary J. Miller. N.d. Pamphlet.
"Speeches at the Constitutional Convention by General Robert Smalls." Compiled by Miss Sarah V. Smalls, Enquirer Print, Charleston, S.C., 1896.
Stackhouse, Mary Fletcher. *A Study of the Boll Weevil Conditions of South Carolina*. Master's thesis, University of South Carolina, June 1922.

Southern Historical Collection, University of North Carolina at Chapel Hill

Grimball Family Papers
Penn School Papers

Newspapers

News and Courier; *Post and Courier*
The State
The Columbia Record

Digitized Census Records

"United States Census, 1880"; "United States Census, 1900"; "United States Census, 1910"; "United States Census, 1920"; "United States Census, 1930"; database with images, *FamilySearch*.
Charleston, South Carolina, U.S., Marriage Records, 1877–1887 [online database]. Provo, Ut.: Ancestry.com, 2007.
South Carolina, U.S., Death Records, 1821–1969 [online database]. Lehi, Ut.: Ancestry.com, 2008.

Published Primary Sources

22 *Years' of Work of Hampton Normal and Agricultural Institute*. Hampton, Va.: Hampton Normal School Press, 1893.

American Forestry Association. *Forestry, Irrigation, and Conservation*. Vol. 14. Washington, D.C.: American Forestry Association, 1908.
Adams, Charles S., ed. *Report of the Jacksonville Auxiliary Sanitary Association of Jacksonville, Florida, covering the work of the association during the yellow fever epidemic, 1888*. Jacksonville, Fla.: Times-Union Print, 1889. Digitized by the New York Public Library.
Barton, Clara. *The Red Cross: A History of this remarkable international movement in the interests of humanity*. Washington, D.C.: American Red Cross, 1898.
———. *A Story of the Red Cross: Glimpses of Field Work*. New York: D. Appleton, 1928.
Bethea, Andrew J., ed. *The Code of Laws in South Carolina*. Vol. 2. Charlottesville: Michie, 1912.
Constitutional Convention of the State of South Carolina, Journal of the Constitutional Convention of the State of South Carolina. Columbia, S.C.: Charles A. Calvo Jr., 1895.
Elkinton, Joseph S. *Selections from the Diary and Correspondence of Joseph S. Elkinton, 1830–1905*. Philadelphia: Press of the Leeds & Biddle Co., 1913.
Fields, Mamie Garvin. *Lemon Swamp and Other Places: A Carolina Memoir*. New York: Free Press, 1985.
Hacker, Debi, and Michael Trinkley. "Chapter 5: History of Kiawah Island." In *The History and Archaeology of Kiawah Island, Charleston County, South Carolina*, edited by Michael Trinkley, 49–113. Columbia: Chicora Foundation, 1993.
Hammond, James Henry "Harry" Jr. *South Carolina. Resources, Population, Institutions, and Industries*. State Board of South Carolina. Charleston, S.C.: Walker, Evans, & Cogswell, Printers, 1883.
Harris, Joel Chandler. "The Sea Island Hurricane: The Devastation." *Scribner's* 15, no. 2 (Feb. 1894), 229–47.
———. "The Sea Island Hurricane: The Relief." *Scribner's* 15, no. 3 (Mar. 1894), 267–84.
Index to the Miscellaneous Documents of the House of Representatives for the First Session of the Forty-Seventh Congress, 1881–'82. Washington, D.C.: Government Printing Office, 1882.
Kiser, Clyde Vernon. *Sea Island to City: A Study of St. Helena Islanders in Harlem and Other Urban Centers*. 1st ed. 1932. Reprint, New York: Atheneum, 1969.
Johnson, Guion Griffis. *A Social History of the Sea Islands, with Special Reference to St. Helena Island, South Carolina*. 1st ed. 1930. Reprint, Chapel Hill: University of North Carolina Press, 2018.
Jones, Katherine M. *Port Royal under Six Flags: The Story of the Sea Islands*. New York: Bobbs-Merrill, 1960.
Leigh, Frances Butler. *Ten Years on a Georgia Plantation Since the War*. London: Richard Bentley & Sons, 1883. Documenting the American South, University of North Carolina at Chapel Hill.
National Conservation Commission. *Report of the National Conservation Commission*. Washington, D.C.: Government Printing Office, 1909.
National Oceanic and Atmospheric Administration. "North Atlantic Storms for August, 1893." *Monthly Weather Review*, 1893, 207.
Oliphant, Albert D. *The Evolution of the Penal System of South Carolina, 1866–1916*. Columbia, S.C.: The State, 1916.

Pringle, Elizabeth. *A Woman Rice Planter.* Edited by Charles Joyner. Columbia: University of South Carolina Press, 1992.
Reports and Resolutions of the South Carolina General Assembly. Columbia: Charles A. Calvo Jr., State Printer, 1893.
Reports and Resolutions of the General Assembly of South Carolina, 1913. Vol. 2. Columbia, S.C.: Gonzales & Bryan, 1913.
Reports and Resolutions of the General Assembly of the State of South Carolina, 1920. Vol. 2. Columbia, S.C.: Walker & Evans, 1921.
Rogers, G. Sherburne. "The Phosphate Deposits of South Carolina." *Contributions to Economic Geology,* no. 1 (1913), 183–220.
Rowe, George C. "The Negroes of the Sea Islands." *The Southern Workman and Hampton School Record* XXIX, no. 1 (January 1990): 709–15.
Satterthwait, Elisabeth Carpenter. *A Son of the Carolinas: A Story of the Hurricane Upon the Islands.* Philadelphia: Henry Althemes, 1898. Online at the Hathi Trust.
Silver, John. *To Live as Free Men.* Produced by Hathen Productions [United States: s.n, 1942], Video. https://www.loc.gov/item/87706576/.
South Carolina General Assembly Reports and Resolutions. Columbia, S.C.: Gonzales & Bryan, 1918.
South Carolina General Assembly Reports and Resolutions. Vols. 1 and 2. Columbia, S.C.: Gonzales & Bryan, 1920.
The Southern Workman and Hampton School Record 29, no. 1. Hampton, Va.: Hampton Normal School Press, 1900.
Watson, E. J. *Handbook of South Carolina, Resources, Institutions and Industries of the State.* Columbia, S.C.: The State, 1907.
Woofter, T. J., Jr. *Black Yeomanry: Life on St. Helena Island.* New York: Henry Holt, 1930.
"Your Servant, Quash": Letters of a South Carolina Freedman. Columbia, S.C.: Chicora Foundation, 1994.

Secondary Sources

Aiken, Charles S. *The Cotton Plantation South since the Civil War.* Baltimore: Johns Hopkins University Press, 1998.
Alexander, Michelle. *The New Jim Crow: Mass Incarceration in the Age of Colorblindness.* New York: New Press, 2012.
Ali, Omar H. *In the Lion's Mouth: Black Populism in the New South, 1886–1900.* Oxford: University Press of Mississippi, 2010.
Arnesen, Eric, ed. *The Black Worker: A Reader.* Urbana: University of Illinois Press, 2007.
Ayers, Edward L. *The Promise of the New South: Life after Reconstruction.* New York: Oxford University Press, 1992.
Bacheler, Nathan M., Kyle W. Shertzer, Robin T. Cheshire, and Jamie MacMahan. "Tropical Storms Influence the Movement Behavior of a Demersal Oceanic Fish Species." *Scientific Reports* 9, no. 1481 (2019). https://doi.org/10.1038/s41598-018-37527-1.
Baird, Keith E., and Mary A. Twining, ed. *Sea Island Roots: African Presence in the Carolinas and Georgia.* Trenton, N.J.: Africa World Press, 1990.

Baker, Bruce E. *This Mob Will Surely Take My Life: Lynchings in the Carolinas, 1871–1947.* New York: Bloomsbury, 2008.

———. *What Reconstruction Meant: Historical Memory in the American South.* Charlottesville: University of Virginia Press, 2009.

Baptist, Edward E. *The Half Has Never Been Told: Slavery and the Making of American Capitalism*. New York: Basic Books, 2014.

Barry, John M. *Rising Tide: The Great Mississippi Flood of 1927 and How It Changed America*. New York: Simon & Schuster, 1997.

Beito, David T. *From Mutual Aid to the Welfare State: Fraternal Societies and Social Services, 1890–1967.* Chapel Hill: University of North Carolina Press, 2000.

Berlin, Ira. *Generations of Captivity: A History of African-American Slaves*. Cambridge, Mass.: Belknap Press, 2004.

———. *The Making of African America: The Four Great Migrations*. New York: Penguin, 2010.

Blackmon, Douglas A. *Slavery by Another Name: The Re-enslavement of Black People in America from the Civil War to World War II*. New York: Doubleday, 2008.

Blakey, Arch Frederic. *The Florida Phosphate Industry: A History of the Development and Use of a Vital Mineral*. Cambridge, Mass.: Harvard University Press, 1973.

Bland, Robert. "'A Grim Memorial of Its Thorough Work of Devastation and Desolation': Race and Memory in the Aftermath of the 1893 Sea Island Storm." *Journal of the Gilded Age and Progressive Era* 17, no. 2 (2018): 297–316. doi:10.1017/S1537781417000846.

Blight, David W. *Race and Reunion: The Civil War in American Memory*. Cambridge, Mass.: Belknap Press, 2001.

Boles, John B., and Bethany L. Johnson, eds. *Origins of the New South Fifty Years Later: The Continuing Influence of a Historical Classic*. Baton Rouge: Louisiana State University Press, 2003.

Bolster, W. Jeffrey. *Black Jacks: African American Seamen in the Age of Sail*. Cambridge, Mass.: Harvard University Press, 1997.

Brown, Ras Michael. *African-Atlantic Cultures and the South Carolina Lowcountry*. Cambridge: Cambridge University Press, 2012.

Bryan, William D. *The Price of Permanence: Nature and Business in the New South*. Athens: University of Georgia Press, 2018.

Burton, Orville Vernon. *In My Father's House Are Many Mansions: Family and Community in Edgefield, South Carolina*. Chapel Hill: University of North Carolina Press, 1985.

Burton, Orville Vernon, and Wilbur Cross. *Penn Center: A History Preserved*. Athens: University of Georgia Press, 2014.

Butler, Christina Rae. *Lowcountry at High Tide: A History of Flooding, Drainage, and Reclamation in Charleston, South Carolina*. Columbia: University of South Carolina Press, 2020.

Campbell, Emory Shaw, Carolyn Grant, and Thomas C. Barnwell Jr. *Gullah Days: Hilton Head Islanders before the Bridge*. Durham, N.C.: Carolina Wren, 2020.

Carlton, David L. *Mill and Town in South Carolina, 1880–1920*. Baton Rouge: Louisiana State University Press, 1982.

Carney, Judith Ann. *Black Rice: The African Origins of Rice Cultivation in the Americas*. Cambridge, Mass.: Harvard University Press, 2001.

Cecelski, David S. *The Waterman's Song: Slavery and Freedom in Maritime North Carolina*. Chapel Hill: University of North Carolina Press, 2001.
Chatelain, Marcia. *South Side Girls: Growing Up in the Great Migration*. Durham, N.C.: Duke University Press, 2015.
Church, Christopher M. *Paradise Destroyed: Catastrophe and Citizenship in the French Caribbean*. Lincoln: University of Nebraska Press, 2017.
Clarke, Erskine. *Dwelling Place: A Plantation Epic*. New Haven, Conn.: Yale University Press, 2005.
Coclanis, Peter A. *The Shadow of a Dream: Economic Life and Death in the South Carolina Low Country, 1670–1920*. Oxford: Oxford University Press, 1991.
Colten, Craig E. *An Unnatural Metropolis: Wresting New Orleans from Nature*. Baton Rouge: Louisiana State University Press, 2005.
———. *Southern Waters: The Limits to Abundance*. Baton Rouge: Louisiana State University Press, 2014.
Cooper, Melissa. *Making Gullah: A History of Sapelo Islanders, Race, and the American Imagination*. Chapel Hill: University of North Carolina Press, 2017.
Cowdrey, Albert. *This Land, This South: An Environmental History*. Lexington: University Press of Kentucky, 1995.
Creel, Margaret Washington. *"A Peculiar People": Slave Religion and Community-Culture among the Gullah*. New York: New York University Press, 1988.
Cronon, William, ed. *Uncommon Ground: Rethinking the Human Place in Nature*. New York: W. W. Norton, 1995.
Cross, Wilbur. *Gullah Culture in America*. New York: John F. Blair, 2008.
Daise, Ronald. *Reminisces of Sea Island Heritage: Legacy of Freedmen on St. Helena Island*. Orangeburg, S.C.: Sandlapper, 1987.
Daniel, Pete. *Dispossession: Discrimination against African American Farmers in the Age of Civil Rights*. Chapel Hill: University of North Carolina Press, 2013.
Davis, Jack E. *Gulf: The Making of an American Sea*. New York: Liverwright, 2017.
Davis, Mike. *Ecology of Fear: Los Angeles and the Imagination of Disaster*. New York: Metropolitan, 1998.
Donegan, Kathleen. *Seasons of Misery: Catastrophe and Colonial Settlement in Early America*. Philadelphia: University of Pennsylvania Press, 2016.
Dorsey, Allison. "'The Great Cry of Our People Is Land!': Black Settlement and Community Development on Ossabaw Island, Georgia, 1865–1900." In *African American Life In The Georgia Lowcountry: The Atlantic World And The Gullah Geechee*, edited by Philip Morgan, 224–52. Athens: University of Georgia Press.
Doyle, Don H. *New Men, New Cities, New South: Atlanta, Nashville, Charleston, Mobile, 1860–1910*. Chapel Hill: University of North Carolina Press, 1990.
Du Bois, W. E. B. *Black Reconstruction in America, 1860–1880*. 1st ed. 1935. Reprint, New York: Free Press, 1998.
Dubcovsky, Alejandra. *Informed Power: Communication in the Early American South*. Cambridge, Mass.: Harvard University Press, 2016.
Dusinberre, William. *Them Dark Days: Slavery in the American Rice Swamps*. Oxford: Oxford University Press, 1996.
Dyl, Joanna L. *Seismic City: An Environmental History of San Francisco's 1906 Earthquake*. Seattle: University of Washington Press, 2017.

Edelson, S. Max. *Plantation Enterprise in Colonial South Carolina*. Cambridge, Mass.: Harvard University Press, 2011.

Edgar, Walter B. *South Carolina: A History*. Columbia: University of South Carolina Press, 1998.

Egerton, Douglas R. *Thunder at the Gates: The Black Civil War Regiments that Redeemed America*. New York: Basic Books, 2016.

Emerson, W. Eric. *Sons of Privilege: The Charleston Light Dragoons in the Civil War*. Columbia: University of South Carolina Press, 2011.

Epstein, Dena J. *Sinful Tunes and Spirituals: Black Folk Music to the Civil War*. Urbana-Champaign: University of Illinois Press, 2003.

Erikson, Kai T. *Everything in Its Path: Destruction of Community in the Buffalo Creek Flood*. New York: Touchstone, 1976.

Ermus, Cindy. *Environmental Disaster in the Gulf South: Two Centuries of Catastrophe, Risk, and Resilience*. Baton Rouge: Louisiana State University Press, 2018.

Faust, Drew Gilpin. *James Henry Hammond and the Old South: A Design for Mastery*. Baton Rouge: Louisiana State University Press, 1985.

Feeser, Andrea. *Red, White, and Black Make Blue: Indigo in the Fabric of Colonial South Carolina Life*. Athens: University of Georgia Press, 2013.

Ferguson, Robert Hunt. *Remaking the Rural South: Interracialism, Christian Socialism, and Cooperative Farming in Jim Crow Mississippi*. Athens: University of Georgia Press, 2018.

Fetig, Barbara. "Pin Point: A Traditional African American Community." In *Coastal Nature, Coastal Culture: Environmental Histories of the Georgia Coast*, edited by Paul S. Sutter and Paul M. Pressly, 138–39. Athens: University of Georgia Press, 2018.

Fields-Black, Edda. *Deep Roots: Rice Farmers in West Africa and the African Diaspora*. Bloomington: Indiana University Press, 2008.

Finnegan, Terence. *A Deed So Accursed: Lynching in Mississippi and South Carolina, 1881–1940*. Charlottesville: University of Virginia Press, 2013.

———. "Lynching in the Outer Coastal Plain Region of South Carolina and the Origins of African American Collective Action, 1901–1910." In *Toward the Meeting of the Waters: Currents in the Civil Rights Movement of South Carolina during the Twentieth Century*, edited by Winifred Moore and Orville Burton, 41–49. Columbia: University of South Carolina Press, 2011.

Foner, Eric. *Nothing but Freedom: Emancipation and Its Legacy*. Lynwood Fleming Lectures in Southern History. Baton Rouge: Louisiana State University Press, 1983.

———. *Reconstruction: America's Unfinished Revolution, 1863–1877*. Rev. ed. New York: Harper, 2014.

Fortenberry, Brent R. "For Refuge and Resilience: The Storm Towers of the Santee Delta." *Arris* 29 (2018): 6–25.

Franklin, John Hope, and Loren Schweninger. *Runaway Slaves: Rebels on the Plantation*. New York: Oxford University Press, 1999.

Fraser, Walter J., Jr. *Lowcountry Hurricanes: Three Centuries of Storms at Sea and Ashore*. Athens: Wormsloe, 2009.

Genovese, Eugene D. *Roll, Jordan, Roll: The World the Slaves Made*. New York: Vintage, 1976.

Giesberg, Judith Ann. *Civil War Sisterhood: The U.S. Sanitary Commission and Women's Politics in Transition*. Boston: Northeastern University Press, 2006.

Giesen, James. *Boll Weevil Blues: Cotton, Myth, and Power in the American South.* Chicago: University of Chicago Press, 2012.

———. "The Truth about the Boll Weevil": The Nature of Planter Power in the Mississippi Delta." *Environmental History* 14, no. 4 (2009): 683–704.

Gilmore, Glenda Elizabeth. *Defying Dixie: The Radical Roots of Civil Rights, 1919–1950.* New York: W. W. Norton, 2008.

———. *Gender and Jim Crow: Women and the Politics of White Supremacy in North Carolina, 1896–1920.* Chapel Hill: University of North Carolina Press, 1996.

Ginzberg, Lori D. *Women and the Work of Benevolence: Morality, Politics, and Class in the Nineteenth-Century United States.* New Haven, Conn.: Yale University Press, 1990.

Giovanni, Nikki. *On My Journey Now: Looking at African American History through the Spirituals.* Somerville, M.A.: Candlewick, 2009.

Glave, Dianne D., and Mark Stoll, eds. *To Love the Wind and the Rain: African Americans and Environmental History.* Pittsburgh: University of Pittsburgh Press, 2006.

Goodwine, Marquetta L. *The Legacy of Ibo Landing: Gullah Roots of African American Culture.* Atlanta: Clarity, 2011.

Graham, Sandra Jean. *Spirituals and the Birth of a Black Entertainment Industry.* Urbana-Champaign: University of Illinois Press, 2018.

Grego, Caroline. "Black Autonomy, Red Cross Recovery, and White Backlash after the Great Sea Island Storm of 1893." *Journal of Southern History* 85, no. 4 (2019): 803–40.

Gregory, James N. *Southern Diaspora: How the Great Migrations of Black Southerners Transformed America.* Chapel Hill: University of North Carolina Press, 2005.

Gutman, Herbert G. *The Black Family in Slavery and Freedom, 1750–1925.* New York: Pantheon Books, 1976.

Hacker, Debi, and Michael Trinkley. "Chapter 5. History of Kiawah Island." In *The History and Archaeology of Kiawah Island, Charleston County, South Carolina*, edited by Michael Trinkley, 48–113. Columbia: Chicora Foundation, 1993.

Hahn, Steven. *A Nation under Our Feet: Black Political Struggles in the Rural South from Slavery to the Great Migration.* Cambridge, Mass.: Belknap Press, 2005.

Haley, Sarah. *No Mercy Here: Gender, Punishment, and the Making of Jim Crow Modernity.* Chapel Hill: University of North Carolina Press, 2016.

Harris, Lynn B. *Patroons and Periaguas: Enslaved Watermen and Watercraft of the Lowcountry.* Columbia: University of South Carolina Press, 2014.

Higgs, Robert. *Competition over Coercion: Blacks in the American Economy, 1865–1914.* Cambridge: Cambridge University Press, 1977.

Hild, Matthew, and Keri Leigh Merritt, eds. *Reconsidering Southern Labor History: Race, Class, and Power.* Gainesville: University Press of Florida, 2018.

Hoffman, Edwin D. "From Slavery to Self-Reliance." *Journal of Negro History* 41, no. 1 (Jan. 1956), 8–42.

Holden, Charles J. *In the Great Maelstrom: Conservatives in Post–Civil War South Carolina.* Columbia: University of South Carolina Press, 2002.

Horowitz, Andy. *Katrina: A History, 1915–2015.* Cambridge, Mass.: Harvard University Press, 2020.

Horowtiz, Andy, and Jacob Remes, eds. *Critical Disaster Studies.* Philadelphia: University of Pennsylvania Press, 2021.

Horsman, Reginald. *Race and Manifest Destiny: The Origins of American Racial Anglo-Saxonism*. Cambridge, Mass.: Harvard University Press, 1981.
Hudson, Janet G. *Entangled by White Supremacy: Reform in World War I–Era South Carolina*. Lexington: University Press of Kentucky, 2009.
Hughes, Patrick. *A Century of Weather Service: A History of the Birth and Growth of the National Weather Service, 1870–1970*. New York: Gordon and Breach, 1970.
Humphries, Margaret. *Malaria: Poverty, Race, and Public Health in the United States*. Baltimore: Johns Hopkins University Press, 2001.
Hunter, Tera W. *To 'Joy My Freedom: Southern Black Women's Lives and Labors after the Civil War*. Cambridge, Mass.: Harvard University Press, 1998.
Inscoe, John C. "Carolina Slave Names: An Index to Acculturation." *Journal of Southern History* 49, no. 4 (Nov. 1983): 537–54.
Irwin, Julia. *Making the World Safe: The American Red Cross and a Nation's Humanitarian Awakening*. Oxford: Oxford University Press, 2013.
Jervey, Theodore D. "The Butlers of South Carolina." *South Carolina Historical and Genealogical Magazine*, no. 4 (Oct. 1903), 296–311.
Johnson, Sherry. *Climate and Catastrophe in Cuba and the Atlantic World in the Age of Revolution*. Chapel Hill: University of North Carolina Press, 2012.
Johnson, Walter. *River of Dark Dreams: Slavery and Empire in the Cotton Kingdom*. Cambridge, Mass.: Belknap Press, 2013.
Jones, Jacqueline. *Labor of Love, Labor of Sorrow: Black Women, Work, and the Family, from Slavery to the Present*. New York: Basic Books, 1985.
———. "My Mother Was Much of a Woman: Black Women, Work, and the Family Under Slavery." *Feminist Studies* 8 (1982): 235–69.
Jones, Marian Moser. *The American Red Cross from Clara Barton to the New Deal*. Baltimore: Johns Hopkins University Press, 2013.
Jones, Norrece T., Jr. *Born a Child of Freedom yet a Slave: Mechanisms of Control and Strategies of Resistance in Antebellum South Carolina*. Hanover, Penn.: Wesleyan University Press, 1990.
Jones-Jackson, Patricia. *When Roots Die: Endangered Traditions on the Sea Islands*. Athens: University of Georgia Press, 1987.
Joyner, Charles. *Down by the Riverside: A South Carolina Slave Community*. Urbana-Champaign: University of Illinois Press, 1984.
Kahrl, Andrew W. *The Land Was Ours: How Black Beaches Became White Wealth in the Coastal South*. Chapel Hill: University of North Carolina Press, 2016.
Kamerling, Henry. *Capital and Convict: Race, Region, and Punishment in Post–Civil War America*. Charlottesville: University of Virginia Press, 2017.
Kantrowitz, Stephen David. *Ben Tillman and the Reconstruction of White Supremacy*. Chapel Hill: University of North Carolina Press, 2000.
———. *More than Freedom: Fighting for Black Citizenship in a White Republic, 1829–1889*. New York: Penguin, 2012.
Karp, Matthew. *This Great Southern Empire: Slaveholders at the Helm of American Foreign Policy*. Cambridge, Mass.: Harvard University Press, 2016.
Kelley, Blair L. M. *Right to Ride: Streetcar Boycotts and African American Citizenship in the Era of Plessy v. Ferguson*. Chapel Hill: University of North Carolina Press, 2010.

Kelley, Robin D. G. *Hammer and Hoe: Alabama Communists during the Great Depression*. 1st ed. 1990. Reprint, Chapel Hill: University of North Carolina Press, 2015.
Kiechle, Melanie. *Smell Detectives: An Olfactory History of Nineteenth Century America*. Seattle: University of Washington Press, 2019.
Kierner, Cynthia A. *Inventing Disaster: The Culture of Calamity from the Jamestown Colony to the Johnstown Flood*. Chapel Hill: University of North Carolina Press, 2019.
Kirby, Jack Temple. *Mockingbird Song: Ecological Landscapes of the South*. Chapel Hill: University of North Carolina Press, 2006.
———. *Rural Worlds Lost: The American South, 1920–1960*. Baton Rouge: Louisiana State University Press, 1986.
Kirby, Rachel C. "Unenslaved through Art: Rice Culture Paintings by Jonathan Green." *Panorama: Journal of the Association of Historians of American Art* 6, no. 1 (Spring 2020), https://doi.org/10.24926/24716839.9889.
Knowles, Scott Gabriel. "Slow Disaster in the Anthropocene: A Historian Witnesses Climate Change on the Korean Peninsula." *Daedalus* 149, no. 4 (202): 192–206.
———. *The Disaster Experts: Mastering Risk in Modern America*. Philadelphia: University of Pennsylvania Press, 2011.
Kytle, Ethan, and Blain Roberts. *Denmark Vesey's Garden: Slavery and Memory in the Cradle of the Confederacy*. New York: New Press, 2018.
La Vere, David. *The Tuscarora War: Indians, Settlers, and the Fight for the Carolina Colonies*. Chapel Hill: University of North Carolina Press, 2013.
LeFlouria, Talitha L. *Chained in Silence: Black Women and Convict Labor in the New South*. Chapel Hill: University of North Carolina Press, 2015.
Lemann, Nicholas. *The Promised Land: The Great Black Migration and How It Changed America*. New York: Vintage, 1992.
Lichtenstein, Alex. *Twice the Work of Free Labor: The Political Economy of Convict Labor in the New South*. New York: Verso, 1996.
Littlefield, Daniel C. *Rice and Slaves: Ethnicity and the Slave Trade in Colonial South Carolina*. Baton Rouge: Louisiana State University Press, 1981.
Litwack, Leon. *Trouble in Mind: Black Southerners in the Age of Jim Crow*. New York: Vintage, 2010.
Lockley, Timothy James, ed. *Maroon Communities in South Carolina: A Documentary Record*. Columbia: University of South Carolina Press, 2009.
Mancini, Matthew. *One Dies, Get Another: Convict Leasing in the American South, 1866–1928*. Columbia: University of South Carolina Press, 1996.
Manigault-Bryant, LeRhonda S. *Talking to the Dead: Religion, Music, and Lived Memory among Gullah/Geechee Women*. Durham, N.C.: Duke University Press, 2014.
Marscher, William and Fran. *The Great Sea Island Storm of 1893*. Macon, Ga.: Mercer University Press, 2004.
Marszalek, John F. *A Black Congressman in the Age of Jim Crow: South Carolina's George Washington Murray*. Gainesville: University Press of Florida, 2006.
Mauldin, Erin Stewart. *Unredeemed Land: An Environmental History of Civil War and Emancipation in the Cotton South*. Oxford: Oxford University Press, 2018.
McAlister, Robert. *The Lumber Boom of Coastal South Carolina: Nineteenth-Century Shipbuilding and the Devastation of Lowcountry Virgin Forests*. Charleston, S.C.: The History Press, 2013.

McCally, David. *The Everglades: An Environmental History*. Gainesville: University Press of Florida, 1999.
McCammack, Brian. *Landscapes of Hope: Nature and the Great Migration in Chicago*. Cambridge, Mass.: Harvard University Press, 2017.
McCandless, Peter. *Slavery, Disease, and Suffering in the Southern Lowcountry*. Cambridge: Cambridge University Press, 2011.
McGerr, Michael. *A Fierce Discontent: The Rise and Fall of the Progressive Movement in America, 1870–1920*. Oxford: Oxford University Press, 2005.
McKinley, Shepherd W. *Stinking Stones and Rocks of Gold: Phosphate, Fertilizer, and Industrialization in Postbellum South Carolina*. Gainesville: University Press of Florida, 2014.
McNeill, John Robert. *Mosquito Empires: Ecology and War in the Greater Caribbean, 1620–1914*. New York: Cambridge University Press, 2010.
Miller, David C. *Dark Eden: The Swamp in Nineteenth-Century American Culture*. Cambridge: Cambridge University Press, 1989.
Miller, Edward A., Jr. *Gullah Statesman: Robert Smalls from Slavery to Congress, 1839–1915*. Columbia: University of South Carolina Press, 1995.
Mizelle, Richard M. *Backwater Blues: The Mississippi Flood of 1927 in the African American Imagination*. Minneapolis: University of Minnesota Press, 2014.
Moore, Winifred, and Orville Vernon Burton, eds. *Toward the Meeting of the Waters: Currents in the Civil Rights Movement of South Carolina during the Twentieth Century*. Columbia: University of South Carolina Press, 2011.
Morgan, Philip, Manisha Sinha, and Patrick Rael, eds. *African American Life in the Georgia Lowcountry: The Atlantic World and the Gullah Geechee*. Athens: University of Georgia Press, 2011.
Morris, Christopher. *The Big Muddy: An Environmental History of the Mississippi and Its Peoples, from Hernando de Soto to Hurricane Katrina*. Oxford: Oxford University Press, 2012.
Mulcahy, Matthew. *Hurricanes and Society in the British Greater Caribbean, 1624–1783*. Baltimore: Johns Hopkins University Press, 2008.
Nash, Linda. *Inescapable Ecologies: A History of Environment, Disease, and Knowledge*. Berkeley: University of California Press, 2006.
Nelson, Lynn A. *Pharsalia: An Environmental Biography of a Southern Plantation, 1780–1880*. Athens: University of Georgia Press, 2007.
Nelson, Megan Kate. *Ruin Nation: Destruction and the American Civil War*. Athens: University of Georgia Press, 2012.
Nixon, Rob. *Slow Violence and the Environmentalism of the Poor*. Cambridge, Mass.: Harvard University Press, 2013.
Oates, Stephen B. *A Woman of Valor: Clara Barton and the Civil War*. New York: Maxwell Macmillan, 1994.
Ogden, Laura. *Swamplife: People, Gators, and Mangroves Entangled in the Everglades*. Minneapolis: University of Minnesota Press, 2011.
Olwell, Robert. *Masters, Slaves, and Subjects: The Culture of Power in the South Carolina Low Country, 1740–1790*. Ithaca, N.Y.: Cornell University Press, 1998.

Ortiz, Paul. *Emancipation Betrayed: The Hidden History of Black Organizing and White Violence in Florida from Reconstruction to the Bloody Election of 1920*. Berkeley: University of California Press, 2005.
Oubre, Claude F. *Forty Acres and a Mule: The Freedmen's Bureau and Black Land Ownership*. Baton Rouge: Louisiana State University Press, 1978.
Pérez, Louis A. *Winds of Change: Hurricanes and the Transformation of Nineteenth-Century Cuba*. Chapel Hill: University of North Carolina Press, 2001.
Perman, Michael. *Struggle for Mastery: Disenfranchisement in the South, 1888–1908*. Chapel Hill: University of North Carolina Press, 2001.
Petty, Adrienne Monteith. *Standing Their Ground: Small Farmers in North Carolina Since the Civil War*. Oxford: Oxford University Press, 2017.
Pietruska, Jamie L. "'A Tornado Is Coming!': Counterfeiting and Commercializing Weather Forecasts from the Gilded Age to the New Era." *Journal of American History* 105, no. 3 (Dec. 2018): 538–62.
Pollitzer, William S. *The Gullah People and Their African Heritage*. Athens: University of Georgia Press, 1999.
Poole, W. *Never Surrender: Confederate Memory and Conservatism in the South Carolina Upcountry*. Athens: University of Georgia Press, 2004.
Postel, Charles. *The Populist Vision*. Oxford: Oxford University Press, 2007.
Prince, Hugh. *Wetlands of the American Midwest: A Historical Geography of Changing Attitudes*. Chicago: University of Chicago Press, 1997.
Prince, K. Stephen. *Stories of the South: Race and the Reconstruction of Southern Identity, 1865–1915*. Chapel Hill: University of North Carolina Press, 2014.
Quintana, Ryan A. *Making a Slave State: Political Development in Early South Carolina*. Chapel Hill: University of North Carolina Press, 2018.
Ramey, Lauri. *Slave Songs and the Birth of African American Poetry*. New York: Palgrave Macmillan, 2008.
Ramsey, William L. *The Yamassee War: A Study of Culture, Economy, and Conflict in the Colonial South*. Lincoln: University of Nebraska Press, 2010.
Rana, Aziz. *The Two Faces of American Freedom*. Cambridge, Mass.: Harvard University Press, 2014.
Reiger, George. *Wanderer on My Native Shore*. New York: Simon and Schuster, 1983.
Remes, Jacob A. C. *Disaster Citizenship: Survivors, Solidarity, and Power in the Progressive Era*. Urbana: University of Illinois Press, 2016.
Ring, Natalie. *The Problem South: Region, Empire, and the New Liberal State, 1880–1930*. Athens: University of Georgia Press, 2012.
Roane, J. T. "Plotting the Black Commons." *Souls: A Critical Journal of Black Politics, Culture, and Society* 20, no. 3 (2018): 239–66.
Rohland, Eleonora. *Changes in the Air: Hurricanes in New Orleans from 1718 to the Present*. New York: Berghahn, 2018.
Rose, Willie. *Rehearsal for Reconstruction: The Port Royal Experiment*. 1st ed. 1964. Repr., Athens: University of Georgia Press, 1999.
Rosén, Hannah. *Terror in the Heart of Freedom: Citizenship, Sexual Violence, and the Meaning of Race in the Postemancipation South*. Chapel Hill: University of North Carolina Press, 2009.

Rosengarten, Theodore. *Tombee: Portrait of a Cotton Planter.* New York: McGraw-Hill, 1987.
Rowland, Lawrence S., Alexander Moore, and George C. Rogers. *The History of Beaufort County, South Carolina: 1514–1861.* Columbia: University of South Carolina Press, 1996.
Rowland, Lawrence S., and Stephen R. Wise. *Bridging the Sea Islands' Past and Present, 1893–2006: The History of Beaufort County, South Carolina, Volume 3.* Columbia: University of South Carolina Press, 2015.
Royster, Charles. *The Fabulous History of the Dismal Swamp Company: A Story of George Washington's Times.* New York: Vintage, 2000.
Salley, A. S., Jr. "Notes and Queries." *The South Carolina Historical and Genealogical Magazine* 1, no. 1 (1900): 98–107.
Saville, Julie. *The Work of Reconstruction: From Slave to Wage Laborer in South Carolina, 1860–1870.* Cambridge: Cambridge University Press, 1996.
Sayers, Daniel O. *A Desolate Place for a Defiant People: The Archaeology of Maroons, Indigenous Americans, and Enslaved Laborers in the Great Dismal Swamp.* Gainesville: University Press of Florida, 2016.
Schick, Tom W., and Don H. Doyle. "The South Carolina Phosphate Boom and the Stillbirth of the New South, 1867–1920." *South Carolina Historical Magazine* 86, no. 1 (Jan. 1985): 1–31.
Schwalm, Leslie. *A Hard Fight for We: Women's Transition from Slavery to Freedom in South Carolina.* Urbana-Champaign: University of Illinois Press, 1997.
Schwartz, Stuart B. *Sea of Storms: A History of Hurricanes in the Greater Caribbean from Columbus to Katrina.* Princeton, N.J.: Princeton University Press, 2015.
Seabrook, Charles. *The World of the Salt Marsh: Appreciating and Protecting the Tidal Marshes of the Southeastern Atlantic Coast.* Athens: University of Georgia Press, 2013.
Shuler, Kristina, and Ralph Bailey Jr. *A History of the Phosphate Mining Industry in the South Carolina Lowcountry.* Mount Pleasant, S.C.: Brockington and Associates, Inc., 2004.
Simon, Bryant. *A Fabric of Defeat: The Politics of South Carolina Millhands, 1910–1948.* Chapel Hill: University of North Carolina Press, 1998.
Sinha, Manisha. *The Counterrevolution of Slavery: Politics and Ideology in Antebellum South Carolina.* Chapel Hill: University of North Carolina Press, 2003.
Shapiro, Karin A. *A New South Rebellion: The Battle against Convict Labor in the Tennessee Coalfields, 1871–1896.* Chapel Hill: University of North Carolina Press, 1998.
Sherr, Evelyn. *Marsh Mud and Mummichogs: An Intimate History of Coastal Georgia.* Athens: University of Georgia Press, 2015.
Smith, Hayden R. *Carolina's Golden Fields: Inland Rice Cultivation in the South Carolina Lowcountry, 1670–1860.* Cambridge: Cambridge University Press, 2019.
Smith, Julia Floyd. *Slavery and Rice Culture in Low Country Georgia, 1750–1860.* Knoxville: University of Tennessee Press, 1985.
Smith, Kimberly K. *African American Environmental Thought: Foundations.* Lawrence: University Press of Kansas, 2007.
Smith, Mark S. *Camille, 1969: Histories of a Hurricane.* Mercer University Lamar Memorial Lectures Series. Athens: University of Georgia Press, 2011.
———. *A Sensory History Manifesto.* State College: Penn State University Press, 2021.
Solnit, Rebecca. *A Paradise Built in Hell: The Extraordinary Communities that Arise in Disaster.* New York: Penguin Books, 2010.

Steeples, Douglas, and David O. Whitten. *Democracy in Desperation: The Depression of 1893*. New York: Praeger, 1998.
Steinberg, Theodore. *Acts of God: The Unnatural History of Natural Disaster in America*. Oxford: Oxford University Press, 2006.
Stewart, Mart A. *"What Nature Suffers to Groe": Life, Labor, and Landscape on the Georgia Coast, 1680–1920*. Athens: University of Georgia Press, 1996.
Strickland, John Scott. "'No More Mud Work': The Struggle for Control of Labor Production in the South Carolina Low Country, 1863–1880." In *The Southern Enigma: Essays on Race, Class, and Folk Culture*, edited by W. J. Fraser and W. B. Moore, 43–62. Westport, Conn.: Greenwood, 1983.
———. "Traditional Culture and Moral Economy: Social and Economic Change in the South Carolina Low Country, 1865–1910." In *The Countryside in the Age of Capitalist Transformation*, edited by Steven Hahn and Johnathan Prude, 141–78. Chapel Hill: University of North Carolina Press, 1985.
Sutter, Paul S., and Christopher J. Manganiello, eds. *Environmental History and the American South: A Reader*. Environmental History and the American South. Athens: University of Georgia Press, 2009.
Sutter, Paul S., and Paul M. Pressly, eds. *Coastal Nature, Coastal Culture: Environmental Histories of the Georgia Coast*. Athens: University of Georgia Press, 2018.
Swanson, Drew. *Remaking Wormsloe Plantation: The Environmental History of a Lowcountry Landscape*. Athens: University of Georgia Press, 2014.
Tetzlaff, Monica Maria. *Cultivating a New South: Abbie Christensen and the Politics of Race and Gender, 1852–1938*. Columbia: University of South Carolina Press, 2002.
Thompson, E. P. "The Moral Economy of the English Crowd in the Eighteenth Century." *Past and Present*, no. 50 (1971): 76–136.
Tindall, George Brown. *South Carolina Negroes, 1877–1900*. Columbia: University of South Carolina Press, 1952.
Tortora, Daniel J. *Carolina in Crisis: Cherokees, Colonists, and Slaves in the American Southeast, 1756–1763*. Chapel Hill: University of North Carolina Press, 2015.
Trinkley, Michael, ed. *The History and Archaeology of Kiawah Island, Charleston County, South Carolina*. Columbia: Chicora Foundation, 1993.
Turner, Lorenzo Dow. *Africanisms in the Gullah Dialect*. Introduction by Katherine Wyly Mille and Michael B. Montgomery. 1st ed. 1949. Reprint, Columbia: University of South Carolina Press, 2002.
Valencius, Conevery Bolton. *The Lost History of the New Madrid Earthquakes*. Chicago: University of Chicago Press, 2013.
Vernon, Amelia Wallace. *African Americans at Mars Bluff, South Carolina*. Baton Rouge: Louisiana State University Press, 1993.
Vileisis, Ann. *Discovering the Unknown Landscape: A History of America's Wetlands*. Washington, D.C.: Island, 1997.
Wayne, Michael. *The Reshaping of Plantation Society: The Natchez District, 1860–1880*. Baton Rouge: Louisiana State University Press, 1983.
White, Monica M. *Freedom Farmers: Agricultural Resistance and the Black Freedom Movement*. University of North Carolina Press, 2021.
Whitnah, Donald R. *A History of the United States Weather Bureau*. Urbana: University of Illinois Press, 1961.

Wilkerson, Isabel. *The Warmth of Other Suns: The Epic Story of America's Great Migration*. New York: Vintage, 2011.
Willis, John C. *Forgotten Time: The Yazoo-Mississippi Delta after the Civil War*. Charlottesville: University of Virginia Press, 2000.
Wilson, Anthony. *Shadow and Shelter: The Swamp in Southern Culture*. 1st ed. Jackson: University Press of Mississippi, 2006.
Williams, Susan Millar, and Stephen G. Hoffius. *Upheaval in Charleston: Earthquake and Murder on the Eve of Jim Crow*. Athens: University of Georgia Press, 2011.
Wise, Stephen R., and Lawrence S. Rowland with Gerhard Spieler. *Rebellion, Reconstruction, and Redemption, 1861–1893*. Columbia: University of South Carolina Press, 2015.
Wood, Peter H. *Black Majority: Negroes in Colonial South Carolina from 1670 through the Stono Rebellion*. New York: Knopf, 1974.
Woodward, C. Vann. *Origins of the New South: 1877–1913*. Rev. ed. Baton Rouge: Louisiana State University Press, 1981.
———. *The Strange Career of Jim Crow*. 1st ed. 1955. Reprint, Oxford: Oxford University Press, 2001.
Yuhl, Stephanie E. *A Golden Haze of Memory: The Making of Historic Charleston*. Chapel Hill: University of North Carolina Press, 2014.

Index

Printed in the USA
CPSIA information can be obtained
at www.ICGtesting.com
LVHW091447230823
756033LV00001B/95